Microcomputer
Applications in Geology, II

COMPUTERS and GEOLOGY

a series edited by Daniel F. Merriam

1976 *Quantitative Techniques for the Analysis of Sediments*

1978 *Recent Advances in Geomathematics*

1979 *Geomathematical and Petrophysical Studies in Sedimentology*
(edited by D. Gill & D. F. Merriam)

1981 *Predictive Geology: with Emphasis on Nuclear-Waste Disposal*
(edited by G. de Marsily & D. F. Merriam)

1986 *Microcomputer Applications in Geology*
(edited by J. T. Hanley & D. F. Merriam)

Professor Merriam also is the Editor-in-Chief of *Computers & Geosciences* — an international journal devoted to the rapid publication of computer programs in widely used languages and their applications.

Other Related Pergamon Publications

Books

HOLLAND
Microcomputers and Their Interfacing
HOLLAND
Illustrated Dictionary of Microcomputers & Microelectronics
MARSAL
Statistics for Geoscientists
NORRIE & TURNER
Automation for Mineral Resource Development

Journals

Acta Geologica Sinica
Acta Seismologica Sinica
Applied Geochemistry
Automatica
Computers & Geosciences
Computer Languages
Information Processing & Management
Information Systems
International Journal of Rock Mechanics and Mining
Sciences & Geomechanics Abstracts
Nuclear Geophysics

Full details of all Pergamon publications/free specimen copy of any Pergamon journal available on request from your nearest Pergamon office.

Microcomputer Applications in Geology, II

edited by

J. THOMAS HANLEY

President, Cape Henry Software, Falls Church, Virginia

and

DANIEL F. MERRIAM

Endowment Association Distinguished Professor of the Natural Sciences, Wichita State University, Wichita, Kansas

PERGAMON PRESS

Member of Maxwell Macmillan Pergamon Publishing Corporation

OXFORD • NEW YORK • TORONTO • SYDNEY
FRANKFURT • TOKYO • SÃO PAULO • BEIJING

U.K.	Pergamon Press plc, Headington Hill Hall, Oxford OX3 0BW, England
U.S.A.	Pergamon Press, Inc., Maxwell House, Fairview Park, Elmsford, New York 10523, U.S.A.
PEOPLE'S REPUBLIC OF CHINA	Pergamon Press, Room 4037, Qianmen Hotel, Beijing, People's Republic of China
FEDERAL REPUBLIC OF GERMANY	Pergamon Press GmbH, Hammerweg 6, D-6242 Kronberg, Federal Republic of Germany
BRAZIL	Pergamon Editora Ltda, Rua Eça de Queiros, 346, CEP 04011, Paraiso, São Paulo, Brazil
AUSTRALIA	Pergamon Press Australia Pty Ltd., P.O. Box 544, Potts Point, N.S.W. 2011, Australia
JAPAN	Pergamon Press, 5th Floor, Matsuoka Central Building, 1-7-1 Nishishinjuku, Shinjuku-ku, Tokyo 160, Japan
CANADA	Pergamon Press Canada Ltd., Suite No. 271, 253 College Street, Toronto, Ontario, Canada M5T 1R5

Copyright © 1990 Pergamon Press plc

First edition 1990

Library of Congress Cataloging in Publication Data
Microcomputer applications in geology, II/edited by
J. Thomas Hanley and Daniel F. Merriam.
p. cm.—(Computers and geology:7)
1. Geology—Data processing. 2. Microcomputers—
Programming.
I. Hanley, J. Thomas. II. Merriam, Daniel Francis.
III. Series: Computers & geology: v. 7.
QE48.8.M532 1989
550'.285'416—dc20 89–22840

British Library Cataloguing in Publication Data
Microcomputer applications in geology, II
1. Geology. Applications of microcomputer systems
I. Hanley, J. Thomas II. Merriam, Daniel F.
(Daniel Francis) III. Series
551'.028'5404

ISBN 0–08–040261–5

Contents

vi

List of Contributors

F. P. Agterberg, Mineral Resources Division, Geological Survey of Canada, 601 Booth Street, Ottawa, Canada K1A OE8

James E. Anderson, Kansas Geological Survey, 1930 Constant Avenue - Campus West, Lawrence, KS 66046 USA

Massimiliano Barchi, Dipartimento di Scienze della Terra, Perugia University, p.za dell'Universita, 06100 Perugia, Italy

Michael J. Bellotti, Olin Chemical Group, Lower River Road, Charleston, TN 37311 USA

J.D. Bliss, U.S. Geological Survey, OMR Arizona Field Office, 210 E. 7th Street, Tucson AZ 85705 USA

Graeme F. Bonham-Carter, Mineral Resources Division, Geological Survey of Canada, 601 Booth Street, Ottawa, Canada K1A OE8

H. Robert Burger, Department of Geology, Smith College, Northampton, MA 01063 USA

John C. Butler, Department of Geosciences, University of Houston, Houston, TX 77004 USA

H.C. Chen, The University of Alabama, Tuscaloosa, AL 354487 USA

Richard Cleave, Historical Productions, Inc., 6 Barrier Strasse, Zug CH-6200, Switzerland

Richard G. Craig, Department of Geology, Kent State University, Kent, OH 44242 USA

J.H. Fang, Department of Geology, The University of Alabama, Tuscaloosa, AL 35487 USA

Maxine V. Fontana, Northern Virginia Community College, Alexandria Campus, Alexandria, VA 22313 USA

D.I. Groves, Key Centre for Strategic Mineral Deposits, Department of Geology, University of Western Australia, Nedlands 6009, Australia

Fausto Guzzetti, IRPI, National Research Council, via Madonna Alta 126, 06100 Perugia, Italy

S.M. Habesch, Poroperm Geochem Group, Chester Street, Saltney, Chester CH4 8RD UK

John K. Hall, Geological Survey of Israel, Marine Geology, Mapping and Tectonics Division, 30 Malchei Israel Street, Jerusalem 95 501 Israel

Ute Christian Herzfeld, Scripps Institution of Oceanography, Geological Research Division, University of California San Diego, La Jolla, CA 92093 USA

A.M. Hittelman, National Geophysical Data Center, NOAA, Boulder, CO 80303 USA

vii

Michael Edward Hohn, West Virginia Geological Survey, P.O. Box 879, Morgantown, WV 26507-0879 USA

Michael M. Kimberley, Department of Marine, Earth, & Atmospheric Sciences, North Carolina State University, Raleigh, NC 27695 USA

Richard Looi, Copland College, Canberra, Australia

D.F. Merriam, Stratigraphic Studies Group, Department of Geology, Wichita State University, Wichita, KS 67208 USA

H. Meyers, National Geophysical Data Center, NOAA, Boulder, CO 80303 USA

Ulf Nordlund, Paleontological Institute, University of Uppsala, Box 558, S-751 22 Uppsala, Sweden

Colin Ong, Water Science Graduate Group, Department of Land, Air & Water Resources, University of California, Davis, CA 95616 USA

R. Poulinet, Key Centre for Strategic Mineral Deposits, Department of Geology, University of Western Australia, Nedlands 6009, Australia

Barry L. Roberts, Department of Geology, Kent State University, Kent, OH 44242 USA

Joseph E. Robinson, Department of Geology, Syracuse University, Syracuse, NY 13210 USA

N.M.S. Rock, Key Centre for Strategic Mineral Deposits, Department of Geology, University of Western Australia, Nedlands 6009, Australia

Eric Rosencrantz, University of Texas Institute for Geophysics, 8701 Mopac Blvd., Austin, TX 78758 USA

Enrico Savazzi, Paleontologisk Institutionen, Box 558, S-751 22 Uppsala, Sweden

Eric Schwartz, formerly of the Environmental Protection Service, Ministry of the Interior, Jerusalem, Israel

J.N. Shellabear, Key Centre for Strategic Mineral Deposits, Department of Geology, University of Western Australia, Nedlands 6009, Australia

A.W. Shultz, The University of Alabama, Tuscaloosa, AL 35487 USA

D.A. Singer, U.S. Geological Survey, 345 Middlefield Road, MS 984, Menlo Park, CA 94025 USA

Mark A. Sondergard, Stratigraphic Studies Group, Department of Geology, Wichita State University, Wichita, KS 67208 USA

John F. Stamm, Department of Geology, Kent State University, Kent, OH 44242 USA

D. Stanley, U.S. Bureau of Mines, Tuscaloosa, AL 35486 USA

John C. Tipper, Department of Geology, Australian National University, GPO Box 4, Canberra, ACT 2601 Australia

Tung Trinh, Lanke Ginninderra College, Canberra, Australia

W. Lynn Watney, Kansas Geological Survey, 1930 Constant Avenue - Campus West, Lawrence, KS 66046 USA

M.R. Wheatley, Western Australian Regional Computing Centre, University of Western Australia, Nedlands 6009, Australia

Jan-Chung Wong, Kansas Geological Survey, 1930 Constant Avenue - Campus West, Lawrence, KS 66046

D. Wright, Shelton State Community College, Tuscaloosa, AL 35405 USA

Availability of Software

Most of the authors in this volume have made their software available through the Computer Oriented Geological Society (COGS) public domain software library. For further information contact COGS directly at the following address:

> COGS Disk Series
> P.O. Box 1317
> Denver, CO 80201-1317

The following authors have contributed their software:

> Barchi and Guzzetti
> Burger
> Hall, Schwartz, and Cleave
> Herzfeld
> Hohn and Fontana
> Kimberley
> Nordlund
> Ong
> Roberts, Craig, and Stamm
> Rosencrantz
> Savazzi
> Singer and Bliss
> Sondergard, Robinson, and Merriam
> Watney, Anderson, and Wong
> Wright, Stanley, Chen, Shultz, and Fang

Other software mentioned in the book generally are available from the authors. Please contact them for further information.

TRADEMARKS

Reference to products with trademarks occur at many places in the book. Where the names are not commonplace, an attempt has been made to include the company name and address.

Preface

The development and utilization of microcomputers is widespread and rapid in all scientific disciplines — geology is no exception. Micros are becoming ubiquitous and indispensable in research and teaching as evidenced in many ways. The success of our previous volume, "Microcomputer Applications in Geology" (this series, 1986), prompted this sequel of Microcomputers II. The diversity of papers presented here on micros will give some indication on the pervasion of their use in the geologic profession.

From the first paper by Bob Burger on the future of geologic education using micros, there is a list of surprising applications as documented in this volume. These applications range from reconstructing fossil shells to reconstructing landscape terraines. In many ways the technology of both hardware and software is outstripping the imaginative and innovative applications that could be made of these advances. As Burger points out, the future generation of geologists will be at home with computers and able to use them to their fullest potential. The papers presented here give evidence of the sophistication as of the end of the decade of the 1980's.

Tom Hanley and I collected the papers on a personal basis — everyone we knew working with micros was invited to contribute. Several papers came to us through submissions to Computers & Geosciences. The results of this solicitation are given here in this volume — a collection of papers on micros by a group of experts in the field.

The mighty micro! The micro has grown in size (capacity) and shrunk in size (physical), so that now even the lap-top computers are as powerful as mainframes just a few years ago. Micros have been configured into workstations so that the working geologist has at his/her fingertips a real mind-extender with limitations only in the mind of the user. Software is one of these limitations — not everyone can create software to do their bidding nor can it be purchased on the open market. However, as more and more software is designed with geologists in mind, this disadvantage will disappear. Until that time, the best source of software and ideas on applications can come from publications such as Micros II and Computers & Geosciences. Many papers being published now in C&G are microcomputer oriented.

Programmers are increasingly aware of the necessity for standards, portability, and ease of use. Thus, the newer programs are user-friendly and adaptable — as availability and ease in use increases, more and more geologists will feel at ease in using their microcomputer. The presentations in this volume are meant to help during this transition time and provide a stimulation for additional and novel use. Many of the programs noted here are available for those interested.

Microcomputers have become a standard work item for the geologist. This involvement is likely to increase during the coming decade — by the turn of the century things certainly will be different. As noted by Burger, "Our science is oriented so visually and depends to such a great extent on communicating graphics information that we truly are in the midst of a revolution." Indeed we are!

D. F. Merriam
Stratigraphic Studies Group Box 153
Department of Geology
Wichita State University
Wichita, Kansas

Introduction

Behold the power of the microcomputer before your very eyes! This book took less than a year to complete from the initial conversations between Dan Merriam and myself to final typeset copy. This would have been impossible when the first volume of this book was published. All of the papers in this volume, with one exception, arrived on floppy disk but in four different disk formats and many wordprocessing formats. These were all handled elegantly by our typesetter and converted into the appropriate Macintosh file format. Even some of the figures were placed in the book electronically. I see this as a great boon to the world of science because of the much shorter time between inception and dispersal of scientific knowledge. As a result of using microcomputers to typeset this book, the information contained in it is current and fresh.

The papers in this book cover a wide range of topics both in geology and computer science. Some papers offer advice to fellow users, others describe a system that the author(s) developed. They come from all over the world. Almost seventy percent of the papers in this volume describe software that is available from COGS (The Computer Oriented Geological Society) through their public-domain software library as stated in the previous section on software availability.

Without further ado, let me introduce our authors. H. ROBERT BURGER (*Geologic education in the 1990's – The impact of personal computers*) speculates on the future uses of microcomputers in geology. Although the applications are from the educational realm, the information presented is of interest to everyone in geology. Dr. Burger explains four categories of software that will be used in education in the future: productivity software, graphics-linked database software, investigatory simulation software, and random-access lecturing software. All of these certainly will be used in geologic education as well as all aspects of geology.

MASSIMILIANO BARCHI and FAUSTO GUZZETTI (*STRANA: A Macintosh computer program for the representation and statistical analysis of orientation data in structural geology*) describe their program, STRANA, which provides a complete procedure for the analysis of orientation data in structural geology. STRANA was written in BASIC and runs on an Apple Macintosh computer. The authors present two examples of the application of the program to structural data.

MICHAEL J. BELLOTTI (*Data and information management for a hydrogeologic study of a waste-disposal site*) explains a strategy for managing data for the hydrogeologic study of a waste-disposal site using a microcomputer with commercial software. He identifies the attendant problems.

G. F. BONHAM-CARTER and F. P. AGTERBERG (*Application of a microcomputer-based geographic information system to mineral-potential mapping*) present the use of a geographic information system in creating gold-potential maps for Nova Scotia. They describe and use the weights of evidence modeling for estimating mineral potential.

JOHN C. BUTLER (*Stimulation via simulation: geochemical modeling*) illustrates how the principles of chemical kinetics and geochemical cycles can be modeled to solve a particular problem. Dr. Butler presents examples of simulations. The software used is STELLA, a dynamic modeling system, which runs on a Macintosh microcomputer.

S. M. HABESCH (*The evaluation of pore-geometry networks in clastic reservoir lithologies using microcomputer technology*) describes the process of evaluating pore-geometry networks to calculate reservoir parameters such as permeability and capillary pressure and to generate lithologic classification schemes.

JOHN K. HALL, ERIC SCHWARTZ, and RICHARD L. W. CLEAVE (*The Israeli DTM (Digital Terrain Map) Project*) describe the production of a new high-resolution digital terrain map of Israel. The DTM was produced with software written in FORTRAN with HALO graphics routines for an IBM-PC AT microcomputer.

UTE CHRISTINA HERZFELD (*Geostatistical software for evaluation of line survey data applied to radio-echo soundings in glaciology*) applies universal kriging to radio-echo soundings in glaciology for evaluating line-survey data. Dr. Herzfeld also describes the software used in this analysis, which is written in FORTRAN 77 for an IBM-compatible microcomputer.

A. M. HITTLEMAN and H. MEYERS (*Regional geophysical data on a compact disk*) describes the Geophysics of North America compact-disk (CD) project. The CD includes topographic, magnetic, gravity, earthquake-seismology, crustal-stress, and thermal-aspect data. They also describe software, written in C, that accesses the CD from an IBM PC.

MICHAEL EDWARD HOHN and MAXINE V. FONTANA (*Dissecting variograms*) present an interactive graphics program that uses advanced geostatistical techniques to identify statistical outliers, detect spatial discontinuities and subpopulations, examine alternative measurements of a regionalized variable, and model variograms. The software is written in Pascal for an IBM-PC AT.

MICHAEL M. KIMBERLEY (*Cross sections and volume measurement of stratigraphic units*) describe a software package that sketches cross sections and calculates either the total or the fluid volume between cross sections. This program can be applied to a wide variety of geologic applications. The code is written in Pascal and runs on an Apple Macintosh microcomputer.

ULF NORDLAND (*A simple Pascal procedure for outline tracing in image analysis*) presents a procedure for outline tracing which is a fundamental procedure in image analysis. This type of procedure is required for different techniques of shape analysis and area computations. The procedure is written in Pascal for an IBM-compatible microcomputer.

COLIN ONG (*CatTrack: A Pascal program to display ternary diagrams on a Macintosh computer*) describes a Pascal program for displaying and printing ternary diagrams on a Macintosh microcomputer. He includes options for three different types of data: major cations, major anions, and soil texture.

BARRY L. ROBERTS, RICHARD G. CRAIG, and JOHN F. STAMM (*A microcomputer reconstruction of paleoclimates*) describe a software system that reconstructs the paleoclimate such as that of the last glacial maximum for the southwestern United States. This program allows the user to compute values of mean monthly maximum daily temperature and total monthly precipitation for each month of the year. The code is written in C and runs on an IBM-compatible microcomputer.

N. M. S. ROCK, J. N. SHELLABEAR, M. R. WHEATLEY, R. POULINET, and D. I. GROVES (*Microcomputers in mineral exploration: A database for modeling gold deposits in the Yilgarn block of Western Australia*) explains the process of setting up a complex database containing geologic, geochemical, and isotopic information. Detailed consideration is given to selecting a computer and the database management system, as well as, determining the structure of the database. The database runs on a Macintosh II using 4th Dimension as its database engine.

ERIC ROSENCRANTZ (*MACS: A Macintosh program for constructing marine magnetic anomaly profile*) describes a program that calculates and displays marine magnetic-anomaly profiles from magnetic-field polarity-reversal sequences provided by the user. The software also allows on-screen editing of magnetic-reversal sequences before calculating the profiles. The program was written in BASIC and runs on a Macintosh microcomputer.

ENRICO SAVAZZI (*Theoretical morphology of shells aided by microcomputers*) describes microcomputer programs that assist in the interactive modeling of shell morphologies and their ontogenetic laws. In his paper, he discusses shell-modeling strategies and some of the problems that are characteristic of the microcomputer environment. The program was written in C and runs on an IBM-compatible microcomputer.

D. A. SINGER and J. D. BLISS (*Program to prepare standard figures for grade-tonnage models on a Macintosh*) describe a program that allows users to plot grade-tonnage distributions. These plots are used in grade-tonnage models of specific types of mineral deposits. This program is written in Pascal and runs on a Macintosh microcomputer.

M. A. SONDERGARD, J. E. ROBINSON, and D. F. MERRIAM (*FILT-PC, a one-dimensional Fourier transform program in FORTRAN for the PC*) present a program that calculates one-dimensional Fast Fourier Transforms. The program is written in FORTRAN 77 and runs on an IBM-PC compatible.

JOHN C.TIPPER, RICHARD LOOI, and TUNG TRINH (*Simulation of sediment-fluid interaction in subsiding basins*) discuss the mechanism of fluid movement in a developing sedimentary basin. A discrete-time, discrete-space simulator is described which allows patterns of diagenetic inhomogeneity in the growing sediment pile to be generated and this mechanism's effect to be investigated.

W. LYNN WATNEY, JAMES E. ANDERSON, and JAN-CHUNG WONG (*Porosity Advisor – an expert system used as an aid in interpreting the origin of porosity on carbonate rocks*) describe Porosity Advisor, which is a rule-based expert system designed to help the user describe the nature, origin, and timing of porosity development and destruction in the Upper Pennsylvanian carbonate rocks in western Kansas. Their paper presents an overview of expert systems, the steps involved in the construction of their system, and results from Porosity Advisor.

D. WRIGHT, D. STANLEY, H. C. CHEN, A. W. SCHULTZ, and J. H. FANG (*A frame-based expert system to identify minerals in thin section*) describe the development of an expert system to identify minerals in thin section using an expert-system shell. The program, XMIN-S, runs on an IBM-compatible microcomputer.

Microcomputers are becoming more prevalent with the passage of time. They are much more powerful today than they were three years ago, when the first volume of this book was put together. The wide variety of applications in this volume shows that the more microcomputers become available to geologists, the more specialized and sophisticated applications will be created, ... and the pace continues to accelerate! In a recent interview with Steve Jobs (Inc. Magazine, April, 1989, p. 116) he told the editors of Inc.:

Let's say that – for the same amount of money it takes to build the most powerful computer in the world – you could make 1,000 computers with one-thousandth the power and put them in the hands of 1,000 creative people. You'll get more out of doing that than out of having one person use the most powerful computer in the world. Because people are inherently creative. They will use tools in ways that toolmakers never thought possible. And once a person figures out how to do something with that tool, he or she can share it with the other 999.

We hope to have accomplished this last thought with this book.

In closing, I thank all the contributor's for their cooperation and assistance. I also thank my wife, Terry, for her great help and encouragement and my children, Catie and Jamie, for their future understanding. Many thanks also go to two companies, KENROB & Associates, for support and sustenance and The Word Cottage, for typesetting this book.

J. Thomas Hanley
Cape Henry Software
7228 Timber Lane
Falls Church, VA 22046

Geologic Education in the 1990's — The Impact of Personal Computers

H. Robert Burger
Smith College

ABSTRACT

The increasing use of personal computers by educators and the evolution of sophisticated graphics capabilities of these machines coupled with the ready availability of commercial software that is both easy to use and is suited to educators' needs are the continuing trends fueling a revolution in the way computers are used in geoscience education. This revolution, yet in its infancy, is supported by the four major categories of software that drive it: (1) productivity software, (2) graphics-linked database software, (3) investigatory simulation software, and (4) random-access lecturing software.

TRENDS AFFECTING INSTRUCTIONAL COMPUTING

In order to set the stage for an appreciation of the categories of instructional software that are beginning to impact instructional computing in the Earth sciences, it is perhaps wisest to first summarize four general trends in instructional computing per se.

One trend is a definite move away from mainframes and minicomputers to personal computers (microcomputers). Just a few years ago virtually all computing tasks performed by students and professors took place on the larger computers. Today, significantly more than one-half the workload has been shifted to personal computers. That this figure is not larger is mainly the result the time involved in transferring programs from mainframes to personal computers and to the limited financial resources available for purchasing new equipment. The first generation of programs used by geologic educators on personal computers was transferred mostly intact from the larger machines and was controlled by their limitations as well as microcomputer limitations. Applications included testing, course management, self-paced instruction, simulations, and standard computationally intensive tasks. As personal computers began to make computer access possible for virtually everyone, word processing became a dominant use and an important tool. Improvements in computational speed, available memory, screen resolution, and graphics capabilities have made personal computers even more powerful, useful, and sought after. As these improvements continue, the move to personal computers will accelerate so that by the early 1990's virtually all instructional computing will take place on desktop machines.

A second trend that began approximately five years ago was the availability of a range of relatively inexpensive commercial software that could be applied to geologic problems. The appearance of the spreadsheet made the personal computer a serious business tool, but it also could be applied to a variety of geologic computational needs. One of the first

1

attempts to integrate computer capabilities into routine classroom work focused on student programming in order to solve typical problems, and, therefore, to improve the independent problem-solving abilities of students. This goal, although laudable, required students to be reasonably competent programmers. As this rarely was realistic, fewer and fewer of these tasks were assigned. Spreadsheets, however, are extremely versatile and fairly easy to master. As such, numerous authors have published papers demonstrating that spreadsheets can be used to solve problems ranging from petrological calculations to deriving Bouguer gravity anomalies.

A third trend, and one that is of great importance for geology, is the increasingly sophisticated graphics capabilities of personal computers. These capabilities exceed by a substantial margin the graphics capabilities of the larger computers at most universities and are one of the main reasons geologists embraced personal computers so early and in such numbers. Once again geologists have a selection of many commercial software programs that aid in creating three-dimensional views of objects, drawing geologic sections, and creating maps that rival those of drafting professionals. In addition, there are programs that graph our data, fit curves to our data, and label axes and legends.

The utility of these programs has taken a quantum leap forward in usefulness with the appearance of the laser printer and the image scanner. The laser printer allows us to prepare diagrams for publication on our computers as well as to create extremely clear and legible course handouts. The scanner permits us to scan just about anything at a resolution comparable to the laser printer. The versatility and usefulness of these devices for our teaching and research is such that both are increasingly present in geology departments and we begin to wonder how we ever got along without them.

The fourth trend can be summarized by the phrase "ease-of-use." The guidelines provided by Apple Computer for developers of Macintosh software stressed a well-defined user interface that was to be followed in all applications. Pull-down menus, mouse-driven events, overlapping and resizeable windows, dialog boxes, scroll bars, text editing, and file handling all work the same in virtually any Macintosh application. The uncharted terrain of a new application always looks somewhat familiar, and all standard operations function in a common manner. Thus, the learning curve for new applications usually is short, and it is straightforward to operate several programs without the usual continual reference to manuals. This standard user interface and the graphics orientation of the Macintosh provides another significant advantage the ability to transfer text and graphics among various applications. Such transfer is enhanced by operating systems that enable several applications to be loaded into memory and provide the user with the capability to move quickly from one application to another. The success of the graphical interface is evident now that the next generation of personal computers being offered by IBM and other manufacturers all are adopting this approach.

It is these four major trends, and others less well defined, working in concert that support the evolution of the educational software that increasingly will impact the way Earth science courses are organized and presented.

EDUCATIONAL SOFTWARE FOR THE FUTURE

Four categories of educational software that will have a tremendous impact on future Earth-science instruction can be summarized as: (1) productivity software, (2) graphics-linked database software, (3) investigatory simulation software, and (4) random-access lecturing software. Examples of each category already exist and a few students now are experiencing a sample of what the future holds for the majority of students. For this vision to become reality, however, the evolving trends in personal computer hardware and software that have occurred during the past five years must continue for at least another five years.

Productivity Software

The "productivity-software" concept meshes ease-of-use, enhanced graphics displays, multitasking, transferability of graphics, text, and data among diverse programs, and a range of utility software (spreadsheet, word processor, CAD, and graphing) to increase the resources available to students to such an extent that course expectations and assignments evolve to a higher level. This concept is dependent heavily on relatively inexpensive commercial software that can be applied to geologic problems.

An excellent example of what is meant by utility software is spreadsheet software that was discussed previously. Because of the versatility of spreadsheets, it is not unreasonable to provide newly declared geology majors with an introduction to spreadsheet use and then to expect them to use this tool in subsequent courses. Figure 1 illustrates one of many applications of spreadsheets in an undergraduate exploration geophysics course. Time-distance field data from a reflection survey are entered into the spreadsheet, and all other parameters are calculated by equations entered into the spreadsheet by students.

Essential to the productivity software concept are the increasingly sophisticated graphics capabilities of personal computers. Figure 2 is a graph of the time-distance data contained in Figure 1. The graphing program has determined the equation of the line of best-fit for the data. These data were copied from the spreadsheet, transferred directly to the graphing program, and plotted by selecting the desired graph type from a menu. Students now can prepare quality graphics so quickly, that such diagrams should be part of any presentation or paper. Figure 3 illustrates a geologic map created in a drawing program as part of a homework assignment in structural geology.

However, these software packages, by themselves, are not sufficient. It is crucial for the software to be easy to use and to follow a well-defined user interface so that a number of different applications can be mastered by the casual user. Finally, the ability to transfer text and graphics among various applications is paramount as is the capability for the user to move quickly from one application to another. Larger memory configurations and the beginning of multitasking capabilities, coupled with ease of use and transferability, combine to create a truly powerful working environment that can be configured to the needs of the individual user.

Although the productivity software concept is applicable equally to professionals, it truly heralds a new era in undergraduate education. In our increasingly complex world, it is more-and-more essential to train problem solvers and good communicators. Geology majors should be trained routinely to use a range of computer software tools: a word-processor, a database program, a spreadsheet program, an extremely powerful and versatile graphing program, and a computer-assisted design (CAD) program. Because of the user-interface concept, students can learn how to use these programs in short order and it will be easy to return to a program they have not used for a period of time. In short, they will view these programs as *resources* (similar to the hand calculator) to be used whenever possible to increase their own productivity.

I now plan courses and assignments so as to take advantage of these resources in helping the students learn how to use them to solve problems on their own and how to use the resources to prepare more professional reports than previously possible. In their wordprocessed reports, these students routinely include computer-generated graphs based on calculations made in a spreadsheet and drawings of geologic relationships produced by a computer drawing program. Students in my undergraduate exploration geophysics course collect seismic refraction and reflection data in the field. They enter the data into a spreadsheet where the necessary computations are performed. The students then switch to a graphing program where the data are plotted, lines fit to data points, and slopes and intercepts determined. Following this step, slopes and intercepts are transferred back to

	1	2	3	4	5	6	7
	Swamp Road						
1	x (m)	t (ms)	t(s)	x-squared	t-squared		
2	--------	--------	--------	--------	--------		slope
3	50	0	0	2500	0		3.24E-07
4	53	0	0	2809	0		
5	56	0	0	3136	0		t-squared at x=0
6	59	54.5	0.0545	3481	0.00297025		0.0018
7	62	55.5	0.0555	3844	0.00308025		
8	65	56.5	0.0565	4225	0.00319225		velocity (m/s)
9	68	57	0.057	4624	0.003249		1757
10	71	57.5	0.0575	5041	0.00330625		
11	74	60	0.06	5476	0.0036		depth (m)
12	77	61.5	0.0615	5929	0.00378225		37
13	80	62.5	0.0625	6400	0.00390625		
14	83	63.5	0.0635	6889	0.00403225		
15							
16							
17	x (m)	t (ms)	t(s)	x-squared	t-squared		
18	--------	--------	--------	--------	--------		slope
19	50	0	0	2500	0		5.94E-07

Figure 1. Spreadsheet utilization in undergraduate exploration geophysics course. Time-distance field data from reflection survey are typed into spreadsheet. All other parameters are calculated by equations entered into spreadsheet by students.

the spreadsheet program where depths and velocities are calculated. After the necessary formulae initially have been placed into the spreadsheet, the complete process from entering data into the spreadsheet to a final solution takes a maximum of five minutes for each data set. Because this process is so rapid and straightforward, students are more willing to be concerned with uncertainties in the data by investigating alternative geologic interpretations and to discuss the limits of analysis in their reports.

Not only has a solution been determined in this short amount of time, but a first-rate data table and graph also are available. The next step is to switch to a drawing program where the geology is diagrammed. Finally, the geologic drawing, time-distance curves, and spreadsheet data are cut and pasted into a word processor where the final report is generated and printed on a laser printer complete with figures that rival those in most journals. This is the essence of productivity software.

Graphics-Linked Database Software

Graphically oriented database programs now are available that allow records to be assigned to graphics objects. It is possible, for example, to enter a geologic map into a database and to plot sample locations or sites where orientation data have been collected. Sample data can be searched for selected characteristics, and samples possessing these characteristics can be highlighted on the map, thereby possibly defining useful trends or patterns. Figure 4 illustrates a diagram of geologic relationships from a local field area created in a database that is capable of representing information in pictures. Students

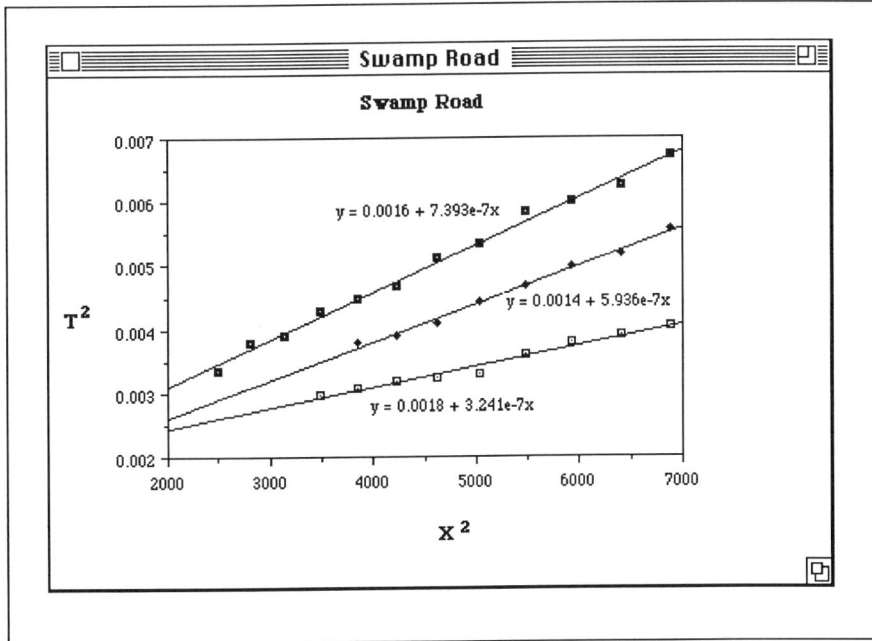

Figure 2. Graph of time-distance data contained in Figure 1. Graphing program has determined equation of best-fit for data.

have plotted sample locations (solid black triangles), stations where numerous joint orientations were measured (thick x's), and sample locations from areas pervaded by shear zones (boxes with zig-zag lines). Each of these locations has associated with it a database record containing all information collected in the field or determined from laboratory measurements. Digitized images (such as a photomicrograph) can be included in the record. Any information type contained in this database can be selected as criteria for a search. Locations satisfying these criteria will be highlighted on the map; all other portions of the graphics image will be presented in a light gray so that the highlighted locations clearly are distinguishable. In today's world of ubiquitous computers, is it not good pedagogy to include database instruction as an integral part of any field methods course? Planning data collecting activities for field and research projects in order to make maximum use of database capabilities should be a skill imparted to all geology majors. In a growing number of geology departments, this skill is viewed to be just as desirable as learning stereographic net techniques and using x-ray diffractometers.

As screen resolution continues to increase and as personal computers become ever more powerful in terms of available memory and database searching capabilities, databases linked directly to graphics images will become the norm in working with field data and in undertaking geologic research. For maximum utility, however, these programs also must function as productivity software. Information gathered from a database search must be transferable to other programs, such as a graphing program, in order to continue the

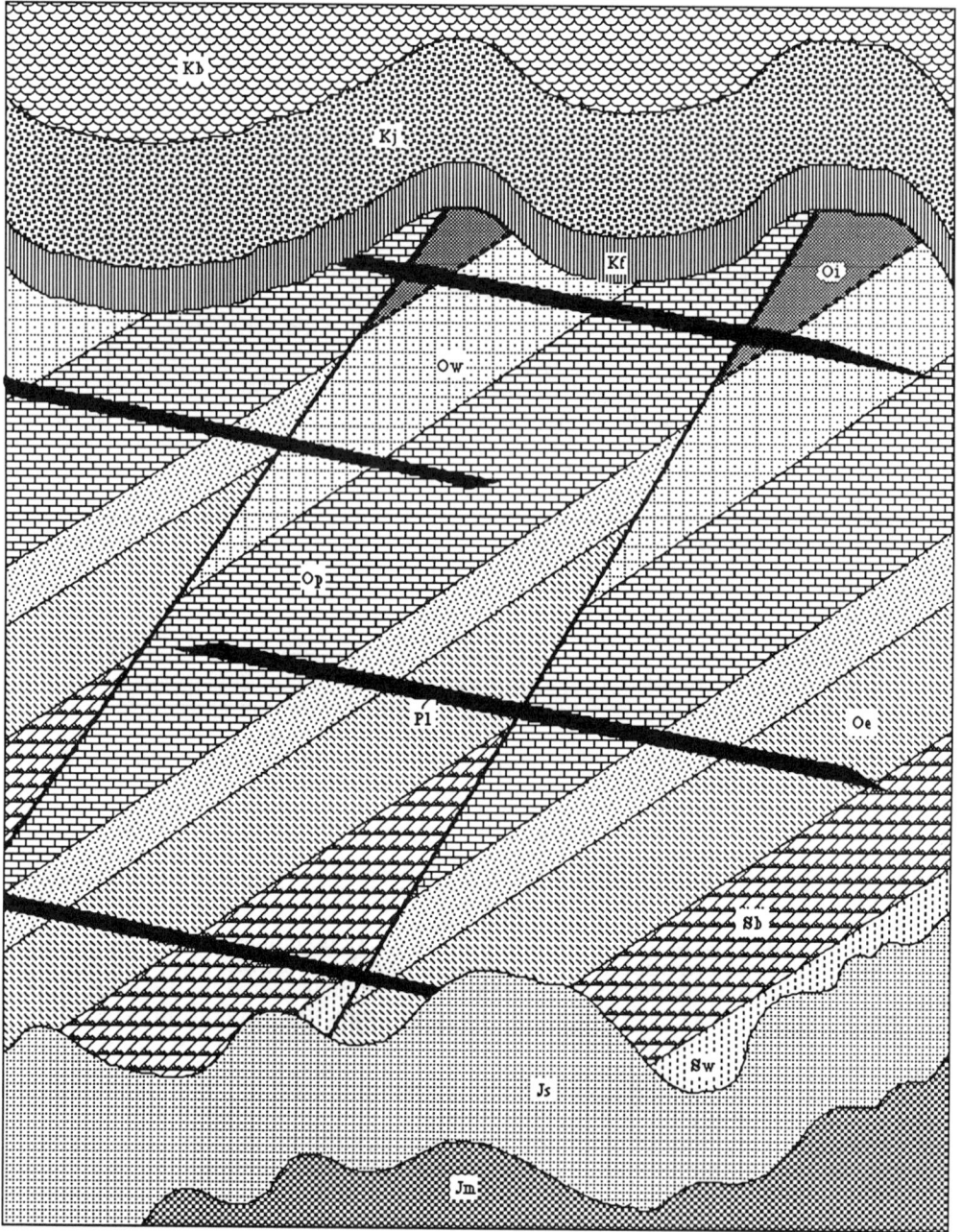

Figure 3. Geologic map created in drawing program as part of homework assignment in structural geology.

analysis. Graphics displays and the subsequent highlighting of information meeting search criteria must be of sufficient quality for laser printing and use in publishing research results.

Investigatory Simulation Software

Possibly the most far-reaching of the four categories of educational software discussed in this chapter is the category of investigatory simulation software. As defined here, this type of simulation is intended to mimic geologic research and field investigations rather than modeling or recreating geologic processes. Of course, the more powerful personal computers now available are capable of handling programs that simulate a specific process, such as the response of an aquifer to multiple wells pumping at differing rates. Such programs are useful and provide detailed graphics displays of computation results. They will grow more powerful and will become more abundant as versions become available that will execute in reasonable times on routinely available machines.

With the increased memory and graphics display capability of personal computers, another type of simulation becomes more attractive for use in geologic education. This type of simulation mimics an actual geologic investigation. The most familiar example is the standard "oil-exploration" program which has been available for a number of years on machines ranging from mainframes to small personal computers. However, this approach now can achieve greater levels of sophistication. Figure 5 is a screen display from such a program that is available through COGS. During the exploration of a selected area, a student is free to employ a wide range of geologic tools including geologic mapping, magnetics, gravity surveying, seismic reflection, and drilling to any desired depth.   The results of selecting two of these tools (drilling and seismic reflection) are illustrated in Figure 5. Additional procedures (Fig. 6) permit data obtained through tool use to be presented for interpretation (the stratigraphic correlation window). Help windows always are available as are information screens summarizing exploration activities, finances, leasing arrangements, and other necessary operations. All operations follow user-interface guidelines. As multiple windows can be present on the screen at one time and can be moved and overlapped, the student has great flexibility in handling data and making decisions. Students especially appreciate the appropriate graphics that appear when a well encounters an oil reservoir.

A wide range of field investigations are amenable to this type of treatment. An even more exciting version of this approach is what could be termed the geologic "adventure game." Most personal computer users are familiar with adventure games and must be impressed with the new breed of these games where a player points and clicks with a mouse in order to investigate objects, collects objects and information about them, and has a seemingly infinite choice about where to go and what to do. Consider a variant of this game where the haunted mansion is a field area, the rooms are outcrops, and the various objects are rock samples, orientation measurements, and mineralogical or chemical analyses. A properly constructed game could introduce a student to important field techniques, organizing principles, and decision-making.

As expert systems become more commonplace in the research environment, the strategies developed for such systems will carry over to the investigatory simulation. Truly sophisticated and intelligent geologic "adventure games" will become available with situations and outcomes dependent upon how an instructor defines a task and upon decisions elected by the student.

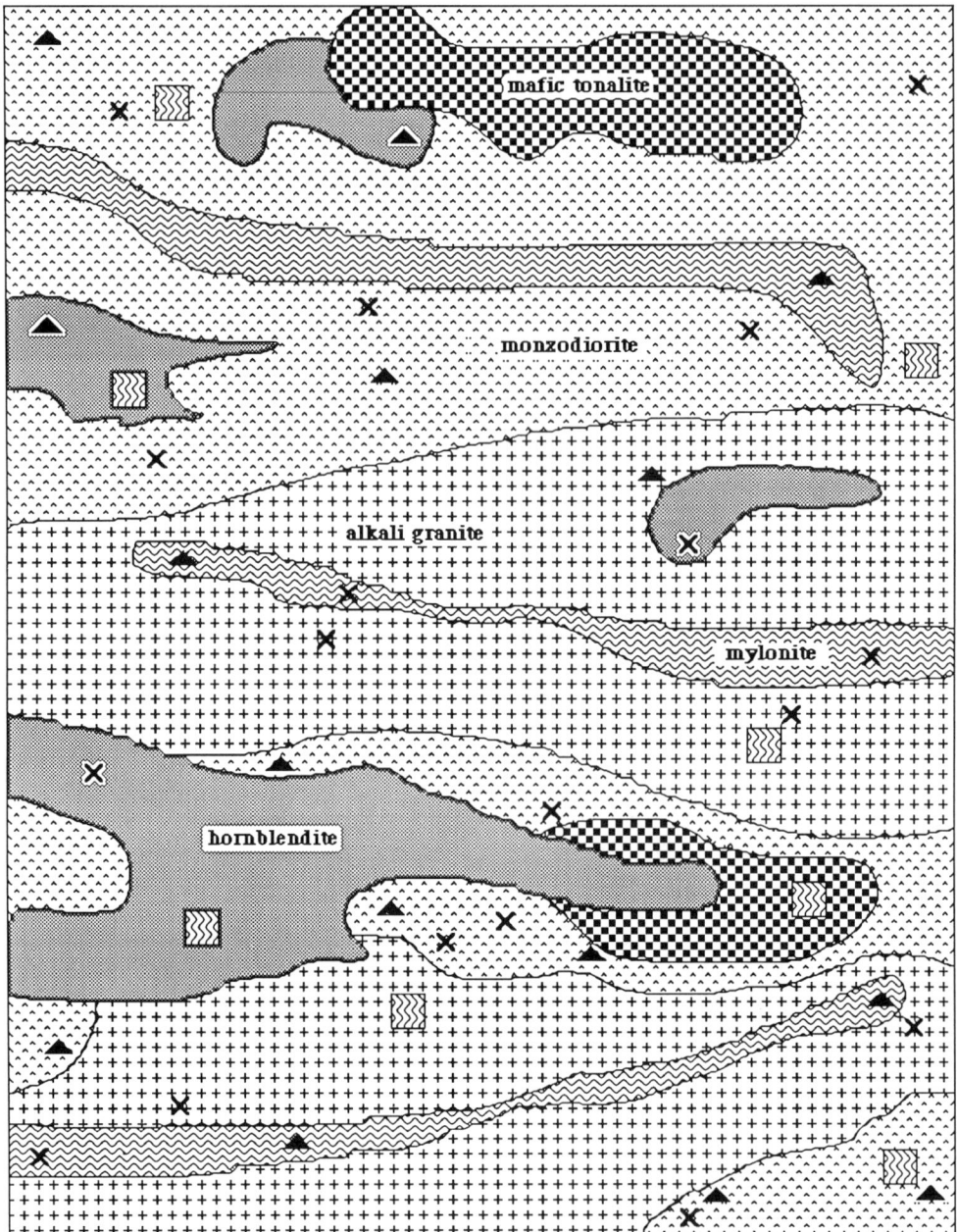

Figure 4. Geologic relationships from field area created in database that is capable of linking information to graphic images. Sample locations (solid black triangles), orientation stations (thick x's), and sample locations from areas pervaded by shear zones (boxes with zig-zag lines) are entered by students.

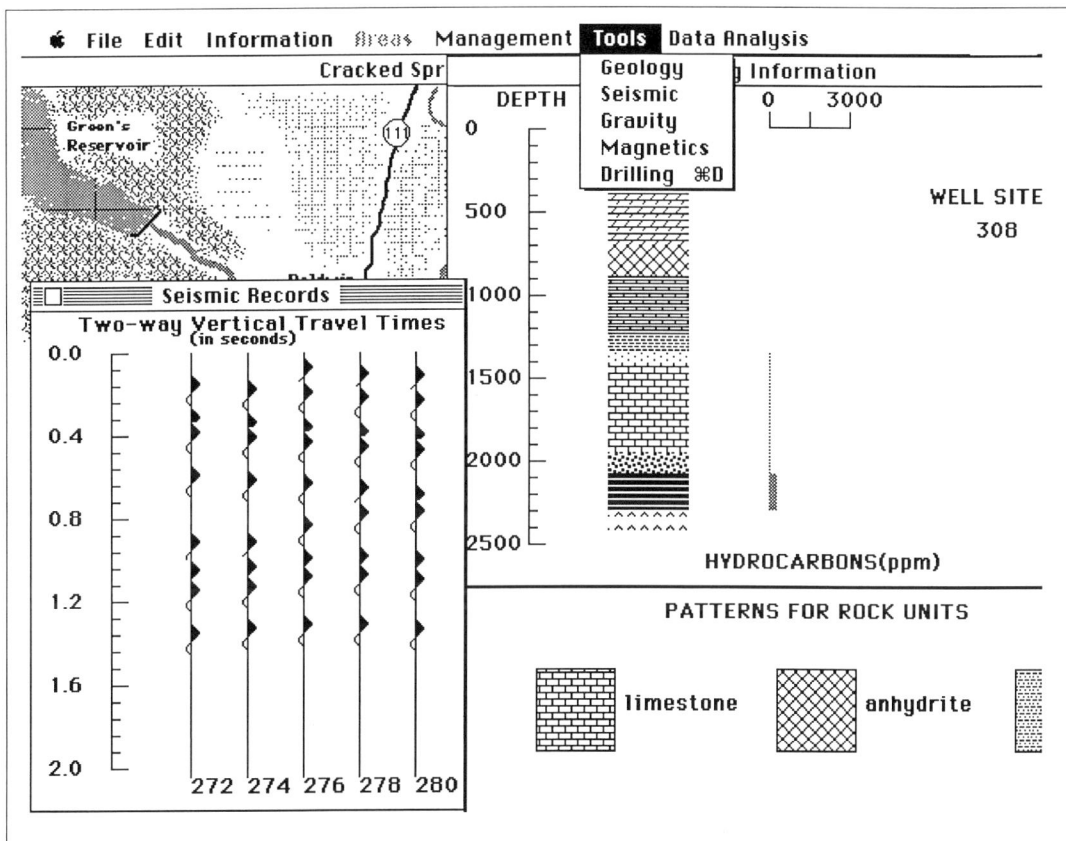

Figure 5. Display from oil-exploration program illustrating sampling of information that can be present on computer screen at one time.

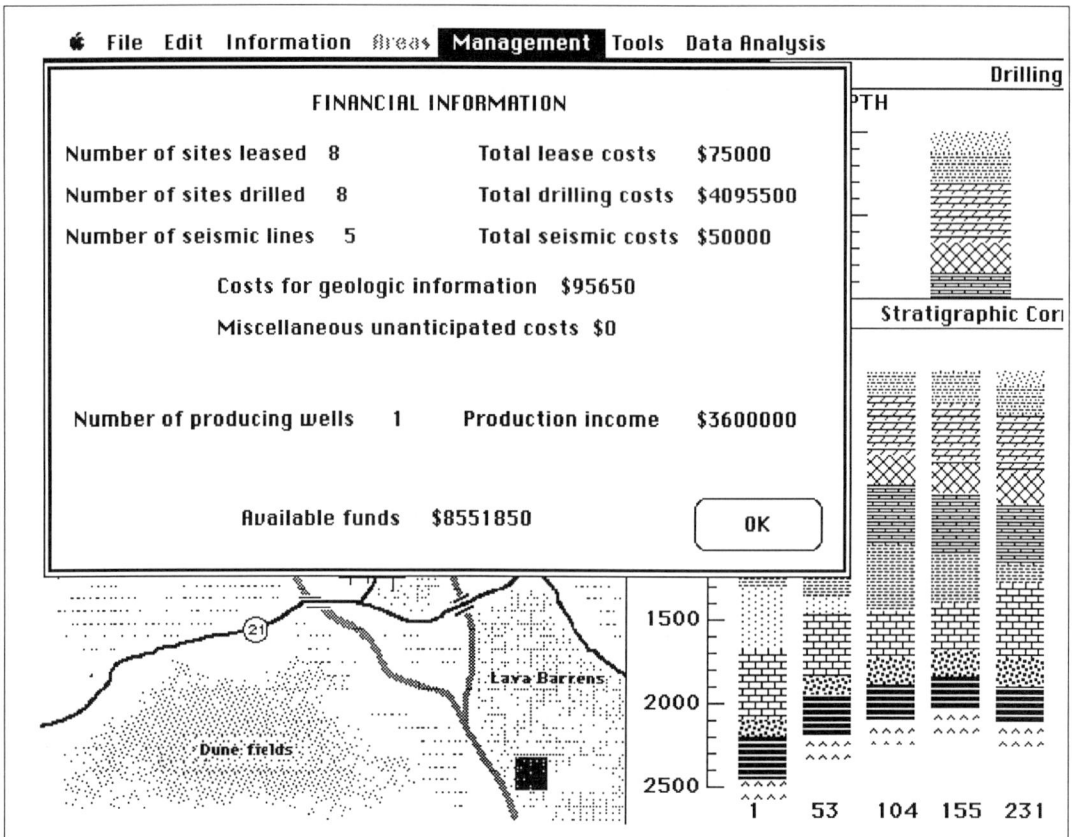

Figure 6. Another display from oil-exploration program illustrating additional
information available in program.

Random-Access Lecturing Software

Perhaps the most exciting aspect about what lies ahead in geologic education is the hardware and software available and being developed that will support what may be categorized as "random-access lecturing." A combination of powerful and versatile outliners, hypertext-like software, interactive videodisks supported by on-line indexes, image-acquisition devices, and multitasking together create a resource base that will permit educators to store a wealth of information pertaining to a specific topic.

For example, a portion of a videodisk would contain material related to earthquakes: slides of faults, slides of various earthquake phenomena (sand blows, building damage, offset features, landsliding, etc.), maps of fault zones, shaking intensity maps, film clips of actual earthquakes, diagrams of subduction zones, seismographs, seismograms, and precursors. All of this and more could be located instantly and projected if needed. Hypertext-like software would provide the way to select a particular visual or subject and would provide a host of other options which support any subject under discussion. A computer using multitasking also would have available powerful outliners to assemble and organize the main points of a discussion under way. In such an environment, the dominant mode of classroom activity no longer will be the formal lecture but rather a free-flowing quest for knowledge, guided by the instructor and supported by a range of organizational and display tools and materials.

Many courses in geology depend heavily on visuals, as they must, but each lecture then is confined to fairly set topics. Hence, there is an inherent resistance to following student insights or questions too far afield from the topic at hand. Such an atmosphere may suppress a free-flowing, intellectually stimulating investigation. In a classroom configured for random-access lecturing, the earthquake discussion can ebb and flow according to student questions, insights, and interests. Visuals always will be available to support the topic of the moment, outliners can help organize what has been discussed, and multitasking and print-spooling can provide an accurate record of the major points covered during discussion. Few limits are present, and realizing this, both students and instructor can approach a given topic from a broader perspective, with fewer constraints, and, not surprisingly, with more interest and enthusiasm.

SUMMARY

What lies ahead? Certainly all the trends mentioned here will continue. Personal computers will assume more-and-more of our computer workload, and the larger machines will be used only for specific tasks. The utility of commercial software will continue to increase. Commercial software now exists that makes it possible to prepare small geologic maps, place the map file on a disk, take the disk to a commercial printer, and have a finished product that is identical in quality to maps in journals. Graphics capabilities will continue to improve. Current color monitors will be supplanted by monitors that display two pages of material in color at high resolution. These will be augmented by laser printers, scanners, and copying machines— all of which will provide color capabilities. Within five years it finally will be possible to use fully the color capabilities of our computers to prepare colored materials for class use and for professional needs. Ease of use surely will continue to evolve to our benefit. Multitasking capabilities will improve and productivity software will become even more useful. Instructors still will engage in lecturing activities, but the technological support will transform the atmosphere and activities in the traditional classroom into a vastly different situation than what is usual today. Hopefully, geologists with programming interests and expertise will organize in order to establish standards and take maximum advantage of these and future opportunities.

The majority of the software mentioned in this chapter is commercial software written for a broad audience but useful for geologists and many other educators. Such software will become more powerful and other useful software not yet imagined will become available and used. However, geologists will have to develop special packages in order to derive maximum benefit from computer capabilities. Investigatory simulations and hypertext packages for use in random-access lecturing must be developed by geologists or by geologists working in concert with programmers. These packages must be at a level of sophistication similar to commercial packages, must follow the same user guidelines, must be able to share information with other applications, and must function in a multitasking environment. This requires considerable programming skills that likely are beyond the majority of most. Agencies providing funds for geologic education must be willing, therefore, to provide such programming support if the possibilities outlined in this discourse are to be realized fully.

Our science is oriented so visually and depends to such a great extent on communicating graphics information that we truly are in the midst of a revolution. Virtually every day brings new announcements of software or hardware advances of one type or another. These increasingly powerful capabilities provide and will continue to provide unlimited possibilities for the way courses are taught and structured. Geoscience education in the 1990's surely will be different than it is today.

STRANA: A Macintosh Computer Program for the Representation and Statistical Analysis of Orientation Data in Structural Geology

Massimiliano Barchi
Perugia University

Fausto Guzzetti
I.R.P.I., National Research Council, Perugia

ABSTRACT

The program STRANA follows, step-by-step, a simple but sufficiently complete procedure for the analysis of orientation data in structural geology, that can be applied easily to various geologic problems. STRANA integrates, in a single program, several different functions for the interactive management of orientation data files and the graphical representation and statistical analysis of these data.

The program is written in Microsoft[1] BASIC for an Apple Macintosh[1] computer and follows entirely the Macintosh style (use of the mouse, menu bars, etc.). It does not require any particular output device: good quality graphic and text output can be obtained on either an Imagewriter[1] or a Laserwriter[1] printer, and assembled in paper-files for the collection and the presentation of data.

Examples of the application of the program to two problems in structural geology also are presented.

INTRODUCTION

Several computer programs are available in literature to perform graphic, geometric, or statistical analysis of orientation data (Schuenemeyer, Koch, and Link, 1972; Williams, 1980; Cooper and Nuttal, 1981; Woodcock and Naylor, 1983; Griffis, Gustafson, and Adams, 1985). The program STRANA, an acronym for STRuctural ANAlysis, is aimed specifically at the analysis of orientation data in structural geology on a personal computer. The program integrates several different functions in a single package intended to:

- interactively manage files of orientation data;
- prepare graphic representations of orientation data on

---

[1] Trade names and trademarks in this paper are for descriptive purposes only, and imply no endorsement by the authors.

equal-area and equal-angle nets, as well as on rose
diagrams;
• perform statistical analyses of a distribution of orientation
data using an eigenvalue method.

The program, written in Microsoft BASIC 3.0 for an Apple Macintosh personal computer
and compiled using Microsoft BASIC Compiler 3.0, follows entirely the user-friendly
Macintosh style (use of windows, menu bars, mouse etc.), and it is completely
error-trapped.

The full listing of the program exceeds 2000 lines, and the compiled version requires 157
Kbytes of memory. STRANA runs on any Macintosh with at least 512 Kbytes and does not
require any particular output device. It will output good quality graphics and texts on
either an Imagewriter or a Laserwriter printer. The program can process a variable
amount of data (up to 10,000) depending on the available computer memory. The
possibility to work with a large number of files is facilitated by the use of an external
drive or a hard disk.

The graphic functions implemented in the program allow the plotting of orientation data
on either equal-area or equal-angle stereonets. Stereographic projections are tools in
several fields of the Earth sciences (e.g. structural geology, mineralogy, seismology,
sedimentology, geomechanics, etc.), and are used to represent easily three-dimensional
data in two dimensions. The literature on the possible applications of stereonets is vast
and it is not the purpose of this report to review it.

Several statistical techniques have been developed to analyze a sample of orientation
data. Probably the most used one is for the construction of contour diagrams. This
method consists of the counting of the number of data that fall on a grid cell of specific
size and shape, moved systematically across an equal-area stereonet. The results are
dependent on the grid size and shape as well as on the contouring technique.

For unimodal and cylindrical distributions the determination of the fabric shape and
strength can be achieved more effectively using an eigenvalue method (Mardia, 1972;
Woodcock, 1977). This technique, based on the assumption that the orientation data are
vectors of unit length in a three-dimensional space, is the basis for all the statistical tests
implemented in the program STRANA.

PROGRAM ORGANIZATION

For the construction of the program STRANA we followed, step-by-step, a simple but
sufficiently complete methodology for the analysis of orientation data in structural
geology (Fig. 1).

The first step in a structural-geology analysis is the definition of the geologic problem
which depends on two main factors: the overall characteristics of the area to be studied
and the goals that have to be pursued. This leads to the definition of the working tools,
and in particular allows one to decide which structural features have to be measured in
the field and their relative importance.

Correspondingly, in the use of the program STRANA, the first step is the preparation of a
number of code files. A code file is a list of up to nine different structural features, whose
order reflects their relative importance and their structural relationships.

A studied planning of the structural features that have to be used is important for the
success of a structural analysis. In our experience a careful definition of one or more code
files is essential for an effective elaboration of the data.

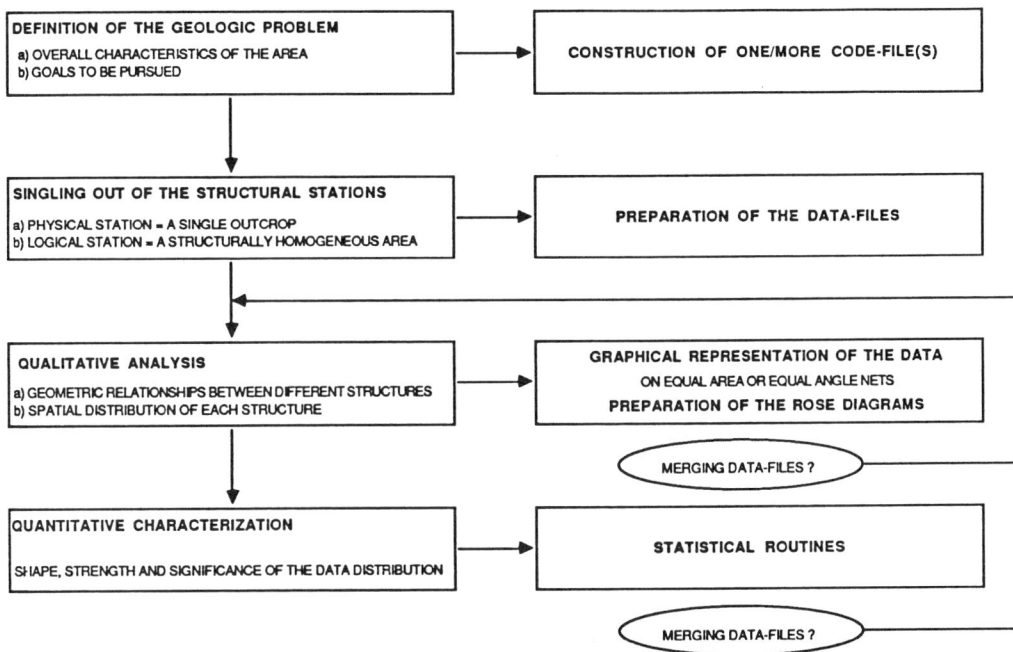

```
┌────────────────────────────────────────┐        ┌────────────────────────────────────┐
│ DEFINITION OF THE GEOLOGIC PROBLEM     │        │                                    │
│  a) OVERALL CHARACTERISTICS OF THE AREA │───────▶│ CONSTRUCTION OF ONE/MORE CODE-FILE(S)│
│  b) GOALS TO BE PURSUED                 │        │                                    │
└────────────────────────────────────────┘        └────────────────────────────────────┘

┌────────────────────────────────────────┐        ┌────────────────────────────────────┐
│ SINGLING OUT OF THE STRUCTURAL STATIONS │        │                                    │
│  a) PHYSICAL STATION = A SINGLE OUTCROP  │───────▶│   PREPARATION OF THE DATA-FILES    │
│  b) LOGICAL STATION = A STRUCTURALLY     │        │                                    │
│     HOMOGENEOUS AREA                     │        │                                    │
└────────────────────────────────────────┘        └────────────────────────────────────┘

┌────────────────────────────────────────┐        ┌────────────────────────────────────┐
│ QUALITATIVE ANALYSIS                    │        │ GRAPHICAL REPRESENTATION OF THE DATA│
│  a) GEOMETRIC RELATIONSHIPS BETWEEN      │───────▶│   ON EQUAL AREA OR EQUAL ANGLE NETS │
│     DIFFERENT STRUCTURES                 │        │ PREPARATION OF THE ROSE DIAGRAMS    │
│  b) SPATIAL DISTRIBUTION OF EACH         │        └────────────────────────────────────┘
│     STRUCTURE                            │              ( MERGING DATA-FILES ? )
└────────────────────────────────────────┘

┌────────────────────────────────────────┐        ┌────────────────────────────────────┐
│ QUANTITATIVE CHARACTERIZATION           │        │                                    │
│ SHAPE, STRENGTH AND SIGNIFICANCE OF THE │───────▶│       STATISTICAL ROUTINES         │
│ DATA DISTRIBUTION                        │        │                                    │
└────────────────────────────────────────┘        └────────────────────────────────────┘
                                                          ( MERGING DATA-FILES ? )
```

Figure 1.   Simple procedure to analyze orientation data in structural geology. On left simple methodology for analysis of orientation data in structural geology; on right implementation of each step of the methodology in program STRANA.

The second step in the structural analysis is the singling out of the structural stations that can be either physical or logical stations. In our view a physical station is a single outcrop, or a well-defined restricted area (Fig. 2B) whereas a logical station is a larger area that is thought to be homogeneous structurally (Fig. 2A).

The user of the program first will prepare several data files, each containing orientation data referring to a single structural station. It is important to use the same code file for data referring to a single work. This will allow the comparison of different structural stations (data files) and eventually their merging in new larger data files.

The third step is a qualitative, structural-geology analysis. In the use of the program STRANA this is obtained by the graphical representation of orientation data on either equal-area (Schmidt) or equal-angle (Wulff) stereonets as well as the preparation of rose diagrams. The visual analysis of these representations allows the user to understand the geometric relationships between different structures and to analyze the spatial distribution of each structure.

The fourth step is a quantitative, statistical characterization of the data distribution. The program offers a series of statistical routines that allow the determination of the principal axes of the ellipsoid of the data distribution, and the performing of different tests intended to identify the shape, the strength, and the degree of significance of each distribution (Flinn, 1962; Woodcock, 1977).

This step allows a statistical quantification of the characteristics of any distribution of orientation data, thus permitting comparisons between different distributions, and eventually the identification of structurally homogeneous areas.

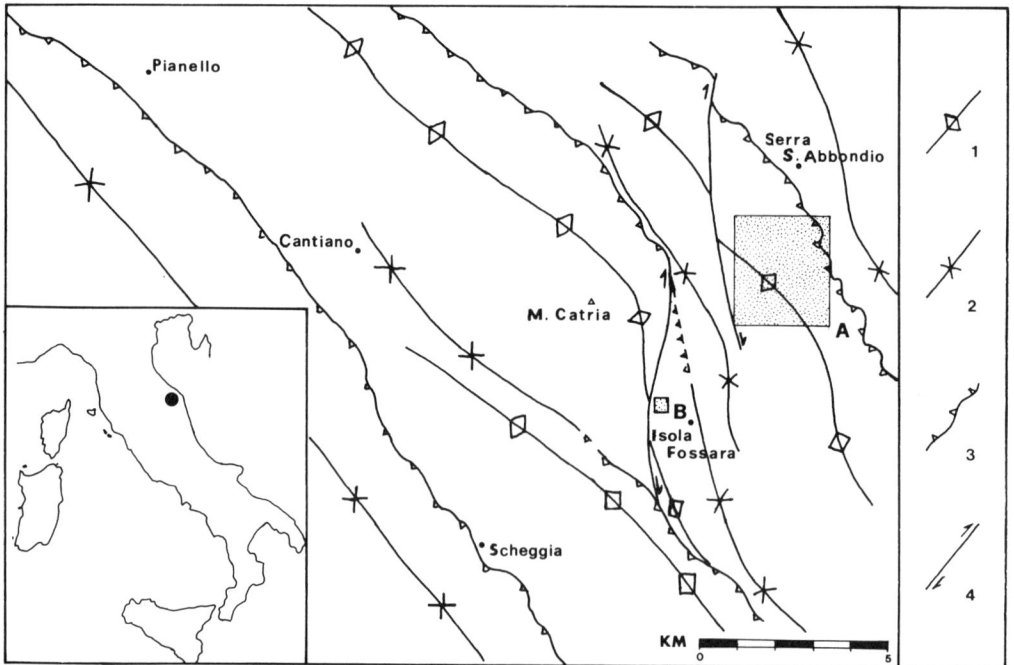

Figure 2. Schematic structural map of Serra S. Abbondio area (northern Apennines - Central Italy) (from Barchi, Lavecchia, and Minelli, 1988, modified). Axes of major folds: 1 - Anticlines; 2 - Synclines; Major shear planes: 3 - Thrusts; 4 - Transpressive Faults. A - Structural station (A) - Logical station: External Anticline (see Fig. 3A) B - Structural station (B) - Physical station: Isola Fossara Shear Zone (see Fig. 3B).

OPTIONS OF THE PROGRAM STRANA

The program STRANA has a modular structure with a main menu that calls several modules that perform different tasks. This structure facilitates the preparation and the correction of the subroutines contained in each module, and makes the implementation of other modules easy. The main menu of the program consists of six options: File, Database, Print, Draw, Utilities, and Statistics.

Under the File option the user can either change the maximum number of data that the program can process (depending on the available computer memory), select the output device, or exit the program and return to the Macintosh desktop.

The Database option allows the interactive management of orientation data files as sequential text files. The user can create and correct either data files or code files, can display data on the screen, and merge different data files. The files created under this option are sequential text files with a header and a body. The header contains general information (i.e. site name, operator, map, lithology, date), the total number of data, and the list of codes. The body of the file contains the list of the orientation data each with its corresponding code. Data are entered from the keyboard. General information and codes are entered as simple strings of predetermined maximum length, whereas orientation data of planar or linear structures are entered as strings of 5 digits: the first 3 for strike and the last 2 for dip.

The Print option allows the printing of data files in tabular form, on the currently selected output device - either an Imagewriter or a Laserwriter printer. All the printouts produced under different options by the program can be used to assemble a paper file containing all the information, the data, and the graphical and statistical elaborations prepared for any particular area.

The Draw option performs graphical and limited statistical analysis of orientation data. The user can plot selected data as poles on an equal-area net (Schmidt) or as planes and poles on an equal-angle net (Wulff) (lower hemisphere projection). It also is possible to prepare full (360°) or half (180°) rose diagrams, selecting the radius of the diagram, the sampling step, and the data dip range. A complete table containing the number and the percentage of data in each sampling class is provided. An optional moving average function also is implemented. Equal-area and equal-angle nets as well as rose diagrams can be sent to the printer with or without the accompanying data table.

The Utility option performs two multipurpose functions used in structural geology: the calculation of the intersection between two planar structures (e.g. the intersection (110) between the bedding plane (S0) and a cleavage (S1)) and the verification of the exact attitude of a linear structure lying on a plane (e.g. a slickenside on a fault plane). In each situation data are read from an original data file and after the calculation, a new file is created. This file contains either the orientation of two planes and the attitude of their intersection, or the corrected attitude of the lineation and its relative plane. The new data files then can be used by any option of the program STRANA.

The Statistics option performs a number of statistical calculations on selected orientation data. Data are first presented to the user on the screen, where those that have to be analyzed can be selected easily using the mouse.

For each data sample the program computes the orientation tensor (Scheidegger, 1965) and defines the ellipsoid of the data distribution using an eigenvalue method (Watson,1966). The best-fit point (or plane) then is identified as the point (or the plane) where the concentration of data is highest: the centroid, or first momentum, of the data sample.

The program also performs a series of auxilliary tests, intended to establish the shape, the degree of clustering, the strength and eventually the randomness of the data sample. These tests are:

   - the Watson test of randomness (Watson, 1966). The test is based on the value of R, the vectorial sum of the orientation data considered as unit vectors. This value is compared against tabled values at 5% and 10% confidence levels to determine the randomness of the data sample (Watson, 1966; Shuenemeyer, Koch, and Link, 1972);

   - the computation of the precision and the radius of the cone of confidence (for 99%, 95%, and 90% confidence levels) parameters (Fischer, 1953; Watson, 1966). These are two different estimators of the tightness of the data sample around the mean value based on the vectorial sum R. Watson and Irving (1957) and Schuenemeyer, Koch, and Link (1972) have shown that these parameters can be used successfully to test if two clusters have the same precision, based on F-test statistics.

It is important to point out that the tests based on the R statistics (i.e. Watson test, precision, and cone of confidence) should be used only for distributions that are clustered around a single point. Distributions that are girdle shaped, or clustered around two points, are misinterpreted by these statistics (Shuenemeyer, Koch, and Link, 1972). Consequently these tests are optional and the user can decide to perform them based on a qualitative visual analysis of the data set.

- the calculation of the parameters for a two axis ratio plot, better known as the Flinn diagram (Flinn,1962; Woodcock,1977). This is a logarithmic eigenvalues ratio plot that provides a convenient representation of the shape and the strength of a sample of orientation data. It is prepared calculating the logarithm of the ratio of the medium to smallest (S2/S3) and largest to medium (S1/S2) normalized eigenvalues. These ratios are used then as x and y on a bilogarithmic graph. Two parameters characterize each distribution:

K = log(S1/S2)/log(S2/S3) is the shape factor, an estimator of the clusterness or girdleness of the sample;

C = log(S1/S3) is the strength factor, an estimator of the degree of tightness of the data sample around the main value, either a point for clusters or a plane for girdles;

-the Woodcock and Naylor test for randomness (Woodcock and Naylor, 1983). The test is based on the value of the parameter C. This value is tested against tabulated data for 99%, 97.5%, 95%, and 90% confidence levels to assess the randomness of the data sample. The program performs this test only for poorly strengthened samples (C<2) and for 1000 data or less.

EXAMPLES OF GEOLOGIC APPLICATION

In this section we present the application of the program STRANA to two problems in structural geology: the determination of the hinge of a major fold and the analysis of the distribution of the structures most frequently present in a shear zone. The examples are taken from an on-going research project of the Structural Geology Group of Perugia University, on the history of the deformation of the Umbria-Marche Apennines (Central Italy) (Barchi, Lavecchia, and Minelli,1988).

The Umbria-Marches Apennines is a thrust and fold belt, nucleated during the Mio-Pliocene. A Meso-Cenozoic sedimentary cover, consisting of massive carbonates (Lower Lias) and pelagic limestones and marls (Middle Lias-Lower Pliocene), overlying incompetent Triassic evaporites, was deformed by buckling, resulting in an en-échelon set of northeast converging anticlines with associated minor folds and extensive axial-plane cleavage. Thrust planes and associated north-south right-lateral and east-west left-lateral shear zones ("lateral ramps") accomplished the progressive deformation. A schematic structural map of the studied area is shown in Figure 2.

The first example (Fig. 3A) shows the application of the program STRANA to the determination of the axial trend of a major anticline.

We first prepare a code file containing the most widespread structural features (i.e. bedding planes, axial plane cleavage, minor fold hinges, etc.) that are measured in the field. With this code file we prepare a data file containing all the orientation data collected in a 5 km^2 structurally homogeneous area (i.e. a logical station).

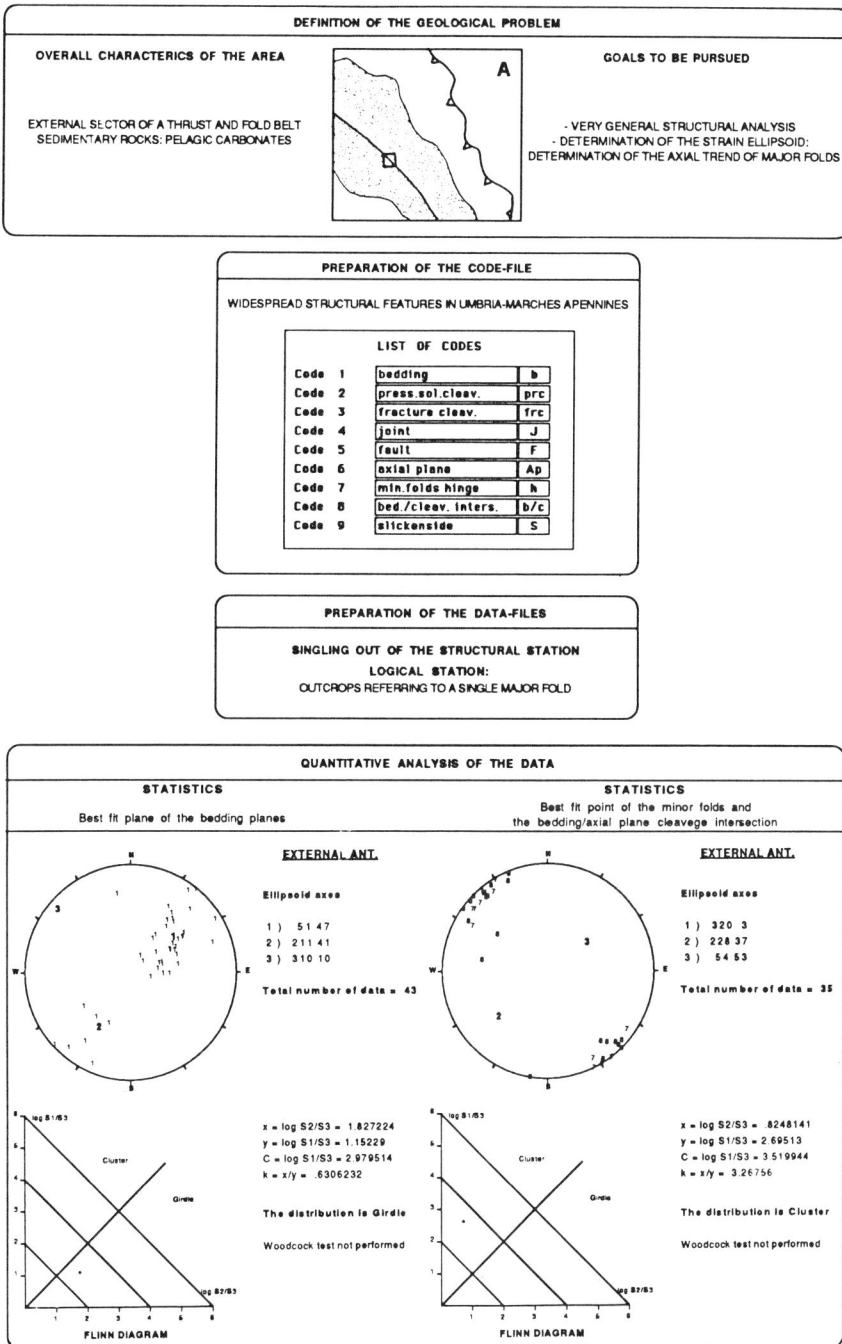

Figure 3A. Application of program STRANA in structural-geology analysis of thrust and fold belt. General purpose analysis of structurally homogeneous area, finalized to determination of axial trend of major fold.

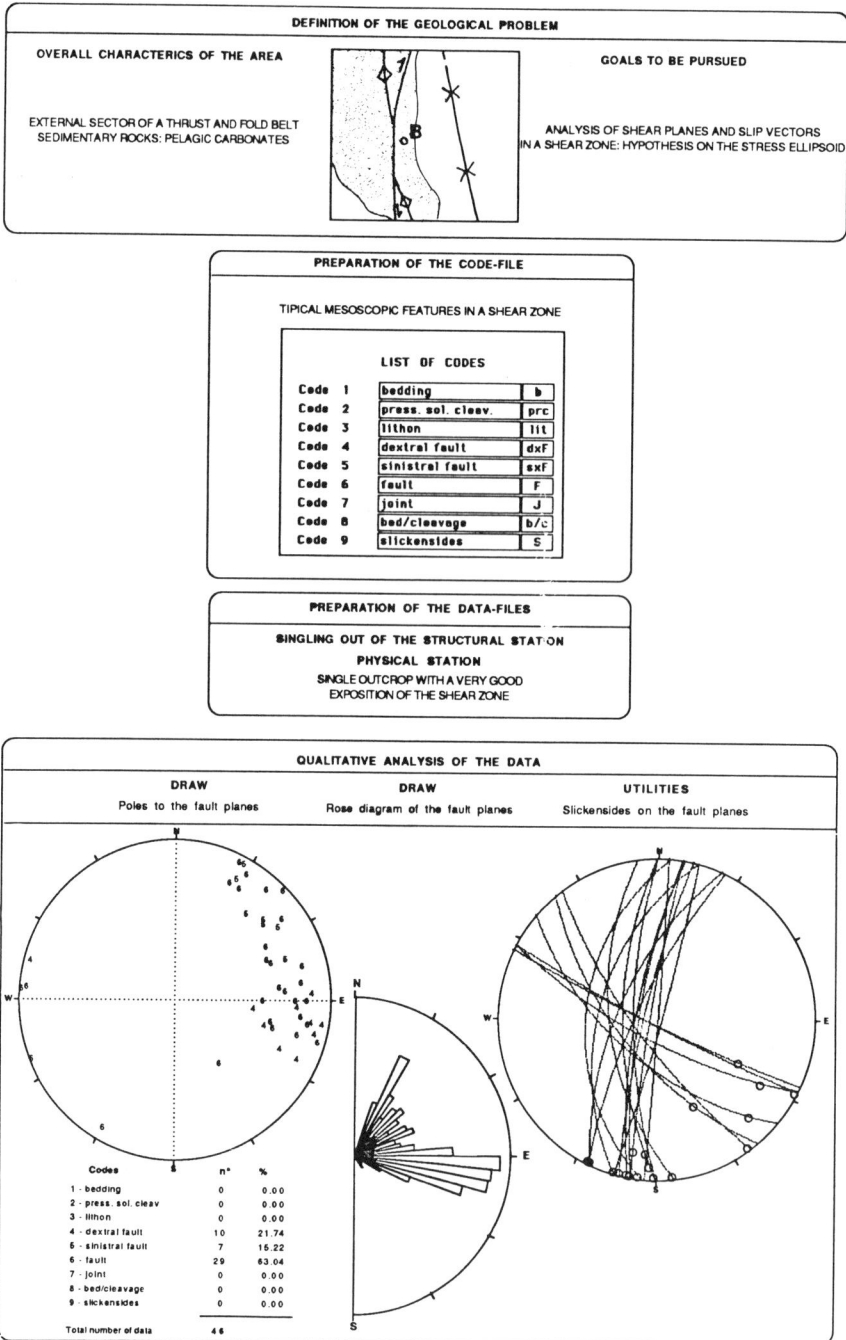

Figure 3B. Different graphical representations of fault planes measured in single outcrop in transcurrent shear zone. All pictures in "windows" of qualitative/quantitative analysis of data are examples of printer outputs produced by program STRANA.

The hinge trend now can be determined in two different ways (Ramsay, 1967):

- as the pole to the best-fit plane of the distribution of the bedding planes, projected as poles on a Schmidt net (π diagram);

- as the best-fit point of all the intersections between bedding planes and axial-plane cleavages.

In the first situation the data distribution is girdle and the hinge trend coincides with the smallest axis (a3) of the distribution ellipsoid. In the second example the data are clustered around two opposite points at the edge of the stereonet, and the hinge trend coincides with the largest axis (a1) of the distribution ellipsoid.

The two different methods give comparable results (N 31010 and N 32003, respectively).

The second example (Fig. 3B) shows the application of the program to the analysis of some of the structural features present in a shear zone. The problem is the determination of the spatial distribution and the geometric relationships of the minor faults, and their associated slickensides, present in a major shear zone.

Once a proper code file has been provided we prepare a data file containing all the structural features (fault planes, shear planes, lithons, slickensides, etc.) that are measured in a well-exposed outcrop (i.e. a physical station) of the shear zone.

We then perform a first qualitative analysis of the data distribution, plotting all the fault planes, as poles, on a Schmidt net. This representation is not clear enough and we therefore prepare a half-circle rose diagram that clearly indicates a bimodal distribution of the attitude of the fault planes, suggesting that two different families of faults were present in the area. Finally, we project both the fault planes and the related slickensides on a single stereonet. The analysis of this last representation shows the existence of a conjugated system of strike slip faults, a right-lateral one striking N10 and a left-lateral one striking N130.

FINAL REMARKS

The program has been developed as a working tool for the Structural Geology Researchers of Perugia University for the effective collection and rapid analysis of a large number of orientation data during the study of the Umbria-Marche fold belt. The program organization therefore is linked strictly to a well-defined methodology for the analysis of orientation data. Nevertheless, we are confident that the program, because of its flexibility and ease-of-use, can be applied successfully in other fields of the Earth sciences where stereographic projections and statistical analysis of orientation data are needed.

The principal advantages and characteristics of STRANA can be summarized as follows:

- the program is written in BASIC and has a modular structure which facilitates the implementation of other functions as well as communication with other programs or databases;

- it is reasonably fast considering the relatively inexpensive hardware requirements. The program can plot, for example, an average of 25 data points per second as poles and 1–2 data points per second as planes;

- the error-trapped input procedure allows the preparation of virtually error-free files that can be used immediately in the graphic and statistical routines;

- it prepares good quality paper outputs.

ACKNOWLEDGMENTS

We wish to thank Prof. Federico Costanzo, Prof. Giusy Lavecchia, and Dr. Giorgio Minelli for reviewing drafts of this report and for their useful comments. We express our sincere thanks to the Structural Geology Research Group of the Earth Sciences Department of the University of Perugia for testing different versions of the program for several months. The research was supported by M.P.I. grants, 40% and 60% (G. Pialli) and by C.N.R., partially under the special project G.N.D.C.I.

REFERENCES

Barchi, M., Lavecchia, G., and Minelli, G., 1988, Sezioni geologiche bilanciate attraverso il sistema a pieghe umbro-marchigiano. 2 - La sezione Scheggia-Serra S. Abbondio: submitted for publication to Boll. Soc. Geol. Italy.

Cooper, M.A., and Nuttal, D.J.H., 1981, GODPP: programs for the presentation and analysis of structural data: Computers & Geosciences, v. 7, no. 3, p. 267-285.

Fisher, R.A., 1953, Dispersion on a sphere: Proc. Royal Soc. London, v. 217, p. 295-305.

Flinn, D., 1962, On folding during three-dimensional progressive deformation: Quart. Jour. Geol. Soc. London, v. 118, pt. 4, p. 385-433.

Griffis, R.A, Gustafson, S.J., and Adams, H.G., 1985, PETFAB: user-considerate FORTRAN 77 program for the generation and statistical evaluation of fabric diagrams: Computers & Geosciences, v. 11, no. 4, p. 369-408.

Mardia, K.V., 1972, Statistics of directional data: Academic Press, New York, 357 p.

Ramsay, J.C., 1967, Folding and fracturing of rocks: McGraw-Hill Book Co., New York, 568 p.

Scheidegger, A.E., 1965, On the statistics of orientation of bedding planes, grain axes, and similar sedimentological data: U.S. Geol. Survey Prof. Paper 525-C, p. 164-167.

Schuenemeyer, J.H., Koch, G.S. Jr., and Link, R.F., 1972, Computer program to analyze directional data, based on the methods of Fisher and Watson: Jour. Math. Geology, v. 4, no.3, p.177-202.

Watson, G.S., 1966, The statistics of orientation data: Jour. Geology, v. 74, no. 5, pt. 2, p. 786-797.

Watson, G.S., and Irving, E., 1957, Statistical methods in rock magnetism: Monthly Notices Royal Astron. Soc., Geophys. Suppl., v. 7, p. 289-300.

Williams, J.D., 1980, ROSENET: a FORTRAN IV program for production of rose diagrams compatible with Gould or Calcomp plotting facilities: Computers & Geosciences, v. 6, no. 1, p. 95-103.

Woodcock, N.H., 1977, Specification of fabric shapes using an eigenvalue method: Geol. Soc. America Bull., v. 88, no. 9, p.1231-1236.

Woodcock, N.H., and Naylor, M.A., 1983, Randomness testing in three-dimensional orientation data: Jour. Struct. Geology, v.5, no.5, p.539-548.

Data and Information Management for a Hydrogeologic Study of a Waste-Disposal Site

Michael J. Bellotti
Olin Chemical Group

ABSTRACT

Hydrogeology is a discipline that usually is applied to environmental investigations at waste-disposal sites. The information objectives from such investigations are diverse and necessitate the development of a variety of data types. The following is an example of the development of such a data-management system, showing how data specific to waste-site investigations can be developed and managed using microcomputer technology and available software.

INTRODUCTION

The Task At Hand

Hydrogeologic data developed in investigations of waste-disposal sites are diverse and span a number of subdisciplines within the field of geology. The geoscientist conducting the site investigation must handle diverse data types such as geologic data, hydrologic data, and water-quality data.

The data diversity is compounded because the mode of data acquisition is varied. Therefore, the format in which data are received is unique to each mode of collection. For example, data can be numerical and differ through space such as measurements of depth to watertable at a series of measurement points; they can be numerical and differ through time such as measurements of watertable fluctuation with time in a drawdown/recovery well test; or they can be nonnumerical such as stratigraphic data from a borehole profile.

Because the data are so diverse, it is crucial that the geoscientist manage his or her data in a systematic and organized manner. The first objective of such data management is to have access to specific segments of the database. Then the data can be transformed, evaluated, and displayed graphically, resulting in a well-reasoned evaluation of the data.

The transformation, evaluation, and display of data is an orderly process, a sequence of logical phases in the progression of a site evaluation. Initially, the data must be processed, that is transformed from the original format at time of collection to a format suitable to the needs of the geoscientist. The data are evaluated then by numerical calculations or by the development of graphic plots representing physical or temporal patterns. Finally, the results of data evaluation are displayed visually to communicate and document the geoscientist's conclusions and complement his/her written evaluation.

The following sections demonstrate how the microcomputer is used to implement a data-management system and facilitate the transformation, calculation, and display phases of a hydrogeologic study of a waste-disposal site.

Approach to Data Development

Data development for a waste-disposal site investigation is a function of the physical geometry of the site. Waste-disposal sites can be regarded as point sources of contamination, although the "point source" is usually an area in which chemical waste or chemical product is contained. This area can range from the size of an underground gas station storage tank to an area of buried waste as large as several acres.

Once the waste area is defined, the geoscientist can proceed with development of a site-study plan and consequently, a strategy for development of a database. Data are developed primarily from discrete points at which boreholes are installed. From these boreholes, the geoscientist must extract the data required to define the hydrogeology at the site of investigation, determine the extent of any contamination that might have entered the groundwater and determine the hydrologic properties of the aquifer so that the direction and rate of migration of any contamination can be determined.

A data point in the form of a borehole yields a variety of data types. As the hole is being drilled, samples of unconsolidated sediments and of the underlying bedrock are collected, logged, and retained for possible analysis later. A monitoring well is installed within the borehole to a depth predetermined by the geoscientist. The well provides data on groundwater quality when a groundwater sample is collected. The sample represents groundwater quality at the time of collection at the location of the well screen within the physical dimensions of the aquifer. Later, the same well can be tested for hydrologic response to pumping and provide data by which the geoscientist can calculate properties of the aquifer such as hydraulic conductivity (permeability) and storage coefficient.

Although the geoscientist has the opportunity to extract a variety of data from a single borehole, he/she must plan the investigation to include an adequate number of data points to characterize the site hydrogeology and the distribution of any contamination moving from it.

The study plan for this type of investigation consists of a number of data points, each yielding the type of data as described. Unlike scientists in other disciplines in which study is conducted on finite, predefined systems, the geoscientist studies the Earth. The geoscientist must study a system which is so large that he/she cannot readily get an overview of it. Therefore, the geoscientist's task is to interpolate between a number of discrete spatial and temporal observations and draw conclusions based on geologic theory and his/her own experience.

Quality Control and Quality Assurance

At the planning stage, any proposed data acquisition should be accompanied by a description of how the data will be evaluated. This compels the geoscientist to think through the program and to eliminate any unnecessary data acquisition. The data acquisition plan should include provisions for quality control (QC) and quality assurance (QA).

Quality control consists of developing a consistency of procedures and protocols during field sampling and measurements, so that data accuracy is maximized. For example, when a monitoring well is sampled, all stagnant water residing within the well casing must be purged before the actual sample is collected. This precaution allows the sample to be drawn from groundwater flowing into the well from the aquifer at the time of collection so that subsequent chemical analyses are indicative of groundwater quality at a given location, at a given time.

Quality assurance consists of designating a representative number of samples to be tested for reproducibility of laboratory procedures. This reproducibility can be checked by designating a percentage of samples to be analyzed in replicate (the number is usually 5 to 10 percent of the total number of samples taken), or, in the situation of more esoteric chemical analyses, by specifying QA checks specific to procedures followed in the analytical laboratory.

Of course, the single most important QA/QC procedure is communication. If the geoscientist is working as part of a team, as is usually the situation, it is imperative that he/she specify, by written procedure or by prefield work meetings, what data are required and the specific methods by which samples are collected or measurements are made. Effective communication with other team members will ensure the geoscientist that the data he/she gets are accurate and are of the correct type.

SEQUENCE AND ORGANIZATION

Data management is accomplished by characterizing data into sequential types and handling each type in a way suitable to its function in the sequence of evaluation, as the hydrogeologic study progresses.

A spreadsheet program rapidly performs numerical transformations of field-generated data to create a formal database. Next, data sorting and repetitive calculations are performed for study objectives such as aquifer permeability. Water quality and piezometric data are tabulated into discipline-specific discrete databases for augmentation with successive sampling or measurement episodes. Creation of these databases allows the user to enter selected blocks of data into applied software or to present appropriate selected data in tabular format suitable for inclusion in a final report. Applied software is used to convert data to traditional geologic tools which the geoscientist can interpret to draw conclusions pertinent to the study. Geologic cross sections are plotted using bar-graph software. Trends in water quality and watertable level are plotted using line-graph software. Piezometric surface and stratigraphic surfaces are plotted by contouring programs. Finally, a wordprocessing program is used to develop and edit preliminary drafts of report text.

DATA TRANSFORMATION

Rationale and Modes of Data Transformation

The process of transforming data requires literally that the data change form. The format of data which the geoscientist receives from the field usually is not in the format which he/she requires for calculations or evaluations. The required transformations of data can be subtle and mundane, but are necessary nevertheless for data comparison and organization. There are a number of reasons why transformation might be required.

Usually, data are developed in the field by people other than the geoscientist. At times the geoscientist must use historic data developed by anonymous persons during, perhaps, a preliminary phase of the investigation or by an agency or firm to which he/she does not have direct technical input. Such data might require transformation as simple as conversion of units. For example, depth to watertable must be measured in a series of wells from the top of the well casing, which has been surveyed as a reference point. These depths to watertable might be taken in feet and inches. The geoscientist usually requires this type of data to be recorded in feet and tenths of feet so that reference to the surveyed reference point at the top-of-well casing is facilitated. Survey elevations are done in feet and decimals thereof.

Although this transformation requires a simple conversion of feet and inches to feet and tenths, conversion of perhaps 50 to 100 measurements for a series of wells is time-consuming and subject to inadvertent error when done by hand calculator. Furthermore, if data are taken during several measurement episodes, the number of calculations increases dramatically. A fast and accurate method of transformation is required.

Different data sources present data of a given type in different formats. Analytical reports from different laboratories usually express their results in different units. Analyses of organic compounds may be expressed in parts per million (ppm) or parts per billion (ppb). Again, unit conversion is required. Furthermore, for analytical data from different sources, the physical page format is rarely the same. The geoscientist usually must transform the data by reorganizing them to a common page format.

For some measurements, the method and units of data recording in the field are selected because of convenience, with knowledge that transformation will be required at a later time. For example, when a pumpdown/recovery test is run to determine hydrologic properties of an aquifer, it is crucial to make a series of water-level measurements corresponding to brief time intervals. The field readings usually are timed by stopwatch, although automated methods are available and will be discussed later. The stopwatch readings are recorded in minutes and seconds. However, the graphic plot which the geoscientist uses to evaluate the aquifer recovery requires a plot in minutes and decimal fractions thereof. Furthermore, a method for evaluation of watertable recovery requires plotting recovery versus a time ratio. Thus, the time measurements taken in the field must be converted to other units and ratios before the geoscientist can evaluate the data.

Nonnumeric or seminumeric data also must be transformed. These data can be stratigraphic in nature. The geoscientist must evaluate drilling logs containing qualitative data and synthesize these data so that they are suitable for quantitative evaluation. These data transformations require the greatest degree of modification and the most subjective judgement. For example, boring logs describe unconsolidated sediments or rock strata in terms of a geologic description and the depth interval at which the sample was taken. The geoscientist must review empirically the field data (i.e.. the drilling logs), note common properties, and identify distinct strata. This is an application of geologic judgement. However, to proceed a step farther to quantitative stratigraphic evaluation, the geoscientist must transform the stratigraphic data by quantifying strata thickness and depth at each boring location. He/she then must summarize these data in spreadsheet form for ready reference later when they are evaluated and graphically represented as cross sections, strata surface contour plots, or isopachous maps.

Approach to Data Transformation

The first step in data transformation is empirical inspection of the "raw" data. This step allows the geoscientist to screen out bad data and determine what type of transformations is required. Bad data are defined as data which are incomplete or which are of quantitative value significantly different from the rest of the data group. Here, significance is defined by the judgement of the geoscientist and his/her experience with the type of data and the specific location. However, the geoscientist must be careful to not eliminate data which do not "seem" right, as these data might be indicative of trends which might be important later in the study.

Bad data may be the result of simple measurement error; inclusion of data from a different subset of measurement points such as a well screened at a different depth than the wells being measured; or to some change in field conditions such as a damaged well casing from which reference measurements are made. Screening out bad data in an early phase reduces the potential for unexplained data anomalies during the evaluation phases of the study.

Organization and Spreadsheet Software

Transformations can be performed with spreadsheet software. Examples are shown on LOTUS 123. LOTUS 123 is available and a powerful tool for implementation of the database management described in this chapter. However, a variety of other spreadsheet software is available.

A working knowledge of the mechanics of the spreadsheet can be developed by investing perhaps a day or two in following the reference manual and tutorial examples. Afterward, specific usages and functions can be learned by consulting the reference manual or others experienced in the software you have selected.

At this point, the geoscientist will develop a system of organization for his/her data which he/she will use for the remainder of the study. It is helpful from an organizational point of view to set up specific files for each data category within the study. For example, such discreet files might consist of piezometric data, water-quality data, stratigraphic data, permeability test data, summary files on monitor-well screening and construction, and other topics specific to the location or type of study.

Methods of Data Transformation

At this point, the geoscientist will have established the type of data that has been collected and the objectives of those data. He/she will have determined the initial format, the format to which the data will be transformed and the calculations required to perform the transformation.

The following are methods for transforming several types of data generated in the field during hydrogeologic investigations at waste-disposal sites.

Piezometric Data

These data are used to determine the direction of groundwater movement, both laterally and vertically, in the aquifer under investigation. The data points are measurements of depth to watertable in monitor wells, measured from a surveyed reference point, usually the top-of-well casing.

Data usually are transformed in two steps. The first step is conversion of units from the units measured in the field to the feet and tenths units comparable to the surveyed casing top elevation. The second step is subtraction of the transformed depth to watertable measurement from the reference elevation to arrive at the elevation of the groundwater table.

A spreadsheet (Table 1) can be organized in columnar form, with well identification and casing-top elevation in the two left columns. The actual field readings then will be entered for each well in the next column(s). An arithmetic conversion will be "programmed" to convert the field readings to feet and tenths, with the resultant data being placed one column to the right of the field data. The "programmed" calculation can be entered for the first line (i.e., the first data point) and then copied to repeat the calculation for the remainder of the lines (i.e., data points). All columns should be labeled appropriately with units specified within the label.

The second transformation then is performed by "programming" the subtraction from reference elevation in a similar manner. The resultant column shows the elevation of the watertable, that is the piezometric level, at each well location. Later, these data will be plotted spatially and contoured. The piezometric data table can be abbreviated for display in a formal report, with several of the calculation columns removed. The "programming" of arithmetic formulae is specific to the commands of the spreadsheet. However, standard protocols and symbols for FORTRAN formulae generally apply.

Table 1. Piezometric data.

```
          SITE NO. 1  PIEZOMETRIC DATA:        SEP 1-88

          PUMP RATE:            7.60 gpm
          PUMP USED:        IW-8 and IW-9
          PUMPING WELL:     IW-11

          WELL      T.O.P            DEPTH TO DEPTH TO     WT
                    ELEV.               WT      WT        ELEV
                    ft-msl              in      ft        ft-msl
          ====      ========         ============================
          GW13       91.01            127.60   10.63       80.38
          GW14       89.19            110.00    9.17       80.02
          GW16       91.56            117.10    9.76       81.80
          GW23       90.93            136.00   11.33       79.60
          B2         98.44            133.80   11.15       77.29
          B3         88.20            111.20    9.27       78.93
          B5         90.39            117.90    9.83       80.57
          B7A        89.03             98.60    8.22       80.81
          B15        90.07              n        n          n
          B17        90.67            112.20    9.35       91.32
          JOB        88.20             dry      dry        dry
          JOD        89.99            126.00   10.50       79.49
          JOE        89.78            121.20   10.10       79.68
          JOF        89.93             dry      dry        dry
          JOG        90.10            122.00   10.17       79.93
          JOH        90.51            131.10   10.93       79.59
          JOI        91.73             dry      0.00       91.73
          IW1        89.28            119.90    9.99       79.29
          IW2        89.60            126.10   10.51       79.09
          IW3        89.35            120.10   10.01       79.34
          IW4        89.6F            118.10    9.84       79.82
          IW6        88.98            111.70    9.31       79.67
          IW7        90.09            123.90   10.33       79.77
          IW8        89.89            125.10   10.43       79.47
          IW9        89.74            122.20   10.18       79.56
          IW10       90.34            127.60   10.63       79.71
          IW11       89.92            139.40   11.62       78.30
          P1D        89.77            122.20   10.18       79.59
          P3         88.94             95.10    7.93       81.02
          P4         88.73            114.80    9.57       79.16
          P5         88.03            116.00    9.67       78.36
          Twelve in  89.78            121.80   10.15       79.63
```

Groundwater Quality Data

Groundwater quality data transformations consist of unit conversions for consistency between lab reports and of format transformations to facilitate data comparisons by location and temporally. Both conversions require empirical inspection of data sheets from lab or field (pH and specific conductance are measured directly in the field for best results) and transposition by the geoscientist to a suitable format. Raw data can be entered directly onto the spreadsheet and ordered by sampling location or date of sampling by sorting functions within the spreadsheet. Sorting can be performed by specifying a data range and subsequent criteria for sorting, such as alphabetically, numerically, etc. The data thus transformed can be used later in the evaluation phase of study by graphing time trends for specific parameters. The data also can be presented in tabular form in a formal report (Table 2).

Stratigraphic Data

Stratigraphic data are transformed by taking information from borelogs and summarizing specific parameters into tabular format in a spreadsheet. Borelogs are a formal documentation of field notes taken at the borehole as drilling takes place. A borelog contains a variety of data and observations concerning the stratigraphy and hydrology of the borehole, and, when a monitor well is installed, the as-built specifications of that monitoring well. Stratigraphic data are developed by sampling discrete intervals of sediment or rock by "split-spoon" soil sampler or by core barrel.

Table 2. Groundwater quality data.

```
                SITE NO. 5  TRENDS IN GROUNDWATER QUALITY

        COMPOUND:        XYLENE (ppb)

        WELL     MAR21-85 JUN26-85 OCT2-85  NOV19-85 MAR18-86 JUN19-86 SEP16-86
        ======== ======== ======== ======== ======== ======== ======== ========
        W1B          3        N       18       ND       ND        N        n
        W2A          4        9        5        9       ND       ND       ND
        W2B          8        5       18       ND       ND       ND       ND
        W3A         ND        5      DRY       ND       ND       ND       ND
        W3B         ND       ND        7       ND       ND       ND       ND
        W4A         ND      DRY      DRY       ND       ND        N       DRY
        W4B       5800     1600     9600        5     2400    10000    57000
        W9A          3       35      DRY       19       ND       ND       ND
        W9B         45       33      265        1       ND       ND       45
        W10A        71       11      DRY       46        3       18       37
        W10B         8       12      228       53       40       31       30
        W11A        ND        2        4       ND       ND       ND       ND
        W11B        ND       ND        7       ND       ND       ND       ND
        W12A        ND      DRY      DRY       ND       ND      DRY      DRY
        W12B         9      170      DRY        2       ND       ND       ND
        W13A        ND      DRY      DRY        8       ND      DRY       ND
        W13B         8       ND      266        2       ND       ND       11
        W14A         N      DRY      DRY      DRY      DRY      DRY      DRY
        W14B         N       75     3900      211     2900       22      240
        W15A         N       ND      DRY        2        4       ND       72
        W15B         N       29       10    18400        2       40       20
        W16A         N    18500    12500        3    27000    23000     6600
        W16B         N     1200    12800    25100    27000    30000    18000
        PW-1
        ND=not detected
        N=not sampled
```

Each discrete sample run is described and characterized by a field geologist on the borelog. The sum total of these descriptions comprise the stratigraphic information in the borelog.

In evaluating these borelogs, the geoscientist interprets and notes strata breaks and changes in lithologic character, and looks for consistencies and correlations between boreholes. This is standard geologic judgement. This judgement can be quantified by noting depths and thicknesses of the strata breaks which the geoscientist deems important, and tabulating them in spreadsheet form (Table 3). These data will be used later in the evaluation phase to plot cross sections and isopachous maps. This type of data transformation is the most subjective and engenders the greatest change in data type, that is from qualitative to quantitative.

Aquifer Characteristic Data

An integral aspect of hydrogeologic studies is the characterization of aquifer properties, specifically the aquifer's capacity to transmit or store groundwater. Permeability and storage coefficient calculations can be made using a response curve of aquifer drawdown or recovery with time. Drawdown/recovery curves are generated by field measurements made at a well during pumpage episodes.

Table 3. Stratigraphic summary data.

```
        SITE NO. 2 STRATIGRAPHIC SUMMARY

        BORING LOCATION: C-1                        wt @ approx 16 ft

        interval (ft)       lithology
        =============       =========
        x 0 to 8            f-vc sand
        x 8 to 20           gravel
        x 20 to 25          bedrock : gneiss
                            competent
                                                          C1
        =================================================================
                                                          C2
        BORING LOCATION: C-2                        wt @ approx 3 ft

        interval (ft)       lithology
        =============       =========
        x 0 to 4            f-m sand
        x 4 to 30           f gravel
        x 30 to 36          c gravel
        x 36 to 52          f gravel
        x 52 to 57          f-c grvl/sand
        x 57 to 62          c grvl/sand
        x 62 to 67          bedrock : gneiss
                            competent
                                                          C2
        =================================================================
                                                          C3
        BORING LOCATION: C-3                        wt @ approx 3 ft

        interval (ft)       lithology
        =============       =========
        x 0 to 3.5          vf sand
        x 3.5 to 6          m-c sand
        x 6 to 22           f-c sand
        x 22 to 28          gravel
        x 28 to 32          f-c sand/tr grvl
        x 32 to 44          f-c sand
        x 44 to 70          f-c sand/tr grvl
        x 70 to 76          f-c sand
        x 76 to 81          bedrock : gneiss
                            competent/several minor wea zones   C3
        =================================================================
```

The data developed during these aquifer tests can be recorded by convenient field methodology, because the tests require a series of rapid measurements with brief time intervals between measurements. The parameters of measurement are time and depth to watertable. The field readings usually are made with watertable probe and stopwatch, so data are recorded as feet to watertable and time in minutes and seconds. However, the response curve for a given aquifer test is a function of feet of drawdown (or recovery) of water level in the well relative to the initial watertable level, and of time in minutes.

For aquifer test data transformations and calculations, a more elaborate spreadsheet must be developed (Table 4). Time in minutes and seconds can be listed in the first two columns and the two columns combined to read as minutes and decimal fractions of minutes by simple arithmetic "programming" of a third column. Depth to watertable is then entered next to the corresponding time of measurement. A simple arithmetic subtraction in an adjacent column transforms this depth data to drawdown data by subtracting the depth from an initial watertable depth reading. Furthermore, if a drawdown test cannot be run to a point at which the outflow pump rate is in equilibrium with the inflow of formation groundwater, recovery is plotted as a function not of time but of a ratio of total time of test to time of recovery (t/t-prime). The recovery ratio is calculated easily using the spreadsheet function of performing multiple calculations. Hence, much tedious calculation by hand calculator is avoided as is the potential for inadvertent error caused by such repetitive calculations. The number of repetitive calculations can range from, as few as perhaps 25 to several hundred for more elaborate aquifer tests which monitor aquifer response in several monitor wells around a pumping well. The specific conditions under which each test is run also should be itemized and noted on the spreadsheet for facility of documentation.

The transformed data, that is time and drawdown, then are used in the evaluation phase of a project to plot aquifer-response curves (Fig. 1 and 2). The slope of these curves is entered into a formula by which permeability is calculated. The formula, although not complex, does have a number of components, each accounting for the field conditions of the aquifer test. The calculations required in implementing this formula, and the necessary units conversions, also can be performed rapidly by "programming" cells within the spreadsheet. The calculations also are presentable as report-quality tables to document the results and methods of the aquifer test (Table 5).

Data collection for aquifer tests has been automated and data can be recorded automatically using pressure transducers lowered into monitor wells. These transducers register differences in head pressure as the overlying column of water changes in height. The data are recorded in the field on a computerized data logger, and transferred to hard or floppy disk later. The data then can be transformed to the appropriate format for response-curve plotting. This equipment increases precision in the field, and could be cost-effective, depending on the frequency that the geoscientist would use it. Software programs currently on the market further automate the development of response curves and more detailed hydrologic formulae. While it is beyond the scope of this article to itemize this software, it should be noted that the geoscientist always should review empirically his/her data and get a "feel" for the data and its variabilities before turning over the process of evaluation to a software package.

Well-Summary Data

A summary of well-construction depths and specifications and the strata screened by each well is a good tool for the geoscientist. As with stratigraphic data, these data are transformed by itemizing information from borelogs. In addition to well construction data, study-specific information useful to the geoscientist also can be recorded. These summary sheets provide a useful overview of the overall monitoring program and are useful for documentation in a final report (Table 6).

Table 4. Aquifer drawdown/recovery data.

AQUIFER TEST DATA: SITE NO. 3 MONITOR WELLS

WELL B4
DEPTH 9.5 ft
LITHOLOGY: CLAY / CHERTY CLAY
PUMP RATE: 0.4 gpm

TIME incr. sec	TIME tot sec	TIME tot min	DEPTH TO W.T. ft	DRAW-DOWN ft	RECOVERY tot sec	RECOVERY tot min	t/tprime
0	0	0	4.075	0			
15	15	0.25	4.65	-0.575			
15	30	0.5	5.2	-1.125			
15	45	0.75	5.7	-1.625			
15	60	1	6.175	-2.1			
15	75	1.25	6.9	-2.825			
15	90	1.5	7.2	-3.125			
30	120	2	7.35	-3.275			
30	150	2.5	7.65	-3.575			
30	180	3	7.9	-3.825			
30	210	3.5	8.15	-4.075			
30	240	4	8.4	-4.325			
30	270	4.5	8.6	-4.525			
30	300	5	8.8	-4.725			
30	330	5.5	9.05	-4.975			
30	360	6	9.25	-5.175			
30	390	6.5	9.45	-5.375			
30	420	7	9.65	-5.575			
30	450	7.5	9.8	-5.725			
30	480	8	10.05	-5.975			
30	510	8.5	10.2	-6.125			
30	540	9	10.4	-6.325			
30	570	9.5	10.55	-6.475			
30	600	10	10.9	-6.825			
30	630	10.5	10.95	-6.875			
30	660	11	11.05	-6.975			
30	690	11.5	11.25	-7.175			
30	720	12	11.4	-7.325			
30	750	12.5	11.6	-7.525			
30	780	13	11.8	-7.725			
100	880	14.66666	11.5	-7.425	100	1.666666	8.8
93	973	16.21666	11.35	-7.275	193	3.216666	5.041450
125	1098	18.3	11.2	-7.125	318	5.3	3.452830
110	1208	20.13333	11.1	-7.025	428	7.133333	2.822429
86	1294	21.56666	11	-6.925	514	8.566666	2.517509
64	1358	22.63333	10.9	-6.825	578	9.633333	2.349480
58	1416	23.6	10.8	-6.725	636	10.6	2.226415
49	1465	24.41666	10.7	-6.625	685	11.41666	2.138686
51	1516	25.26666	10.6	-6.525	736	12.26666	2.059782
48	1564	26.06666	10.5	-6.425	784	13.06666	1.994897
45	1609	26.81666	10.4	-6.325	829	13.81666	1.940892
40	1649	27.48333	10.3	-6.225	869	14.48333	1.897583
42	1691	28.18333	10.2	-6.125	911	15.18333	1.856201
38	1729	28.81666	10.1	-6.025	949	15.81666	1.821917
42	1771	29.51666	10	-5.925	991	16.51666	1.787083
40	1811	30.18333	9.9	-5.825	1031	17.18333	1.756547
44	1855	30.91666	9.8	-5.725	1075	17.91666	1.725581
51	1906	31.76666	9.7	-5.625	1126	18.76666	1.692717
52	1958	32.63333	9.6	-5.525	1178	19.63333	1.662139
54	2012	33.53333	9.5	-5.425	1232	20.53333	1.633116
54	2066	34.43333	9.4	-5.325	1286	21.43333	1.606531
49	2115	35.25	9.3	-5.225	1335	22.25	1.584269
92	2207	36.78333	9.1	-5.025	1427	23.78333	1.546601
120	2327	38.78333	8.9	-4.825	1547	25.78333	1.504201
126	2453	40.88333	8.7	-4.625	1673	27.88333	1.466228
130	2583	43.05	8.5	-4.425	1803	30.05	1.432612
146	2729	45.48333	8.3	-4.225	1949	32.48333	1.400205
178	2907	48.45	8.1	-4.025	2127	35.45	1.366713
165	3072	51.2	7.9	-3.825	2292	38.2	1.340314
178	3250	54.16666	7.7	-3.625	2470	41.16666	1.315789
186	3436	57.26666	7.5	-3.425	2656	44.26666	1.293674
189	3625	60.41666	7.3	-3.225	2845	47.41666	1.274165
198	3823	63.71666	7.1	-3.025	3043	50.71666	1.256325
64	3887	64.78333	6.9	-2.825	3107	51.78333	1.251046
68	3955	65.91666	6.7	-2.625	3175	52.91666	1.245669
85	4040	67.33333	6.5	-2.425	3260	54.33333	1.239263
80	4120	68.66666	6.3	-2.225	3340	55.66666	1.233532
69	4189	69.81666	6.1	-2.025	3409	56.81666	1.228806
81	4270	71.16666	5.9	-1.825	3490	58.16666	1.223495
88	4358	72.63333	5.7	-1.625	3578	59.63333	1.217998
121	4479	74.65	5.5	-1.425	3699	61.65	1.210867
134	4613	76.88333	5.3	-1.225	3833	63.88333	1.203495
158	4771	79.51666	5.1	-1.025	3991	66.51666	1.195439
190	4961	82.68333	4.9	-0.825	4181	69.68333	1.186558
215	5176	86.26666	4.7	-0.625	4396	73.26666	1.177434
235	5411	90.18333	4.5	-0.425	4631	77.18333	1.168430

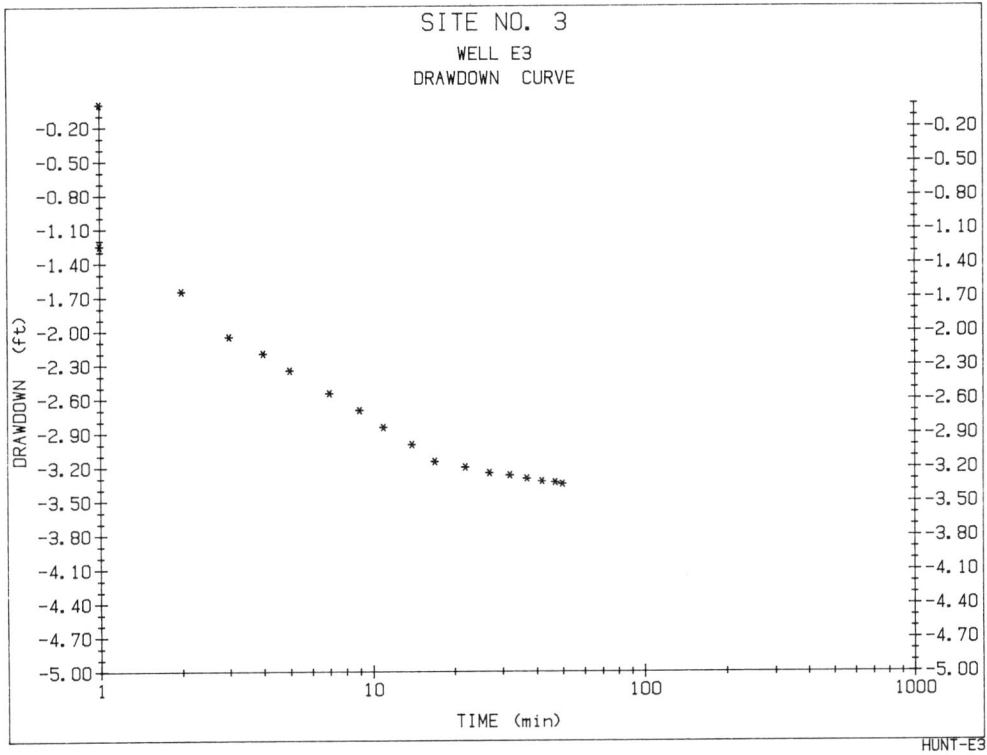

Figure 1. Aquifer drawdown curve.

Table 5. Permeability calculations.

SITE NO. 3
PERMEABILITY AND TRAVEL TIME CALCULATIONS

WELL	pump rate	delta s	Transmis -sivity	saturated thickness	PERMEABILITY			gradient at creek max	porosity	VELOCITY	lithology	WELL
	gpm		gpd/ft	ft	gpd/ft2	cm/sec	ft/day	ft/ft		ft/year		
A4	2.0000	12.2500	43.1020	14.0000	3.0787	0.0001	0.3965	0.0020	0.2000	1.4474	silty clay	A4
A4R	4.0000	1.0000	211.2000	5.0000	42.2400	0.0019	5.4405	0.0020	0.3000	13.2386	weathered limestone	A4R
B4	0.4000	15.0000	7.0400	8.0000	0.8800	0.0001	0.2800	0.0010	0.2000	0.5110	clay/ cherty clay	B4
B4R	0.5000	1.2500	105.6000	5.0000	21.1200	0.0010	2.7203	0.0010	0.3000	3.3096	limestone	B4R
C4	0.4000	7.0000	15.0857	13.0000	1.1604	0.0001	0.1495	0.0030	0.2000	0.8183	clay/ wea. limestone	C4
C4R	0.4000	6.2500	16.8960	5.0000	3.3792	0.0002	0.4352	0.0030	0.3000	1.5886	limestone	C4R
D3	10.0000	0.5000	264.0000	20.0000	13.2000	0.0006	1.7002	0.0100	0.2000	31.0279	fine-med sand	D3
E3	0.6000	1.7500	90.5143	25.0000	3.6206	0.0002	0.4663	0.0050	0.2000	4.2553	cherty clay	E3
E3R	0.4000	5.7500	18.3652	5.0000	3.6730	0.0002	0.4731	0.0050	0.3000	2.8780	limestone	E3R

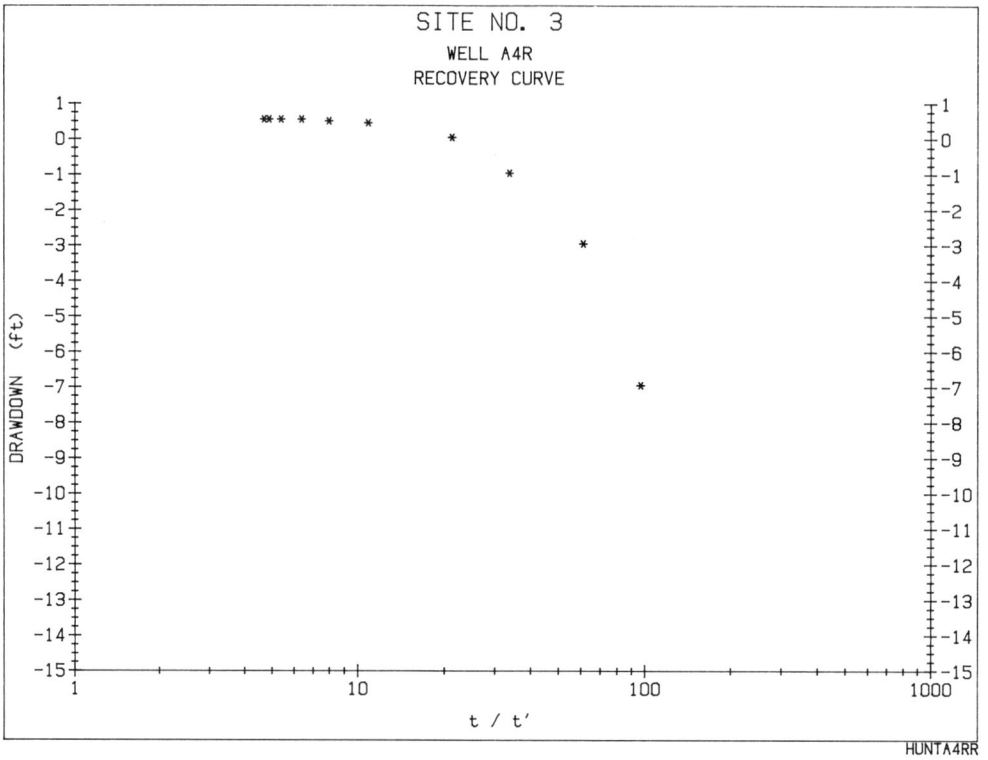

Figure 2. Aquifer recovery curve.

Table 6. Well summary data.

```
                    SITE NO. 4   WELL DETAILS


        WELL      DEPTH     SCREEN    PIPE      GROUND    BOTTOM
                            LENGTH    LENGTH    ELEV.     ELEV.
                  ft        ft        ft        ft-msl    ft-msl

        ========  ========  ========  ========  ========  ========
        P1              25        20         5       7.1     -17.9
        P2               8         5         3       6.4      -1.6
        P3               8         5         3       7.5      -0.5
        P4               8         5         3       6.2      -1.8
        P5               8         5         3       6.5      -1.5
        P6              13        10         3       9.6      -3.4
        P7              10         5         5       6.4      -3.6
        P8             8.8         5       3.8       6.4      -2.4
        P9               8         5         3       6.9      -1.1
        P10             13        10         3       6.5      -6.5
        P11           13.9        10       3.9       7.2      -6.7
        R1              11       7.5       3.5       8.7      -2.3
        R2              10       6.5       3.5       7.9      -2.1
        R3              10       6.5       3.5         6        -4
        R4              10       6.5       3.5       7.9      -2.1
```

Geophysical Data

Various geophysical techniques are used to aid in stratigraphic evaluation of site-specific data or to trace the lateral extent of highly ionized contaminant "plumes." Soil-resistivity measurements, for example, can be facilitated by use of spreadsheets, as they require repetitive calculations and transformations of field measurements to a usable database. Analogous to procedures as described, the database, in turn, is applied to a variety of methods to plot curves of resistance vs. depth below ground. Breaks in the curves can indicate changes in strata or in groundwater quality.

CALCULATIONS AND DATA EVALUATION

As noted in the previous section, calculations can be performed by "programming" formulae into cells of the spreadsheet. Sequences of calculations usually must be performed to synthesize data and arrive at conclusions pertinent to the site study.

Mass Loading

For example, mass-loading calculations are required to determine the degree of contamination moving in groundwater from a waste-disposal site. Mass-loading calculations use data from several subdisciplines used in the site study, and therefore require a synthesis of data types to arrive at a numerical estimate of contaminant degree.

It should be noted that any mass-loading calculations performed on a spreadsheet (or by any method for that matter) are as accurate as the component with the greatest degree of variability. In the situation of mass-loading calculations, permeability is the limiting measurement, as permeability calculations are accurate to within roughly one-half an order of magnitude. This variability is the result partly of measurement error, but more to the uncertainty of how representative are the locations of the test wells within the aquifer. Geologic formations, by nature, are not homogeneous and nonisotropic. Therefore, measurements at one location will have some variability relative to the aquifer as a whole.

The initial step in mass-loading calculations is calculation of groundwater flow volume by Darcy's law. The flow is a function of the aquifer's permeability, the head imposed by the prevailing piezometric gradient and the physical dimensions of the aquifer (Table 7). Aquifer dimensions can be subdivided for a more precise mass-loading breakdown. A lateral flow net can be developed, with flow lanes defined to include locations of monitor wells. Thus, mass loadings are calculated for each flow zone, based on that flow zone's physical dimensions and contaminants identified specifically within that zone.

The three variables of flow volume have been quantified by data developed in the field and in the laboratory. Permeability has been calculated by drawdown/recovery tests and the subsequent response curves and calculations; the piezometric gradient has been determined by watertable-depth measurements and their transformation to piezometric-head measurements and gradient isograds; and the aquifer dimensions have been determined both laterally and vertically by development and quantification of stratigraphic data and cross-section plotting.

The mass loading is the product of the groundwater flow across the site and the contamination detected in groundwater at the site perimeter. Of course, appropriate unit conversion is required (Table 8).

Table 7. Aquifer flow calculations.

```
                              SITE NO. 1
                     AQUIFER FLOW VOLUME CALCULATIONS


UPPER FLOW ZONE : SAND  - PERM = 10 -4 CM/SEC
```

UPPER FLOW ZONE : SAND - PERM = 10 -4 CM/SEC

FLOW LANE	PERM ft/day	GRAD ft/ft	WIDE lat ft	THICK vert ft	AREA ft 2	FLOW ft3/day	FLOW gpd
1	0.28	0.007	200	5	1000	1.96	15.07692
2	0.28	0.007	300	5	1500	2.94	22.61538
3	0.28	0.003	200	5	1000	0.84	6.461538
4	0.28	0.002	200	5	1000	0.56	4.307692
5	0.28	0.02	100	5	500	2.8	21.53846

```
                                   SUBTOTAL: SAND ZONE            70
```

LOWER FLOW ZONE : GRAVELLY 'TILL' - PERM = 10 -2 CM/SEC

FLOW LANE	PERM ft/day	GRAD ft/ft	WIDE lat ft	THICK vert ft	AREA ft 2	FLOW ft3/day	FLOW gpd
1	28	0.015	200	10	2000	840	6461.538
2	28	0.01	300	10	3000	840	6461.538
3	28	0.004	200	10	2000	224	1723.076
4	28	0.005	200	10	2000	280	2153.846
5	28	0.01	100	10	1000	280	2153.846
6	28	0.009	200	10	2000	504	3876.923
7	28	0.009	150	10	1500	378	2907.692

```
                                  SUBTOTAL: GRAVEL ZONE       25738.46

                                  TOTAL OFFSITE GW FLOW:      25808.46
```

Table 8. Mass-loading calculations.

```
                     SITE NO. 1  CONTAMINANT LOADING

                     CHROMIUM, TOTAL : JULY, 1986
```

FLOW LANE	WELL	FLOW gpd	CONC ppm	CONC lb/gal X10 -8	MASS lb/day	MASS lb/yr
1	GW3S	15	0.006	4.98	0.000000	0.000272
1	GW3D	6500	0.022	18.26	0.001186	0.433218
2	GW17S	23	0.007	5.81	0.000001	0.000487
2	GW17D	6500	2.3	1909	0.124085	45.29102
3	GW27S	7	0.008	6.64	0.000000	0.000169
3	GW27D	1700	120	99600	1.6932	618.018

```
                                   SUBTOTAL  663.7431 EAST
```

FLOW LANE	WELL	FLOW gpd	CONC ppm	CONC lb/gal X10 -8	MASS lb/day	MASS lb/yr
4	GW19S	6	0.012	9.96	0.000000	0.000218
4	GW19D	2700	0.031	25.73	0.000694	0.253569
5	GW30S	32	0.059	48.97	0.000015	0.005719
5	GW30D	3200	1500	1245000	39.84	14541.6
6	GW28D	3877	0.11	91.3	0.003539	1.291990
7	GW21D	2908	0	0	0	0

```
                                   SUBTOTAL 14543.15 WEST, SW

            TOTAL CHROMIUM MASS LOADING:          15206.89 OFFSITE
```

The spreadsheet facilitates the multiple calculations required by performing sequential calculations and finally summing the loadings for each flow zone. Flow and loadings can also be calculated for upper and lower portions of an aquifer in which the hydrogeologic character changes with depth.

Testing Assumptions

Spreadsheet calculations can be used in planning for site remediation by facilitating the evaluation of alternatives. For example, should site clean-up require groundwater containment by pumpage and treatment of groundwater, capture-zone calculations can be used to evaluate pumping strategy alternatives. Capture zones can be calculated for a variety of wells and interwell spacing to ensure that cones of influence from any of a series of pumping wells overlap and intersect, thereby precluding groundwater flow between the wells. A formula with appropriate variables is "programmed" into a series of spreadsheet cells, and variables are altered progressively to determine the optimum well array (Table 9).

Predictive modeling has become commonplace in hydrogeologic studies. These models mathematically simulate hydrologic properties of an aquifer, then impose hypothetical pumpage patterns on that simulated system to predict the amount and location of groundwater withdrawal required to contain or control groundwater flow at the site. Although it is beyond the scope of this paper, it should be noted that the data entered into hydrogeologic models are geologic and hydrologic data developed and handled in the databases described herein. It is important that the informational input to the predictive models be accurate to ensure accurate results.

Table 9. Assumption testing spreadsheet.

```
          SITE NO. 1 CAPTURE ZONE CALCULATIONS WITH VARIED PUMP RATE


                  PUMP RATE                 PUMP RATE
      parameter     gpm                        gpm
      ========= ========                    =========
      Q            4.50    824.00 ft3/day     12.50 18000.00 ft3/day
      h                     14.00 ft                  14.00 ft
      n                      0.20                       0.20
      v @ 10 -2 cm/s         1.40 ft/day                1.40 ft/day

      r                     33.00 ft                  89.00 ft
      CAPTURE ZONE         104.00 ft                 290.00 ft



                  PUMP RATE                 PUMP RATE
      parameter     gpm                        gpm
      ========= ========                    =========
      Q            4.50    824.00 ft3/day     12.50 18000.00 ft3/day
      h                     14.00 ft                  14.00 ft
      n                      0.30                       0.30
      v @ 10 -2 cm/s         1.40 ft/day                1.40 ft/day

      r                     22.00 ft                  69.00 ft
      CAPTURE ZONE          70.00 ft                 188.00 ft


      r =     distance from well to downgradient 'stagnation point'
      C =     pi x r = radius of lateral 'capture zone' of well

      Q = pumping rate, in ft3/day
      h = saturated thickness of aquifer
      n = porosity
      v(nat) = gw velocity under natural gradient and conductivity

      r = Q / * 2 * pi * h * n * v
```

GRAPHIC REPRESENTATION

Graphic representation of data can be used to display trends and results and to evaluate
data by inspecting visually temporal and spatial trends in parameters under study. Thus,
traditional geologic methods of data evaluation are facilitated. Data used in graphic plots
are data which have been transformed and which have been incorporated into the study
database. Such plots can be created by printer or by plotter.

Trend Plots

Trend plots have been demonstrated as an evaluation tool when used in
drawdown/recovery curve plotting. However, by plotting trends of contamination with
time and of contamination as a function of watertable level the geoscientist can observe
trends empirically and note similarities and anomalies. Such graphic plots (Fig. 3) also
are effective communicative tools when included in a hydrogeologic report.

Cross Sections

Graphic representation of data can facilitate the most traditional evaluation tool, the
cross section. Geologic cross sections can be drawn to differing degrees of detail by
manipulating a simple bar-graph software package. The software can be "tricked" into
representing stratigraphic data in the form of a cross section, as described next.

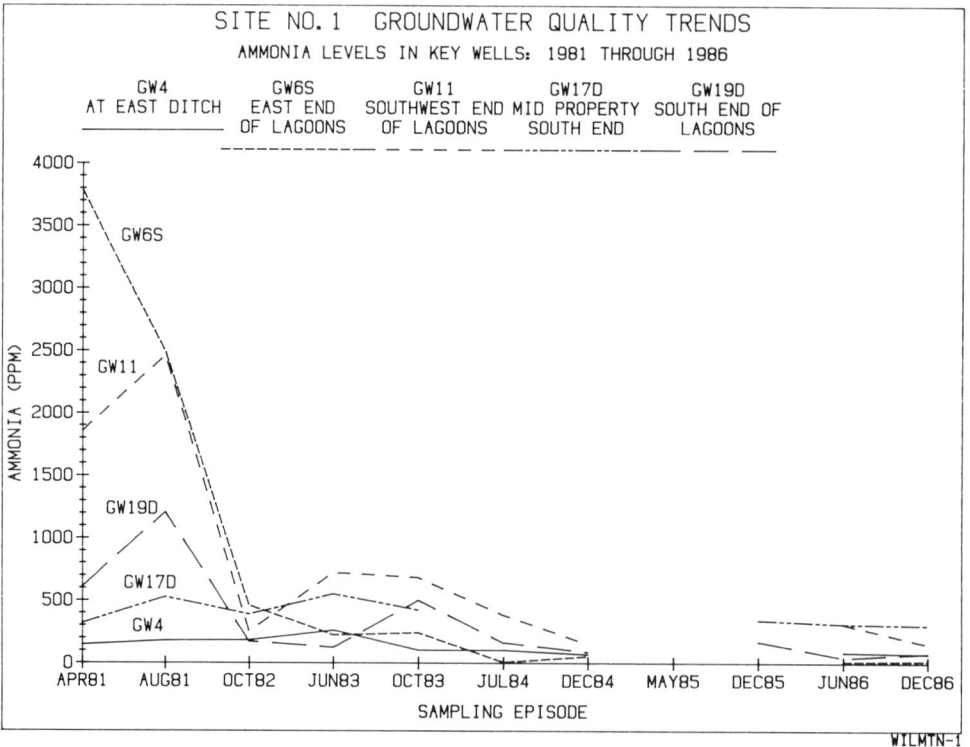

Figure 3. Time trend plot.

There are some cross-section software packages available commercially. However, in lieu of specific software, a bar-graph software package is versatile, allows a great degree of creativity in the placement of labels, and explanations and affords the user the luxury of experimenting with the vertical depth scale. Further, cross sections created using bar-graph software allow a surprising degree of detail.

To develop a cross section using bar-graph software, stratigraphic data are developed in tabular format, as they ordinarily would be organized in the preparation of a cross section. Each discrete stratum is tabulated by its thickness and described by its texture and geologic characteristics. The ground surface elevation of each borehole also is listed for elevation control. The data are entered into a spreadsheet as they are transformed from the "raw" information on individual borelogs. The spreadsheet data are transferred to the bar-graph software either electronically or manually, depending on the hardware available and the size of the database. An example spreadsheet is shown in Table 10.

Table 10. Cross section data spreadsheet.

```
              SITE NO. 4   CROSS SECTION DATA SPREADSHEET


    WELL              depth to depth to stratum  top of   bottom of
                      top of   bottm of thk.     stratum  stratum  grade
    P1                stratum  stratum           elev.    elev.    elev.
                        ft       ft       ft     ft-msl   ft-msl   ft-msl
    ========          ======== ======== ======== ======== ======== ========
    fill gr/sd/brk/wd       0     9.75     9.75      7.1    -2.65      7.1
    silty sand           9.75       12     2.25    -2.65     -4.9      7.1
    gravel                 12       14        2     -4.9     -6.9      7.1
    silty sand             14       18        4     -6.9    -10.9      7.1
    gravel                 18       25        7    -10.9    -17.9      7.1

    *********************************************************************


    WELL              depth to depth to stratum  top of   bottom of
                      top of   bottm of thk.     stratum  stratum  grade
    P2                stratum  stratum           elev.    elev.    elev.
                        ft       ft       ft     ft-msl   ft-msl   ft-msl
    ========          ======== ======== ======== ======== ======== ========
    fill gr/sd              0        4        4      6.4      2.4      6.4
    marsh organics          4      7.8      3.8      2.4     -1.4      6.4
    silty sand            7.8      8.1      0.3     -1.4     -1.7      6.4
    clayey silt           8.1       10      1.9     -1.7     -3.6      6.4

    *********************************************************************


    WELL              depth to depth to stratum  top of   bottom of
                      top of   bottm of thk.     stratum  stratum  grade
    P3                stratum  stratum           elev.    elev.    elev.
                        ft       ft       ft     ft-msl   ft-msl   ft-msl
    ========          ======== ======== ======== ======== ======== ========
    fill gr/sd              0      4.6      4.6      7.5      2.9      7.5
    marsh organics        4.6      8.2      3.6      2.9     -0.7      7.5
    clayey silt           8.2       10      1.8     -0.7     -2.5      7.5

    *********************************************************************


    WELL              depth to depth to stratum  top of   bottom of
                      top of   bottm of thk.     stratum  stratum  grade
    P4                stratum  stratum           elev.    elev.    elev.
                        ft       ft       ft     ft-msl   ft-msl   ft-msl
    ========          ======== ======== ======== ======== ======== ========
    fill gr/sd  -oil-       0        7        7      6.2     -0.8      6.2
    marsh organics          7      8.3      1.3     -0.8     -2.1      6.2
    clayey silt           8.3       10      1.7     -2.1     -3.8      6.2
```

Next, the bar-graph software is "programmed" to accommodate the stratigraphic data and the desired format.

Data from each borehole are represented by a series of vertically occurring strata which, in turn, are represented by a series of bars. The bar lengths are proportional to the vertical thickness of the strata which they represent. The bars are formatted as "stacks" in lieu of "clusters." The space between bars is formatted as zero, allowing the stacked bars to touch, that is, allowing each stratum to contact the stratum above and below it. Specific lithologic units can be identified by patterns superimposed on each bar, as shown in Figure 4. Bar widths can be changed to show representative patterns more fully, as per the preference of the user. Each borehole thus is represented by stratigraphic data transformed to bars and patterns. Strata then can be correlated between boreholes as per the judgement of the geoscientist. Continuous strata can be delineated graphically by lines connecting each stratigraphic horizon. This delineation requires that the elevations of each stratum break be added to the database.

In developing more complex cross sections, the geoscientist may prefer to plot the stratigraphic data solely in bar form and draw the stratigraphic correlations by hand on the resulting plot. Furthermore, more complex cross sections can be represented by all blank bars, allowing the geoscientist to interpret the presence of lenses or discontinuous strata and manually draw them onto the cross section, as shown in Figure 5.

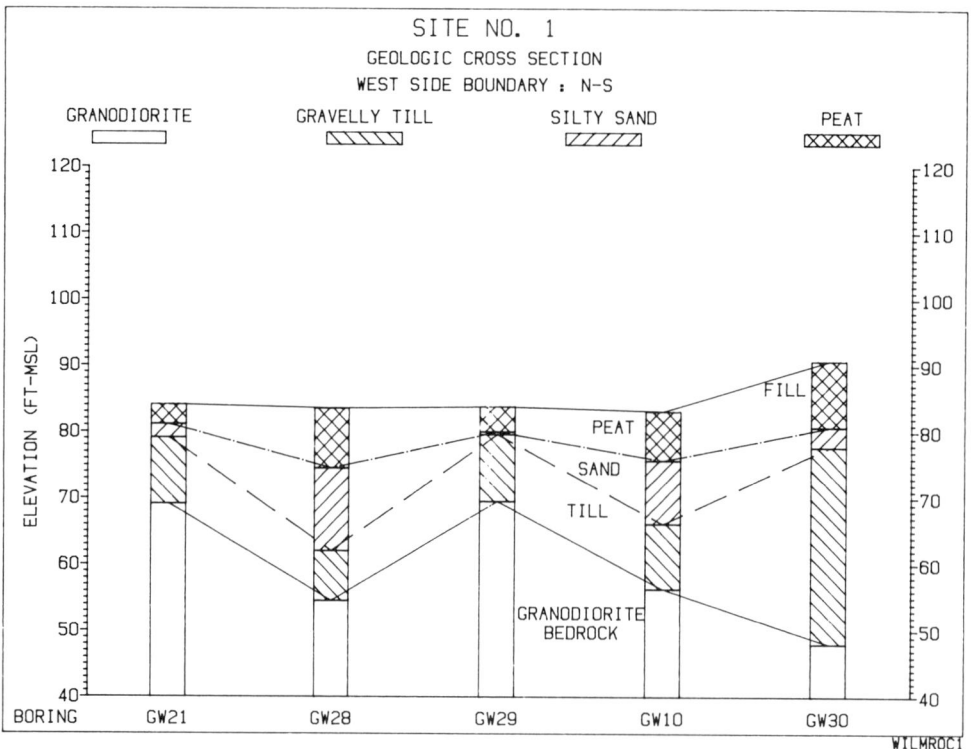

Figure 4. Cross Section 1.

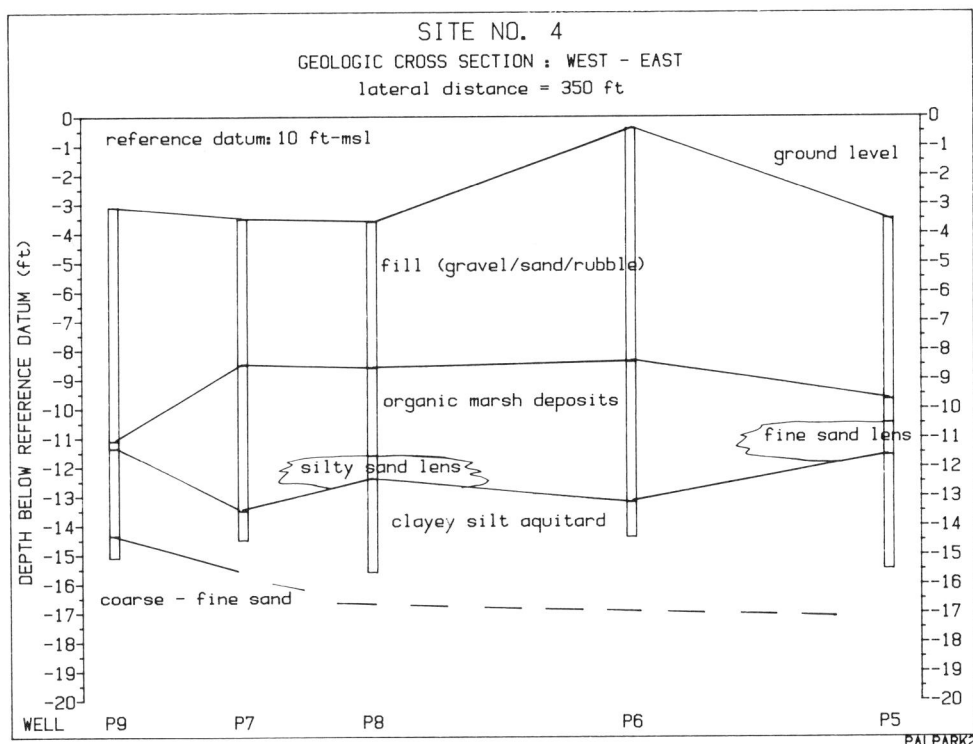

Figure 5. Cross Section 2.

Several notes are offered regarding the formatting of the bar-graph software.

The x-axis of the bar-graph indicates locations of the boreholes in relation to each other. The y-axis indicates relative elevation or depth below ground. Relative distances between boreholes that are not spaced equally are represented proportionally by blank data rows between data rows containing stratigraphic data.

For more simplified cross sections, such as those containing data from boreholes drilled at a uniform elevation, borehole profiles can be represented in terms of their depth below ground. Thus, stratigraphic data can be entered from youngest to oldest, with a negative sign indicative of depth below ground level. The vertical axes are represented as depth in feet. This type of cross section is shown in Figure 6.

When relative elevation data are used, as when boreholes are drilled at differing elevations, borehole profiles can be represented in terms of relative elevation (feet above mean sealevel) The vertical axes are represented by surveyed elevation. The lowermost bar will be assigned a value of the elevation of the bottom of the borehole and will be blanked so it does not appear on the final plot. Subsequent overlying strata thickness data are entered above the blanked bar from oldest to youngest, as shown in Figure 7.

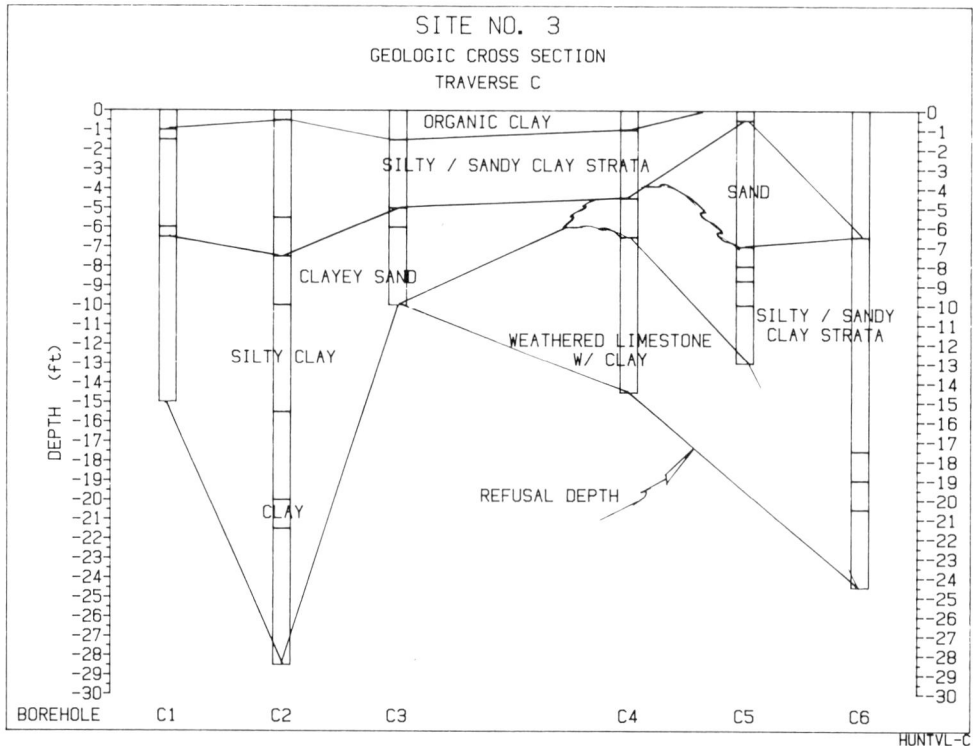

Figure 6. Cross Section 3.

Similarly, profiles of well placement can be developed to represent vertical zones within the aquifer which are screened for groundwater-quality sampling. Here, well screen lengths and standpipe lengths replace stratigraphic data and graphically display a profile of screened zones, as shown in Figure 8.

Contouring/Data Smoothing

Automated contouring packages are used to develop precise piezometric-gradient maps and isograds of contours representing groundwater-quality parameters. The user has options, for example, to determine the number of data points used in interpolation to establish the exact direction of a contour line, and the degree of "smoothing" of isograd lines on the final plot. Statistical smoothing of the areal distribution of groundwater-quality data also is possible by weighted averaging methods such as kriging.

Although the geoscientist should take advantage of these tools, for purpose of evaluation of data and for purpose of graphic display and communication of results, a word of caution is offered. All data that are plotted and contoured, in a preliminary sense, are controlled by the geology and hydrology of the study area. For example, groundwater flow, as represented by contour lines of piezometric gradient, may be controlled by geologic boundaries of the study area such as by a basal clay or rock stratum or by a surface-water discharge point. The contour program has no way of "knowing" these site-specific hydrogeologic details.

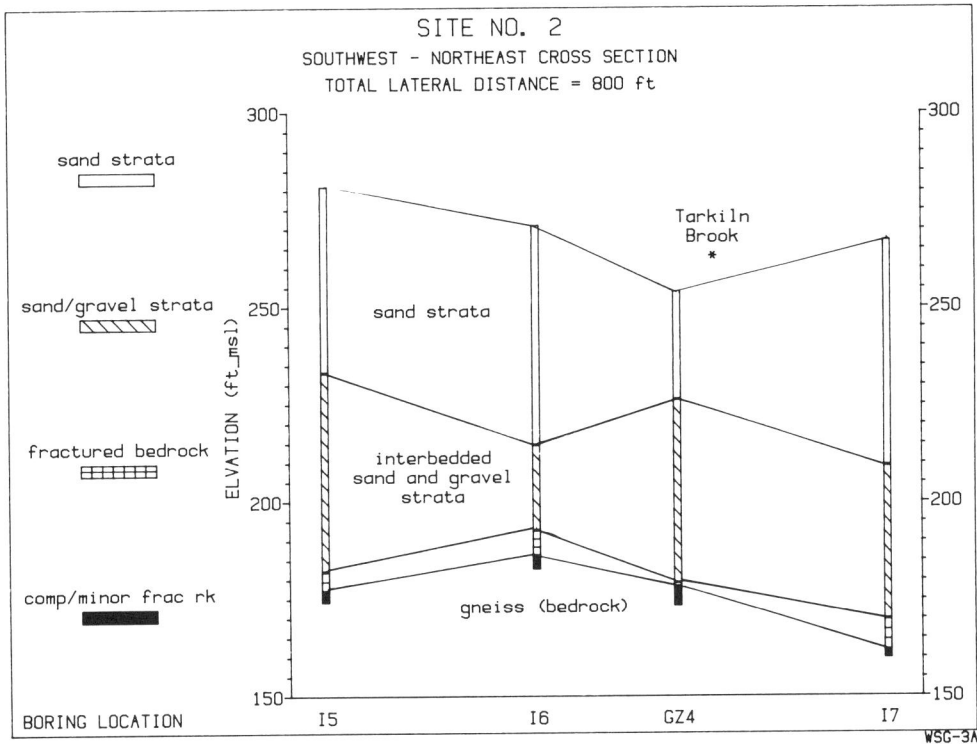

Figure 7. Cross Section 4.

However, the geoscientist incorporates all site-specific knowledge into contour plots which he/she generates by manual plot. Therefore, the geoscientist must develop a first-hand knowledge of the site by preliminary manual plots and data inspection before submitting his/her data to an automated contouring program. After a contour plot is developed, the geoscientist must inspect the plot to ensure that it reflects the site's hydrogeologic constraints as he/she understands them. Only then can the current and future plots be considered reliable evaluation tools. An example of a contoured plot is shown in Figure 9, an isopachous contour plot of a stratum of "fill" soil atop a naturally occurring stratum.

All graphic plots and cross sections shown in this chapter were developed using Picture Perfect software. Contouring plots were developed using Surfer software.

WORDPROCESSING

Wordprocessing software has been used for years as a secretarial and clerical aid in office work. Such software usually is available for use with microcomputers. The geoscientist with rudimentary typing skills can use this "electronic typewriter" to develop and edit drafts of reports. The geoscientist can review and quickly change drafts of text, as he/she evaluates report format, length, and wording. If a near-letter-quality printer is available, final reports can be generated on the microcomputer.

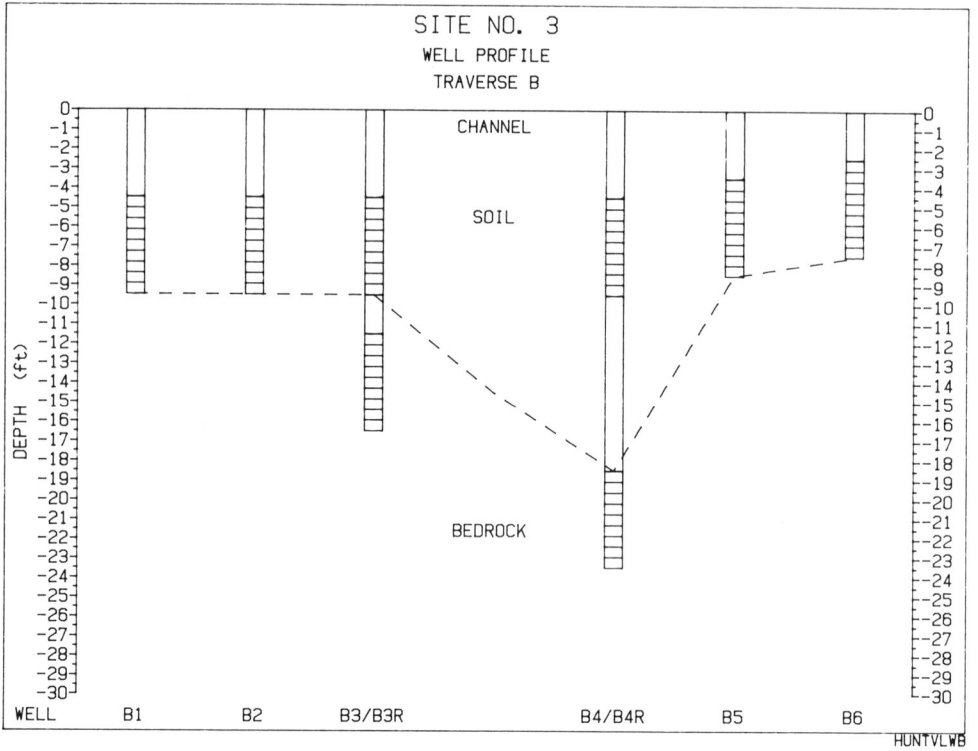

Figure 8. Monitor well profile.

SITE NO. 4   FILL ISOPACH (FT)

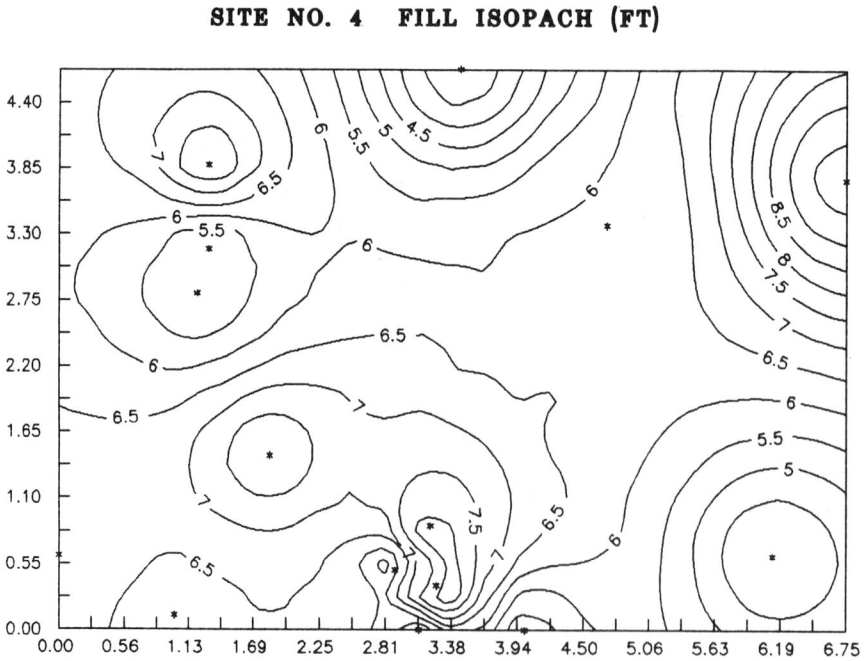

Figure 9. Isopachous contour plot.

CONCLUSION

As demonstrated here, hydrogeologic studies at waste-disposal sites require a synthesis of data and knowledge from a variety of disciplines within the science of geology. Furthermore, these studies require management and transformation of data through a progression of stages from "raw" data to conclusions. If a hydrogeologic study can be likened to an alluvial aquifer, it may be said that the sedimentary particles comprising the aquifer are individual data points, the groundwater flowing through the aquifer is information, and the cementing agent of the geologic system is the personal computer.

Application of a Microcomputer-based Geographic Information System to Mineral-Potential Mapping[1]

G.F. Bonham-Carter and F.P. Agterberg
Geological Survey of Canada

ABSTRACT

Recent advances in technology and methods of integrating map data are making the task of mapping mineral potential by computer more appealing to the practicing geologist. Micro-based Geographic Information Systems (GIS) make the task of building a database from diverse maps relatively straightforward, even for those who are not computer specialists. Efficient commercial software packages, hierarchical data structures such as quadtrees, low-cost disk storage, and faster micros make multiple map analysis by computer easier than manual procedures, such that the light table for overlaying maps which may soon become a thing of the past.

A recent development in methodology, termed here "weights of evidence" mapping, uses an approach for combining evidence from several predictor maps to estimate probability of mineral occurrence. Not only does the GIS aid in capturing and coregistering the input maps for this task, it also facilitates the modeling and visualization of results. In the simplest situation, each input map is converted to binary form, and a pair of weights, W^+ and W^-, are determined corresponding to the two map classes. The posterior log odds of a mineral occurrence within a unit cell is calculated by adding the weights of evidence (W^+ and W^-) from the input maps to the prior log odds. The weights of evidence are determined using the distribution of known mineral occurrences with respect to the input maps. The method assumes that the input maps are conditionally independent of one another with respect to the points, and tests are made to verify this assumption. The uncertainty of the probability estimates, because of both the variances of the weights of evidence and incomplete or missing data, also can be mapped.

A small worked example illustrates the calculations involved. The method also is applied to gold potential mapping in Nova Scotia. Maps showing rock type, regional geochemistry, proximity to structures, formation contacts, and fold axes are used to predict gold mineralization. The proximity maps are generated by dilating linear features, a useful

---

[1] Geological Survey of Canada Contribution Number 47488

function of GIS's for mineralization modeling. Hardcopy of the output maps can either be made by photographing the color monitor or by using raster plotting devices. The GIS map-query function provides a powerful tool to explore the output maps interactively with the cursor, simultaneously interrogating the "map stack" to aid in understanding the relationships between input and output.

INTRODUCTION

The task of integrating diverse types of map data in order to produce a new map showing areas favorable for mineralization traditionally has been carried out by hand, or by overlaying maps on a light table. At a regional scale, this process leads to maps of mineral potential, or mineral favorability; at a local scale, it may involve target selection. Considerable effort during the past two decades has been directed towards methods of quantifying and modeling the mapping of mineral potential. This work generally has involved multivariate statistics to predict areas with a linear combination of predictor variables that discriminate between mineralized and nonmineralized regions, for example, Agterberg and Franklin (1987), Harris (1984), and many other authors. More recently, expert systems have been used to combine predictor maps, using the knowledge of an expert to estimate probabilities and define rules for combining evidence, for example, Campbell, Hollister, and Duda (1982). With the multivariate statistical approach, a training area is required, where exploration is at a relatively mature stage, and there are a sufficient number of known mineral occurrences to obtain reliable estimates of regression coefficients. For the expert-system approach, no knowledge of mineral-occurrence locations is required for the study area, and a variety of rules, including Bayes' rule, are used to model the reasoning of the expert for combining evidence.

The acceptance of these quantitative approaches by geologists has been limited, partly because of difficulties in capturing as digital maps the variables important for mapping mineral potential, and partly because the methodology has been rather difficult to understand, to implement, and to interpret.

Some recent work by Agterberg (1989), Agterberg, Bonham-Carter, and Wright (1988) and Bonham-Carter, Agterberg, and Wright (1988) has attempted to overcome some of these problems. Using methodology developed in the field of quantitative medical diagnosis (Spiegelhalter, 1986), which employs additive weights of evidence for combining knowledge of symptoms to estimate the probability of disease, geoscience maps are weighted to estimate probability of mineralization. In the simplest situation, maps are first converted to binary form, each one showing areas that are either on the binary pattern (score = +1) or not on the pattern (score = –1). In some situations, the pattern may be unknown (score = 0). A pair of weights is computed for each map, positive weights (W^+) for pattern present and negative weights (W^-) pattern not present. The posterior probability of mineral occurrence is determined from a prior probability and the summed weights of evidence, calculated for each area of the map where a unique overlap combination of the binary input maps occurs.

This approach is somewhat similar to an expert system: Bayesian probability is used to combine evidence for mineralization, and it is assumed that the predictor maps are conditionally independent with respect to mineral occurrences. However, instead of using expert judgement to estimate the weights of evidence, the proportion of known mineral occurrences falling on the binary pattern area to total area, are used to estimate W^+ and W^- for each predictor map. These weights are usually straightforward to interpret, similar to regression coefficients which usually have relatively large variances in mineral-resource evaluation applications. The weights of evidence methodology is appealing because it resembles the approach of the practicing geologist.

The development of Geographic Information Systems (GIS) also is playing an important role in making quantitative methods of spatial data integration easier and more acceptable to the geologist. This is because the digital capture and coregistration of maps from diverse scales and projections is now straightforward. Several GIS systems, such as SPANS (TYDAC, 1987) used for this study, operate on MS–DOS based microcomputers, and offer speed and functionality previously not available on low-cost equipment. The ability to generate maps showing distance to linear features such as contacts, fold axes, and lineaments, is useful particularly for mineral-potential studies. These maps capture proximity information at a detailed spatial resolution, providing a method to quantify factors that are essential for many mineral-deposit models. Combining large maps using models can be a computationally expensive task, but a new approach of using a "unique-conditions" map makes these computations efficient by collapsing the combined map into a relatively small number of classes. For each class, a unique combination of the predictor maps occurs, and model calculations need only be carried out for each class, instead of being repeated for each pixel in a raster.

MICRO-BASED GEOGRAPHIC INFORMATION SYSTEM

The mineral-potential mapping described here is being carried out using SPANS (TYDAC, 1987), a GIS operating on an Intel 80386-based microcomputer equipped with 640 Kb main and 2 Mb extended memory, a Number Nine Corporation graphics card (512 x 512 x 32) and 20" color monitor. The GIS software is raster-based, using quadtrees for storing map data (Samet, 1984). A 48" x 36" digitizing table, and TYDIG — an associated digitizing package — is being used to capture point, line, and polygon maps. Remote sensing and airborne geophysical data, available as raster images, are read directly into the GIS, assuming the data is georeferenced with information about the projection, location of the origin, and the extents of the digital image. Point data, with associated geographic coordinates and nonspatial attributes (geochemical data, mineral-deposit data, for example) are imported directly from text files, and either converted to maps using tesselation or interpolation routines, or left in point form. A wide variety of input formats are possible, allowing import of map data from vector and raster-based systems. Thus data can be gathered from a variety of digital sources, originally captured from diverse scales and projections.

Having established a study area defined by the geographic corner points, and having selected a suitable projection for manipulation and display, each component map is imported, and automatically registered and stored using a pixel resolution appropriate to the map. In the quadtree system, the width of the study area in pixels depends on the quadlevel, which changes as a power of 2 from 0 to 15, that is, from 1 to 32,768 pixels wide. In practice the quadlevel for most maps in our work ranges between 9 and 12, that is, between 512 and 4,096 pixels wide. The quadtree results in considerable space saving for many thematic maps. For example, a 1,024 x 1,024 x 8-bit raster image in normal "expanded" form requires a megabyte of storage space, whereas the equivalent map quadded at level 10 usually require only 100 x 300 K bytes, depending on the nature of the map. The quadtree representation not only cuts down considerably on storage requirements, but access time and therefore processing efficiency is improved over conventional image-analysis systems. Maps in a quadtree database can be at various quadlevels, that is, a remote-sensing image at level 13 can be in the same database with a geologic map at quadlevel 10.

A powerful feature of many GIS's is the ability to dilate linear features to produce corridors or buffer zones. In SPANS, lines can be imported and stored as vectors. These lines can be overlain for display purposes on any map, or converted to maps themselves by generating one or more corridors at specified widths. We have used this facility to build maps showing proximity to fold axes, to lineaments or faults, and to geologic contacts. The resulting raster images as quadtree data structures provide compact yet finely resolved structural information invaluable for mineral-potential mapping.

The ability to combine two or more maps using simple overlay primitives is enhanced by flexible modeling statements. Such operations can be carried out remarkably rapidly, considering that the hardware platform is a microcomputer. When modeling is being performed, a unique-conditions map is first created. For a simple example, suppose two binary maps are to be combined using an equation, the following unique conditions would be possible (1, 1), (1, 0), (0, 1), and (0, 0). The unique-conditions map therefore would show only four classes, one for each unique condition (if it actually exists on the map). The subsequent modeling operation need be carried out only four times, and each unique-condition class is assigned the appropriate result. Suppose the original maps are 1,024 x 1,024 images, the model calculations normally would be repeated 1,048,576 times if carried out on the expanded image, versus four times if carried out on the unique-conditions image. This leads to rapid interactive modeling even on large numbers of spatially detailed maps.

In the next sections we describe some details of the weighted-evidence methodology, show how the calculations are made on a simple artificial example, and discuss results from a study of gold potential in Nova Scotia.

WEIGHTS OF EVIDENCE MODELING

Suppose that a mineral-deposits geologist has a number of coregistered maps, each one showing two classes: pattern present, pattern absent. These are typically rock types, geophysical or geochemical anomalies, fault or contact neighborhoods, and other information diagnostic for a particular type of mineralization. Assume also that a point map is available, showing locations of known mineral occurrences. The object is to determine weights of evidence for each input map that reflect the correlation with the occurrences, and to produce an output map showing the probability of occurrence based on the accumulated evidence from the input maps. Additionally, a map showing the uncertainty of the probability estimate is desired.

Following the work summarized by Speigelhalter and Knill-Jones (1984) and Spiegelhalter (1986) applied to medical diagnosis, Agterberg, Bonham-Carter, and Wright (1988) and Agterberg (1989) show how these calculations can be used in a spatial context. One major assumption must be made: that the input maps are conditionally independent of the points, but tests are available for this, as will be described next. The sequence of calculations can be broken into the following steps:

(1) Calculate a prior probability of occurrence by dividing the study area into unit cells of fixed area. As will be seen, the selection of unit-cell size is not important to the final outcome. However, it does effect the prior probability, which usually is selected as the unconditional probability of locating an occurrence in a unit cell selected at random. Given total area as t km^2, and area of unit cell u km^2, then T = t/u is the total number of unit cells in the study area. Let D be the number of unit cells containing an occurrence, equal to the number of occurrences if u is small

enough. Then P(D) = D/T is the prior probability and is assumed to be constant throughout the whole area.

(2) Transform this prior probability to prior odds, O(D)

$$O(D) = \frac{P(D)}{1 - P(D)}$$

(3) For the j-th binary-input map, let $B_j = b_j/u$ be the area where the binary pattern is present, expressed in unit cells. Then $B_j \cap D$ is the area of intersection, or the number of occurrences falling on binary pattern j; $B_j \cap \bar{D}$ is the number of unit cells on the pattern not occupied by an occurrence; and $\bar{B}_j \cap D$ and $\bar{B}_j \cap \bar{D}$ are the remaining intersection areas as shown in Figure 1. The weights of evidence for the j-th map are defined by

$$W_j^+ = \log_e \frac{P\left(B_j \mid D\right)}{P\left(B_j \mid \bar{D}\right)}$$

$$W_j^- = \log_e \frac{P\left(\bar{B}_j \mid D\right)}{P\left(\bar{B}_j \mid \bar{D}\right)}$$

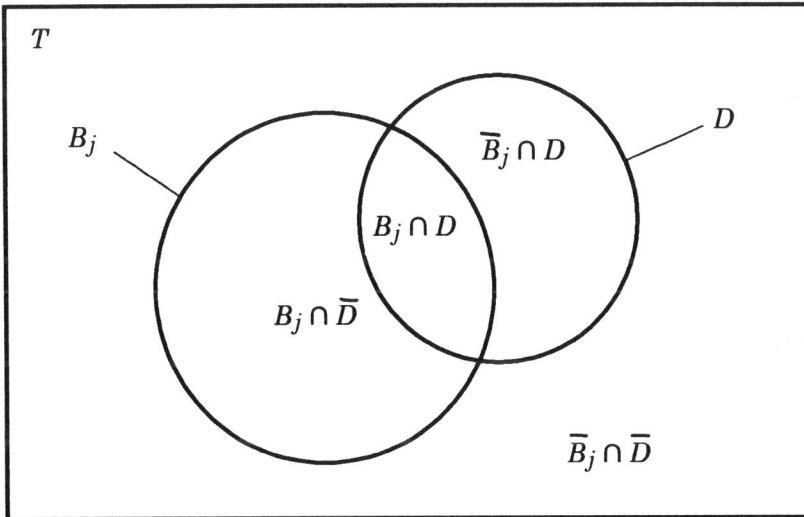

Figure 1. Venn diagram showing overlap of binary pattern, B, with deposits, D.

The positive weight of evidence, W_j^+, is used where the binary pattern is present, W_j^- where it is absent. The conditional probabilities are determined from the overlap areas, for example

$$P(B_j \mid D) = \frac{B_j \cap D}{D} \quad \text{and} \quad P(B_j \mid \bar{D}) = \frac{B_j \cap D}{\bar{D}}$$

In this step, notice that the weights W_j^+, W_j^- are determined independently for each binary map.

(4) The variances of the weights are calculated from

$$\sigma^2(W_j^+) = \frac{1}{B_j \cap D} + \frac{1}{B_j \cap \bar{D}}, \quad \text{and}$$

$$\sigma^2(W_j^-) = \frac{1}{\bar{B}_j \cap D} + \frac{1}{\bar{B}_j \cap \bar{D}}$$

using an asymptopic result (Bishop, Fienberg, and Holland, 1975), which assumes that the number of unit cells containing an occurrence is large, as discussed in Agterberg, Bonham-Carter, and Wright (1988). These expressions are approximate where a small number of occurrences are used or available.

(5) A convenient measure of correlation between the j-th map and the occurrences can be defined as the contrast $C_j = W_j^+ - W_j^-$. The variance of C is $\sigma^2(C_j) = \sigma^2(W_j^+) + \sigma^2(W_j^-)$.

(6) The log posterior odds for any overlap combination (unique condition) of m input maps is given by

$$\log_e O(D \mid B_1^k \cap B_2^k \cap B_3^k \ldots \cap B_m^k) = \log_e O(D) + \sum_{j=1}^{m} W_j^k$$

where k refers to presence or absence of binary pattern: thus $B_j^k = B_j$ for pattern j present, $B_j^k = B_j$ for pattern absent. Similarly

$$W_j^k = \begin{cases} W_j^+ & \text{if } B_j^k = B_j, \\ W_j^- & \text{if } B_j^k = \bar{B}_j, \\ \varnothing & \text{if } B_j^k \text{ is unknown.} \end{cases}$$

Then the posterior probability is

$$P_{post} = \frac{O_{post}}{1 + O_{post}}$$

which is mapped for each unique condition.

(7) The variance of P_{post} due to variances of the weights also can be mapped for each unique condition. The variances of the weights are summed for each map and added to the variance of the prior log odds. Following the discussion of Agterberg, Bonham-Carter, and Wright (1988), and assuming that the prior probability is D/T, the variance of the prior log odds can be shown to be 1/D. Then

$$\sigma^2 (\log_e O_{post}) = \frac{1}{D} + \sum_{j=1}^{m} \sigma^2 (W_j^k)$$

where k signifies + or – as before, and

$$\sigma^2 (P_{post}) = \sigma^2 (\log_e O_{post}) * P_{post}^2$$

(8) Where one map is incomplete, as may happen in practice so that whether a binary pattern is present is unknown, another component of variance of the posterior probability can be calculated (Agterberg, Bonham-Carter, and Wright, 1988). For the situation of missing data from the j-th input map, this variance component is

$$\sigma_1^2 (P_{post}) = \{ p(D|B) - P(D) \}^2 P(B) + \{ P(D|\bar{B}) - P(D) \}^2 P(\bar{B})$$

where P (D) is now the posterior probability calculated for a region where B_j is

missing, and P (D I B) and P (D I $\bar{B}$) are updated posterior probabilities calculated as if the missing pattern actually was known. This variance component is zero where the j-th pattern is known. It therefore is not a constant in the unknown areas but will differ depending on the unique overlap conditions of the other input maps. Notice that the variance is the average squared difference between the posterior probability and the updated posterior probabilities, weighted by the known area components of the missing pattern.

(9) A combined variance map can be calculated by summing the variances because of uncertainty of the weights with uncertainty because of missing data.

(10) In order to test the assumption of conditional independence, either a chi-square test can be applied to compare observed and expected frequencies of deposits over all the unique conditions on the map, as described by Agterberg, Bonham-Carter, and Wright (1988), or a Kolmogorov–Smirnov test can be made which is similar, but avoids the requirement of binning the data into classes required by the chi-square test.

(11) Finally, a pairwise test of conditional independence can be applied, showing the degree with which each pair of input maps satisfies the conditional independence

assumption, thereby highlighting particular maps where the assumption may be weak or violated.

In order to make this test, the observed and expected number of unit cells are computed for the following table, given two input maps B_1 and B_2.

	$B_1 \cap B_2$	$B_1 \cap \overline{B}_2$	$\overline{B}_1 \cap B_2$	$\overline{B}_1 \cap \overline{B}_2$
D	$P(D\,B_1 B_2)$	$P(D\,B_1 \overline{B}_2)$	$P(D\,\overline{B}_1 B_2)$	$P(D\,\overline{B}_1 \overline{B}_2)$
$\overline{D}$	$P(\overline{D}\,B_1 B_2)$	$P(\overline{D}\,B_1 \overline{B}_2)$	$P(\overline{D}\,\overline{B}_1 B_2)$	$P(\overline{D}\,\overline{B}_1 \overline{B}_2)$

For each class in this table, let x be the observed area (frequency of unit cells) and m be the frequency estimated from the model, using the weights for B_1 and B_2 and the four unique conditions expressed by the four columns.

$$m = P(D \mid B_1 B_2)\, P(B_1 B_2) T$$

gives the expected frequency for the area of overlap between occurrences and both binary patterns. The term $P(D \mid B_1 B_2)$ is obtained from the model, and $P(B_1\,B_2)$ is the measured area of overlap of the two binary patterns as a proportion of the total area. The other terms in the table are calculated similarly and then

$$X^2 = \sum_{i=1}^{8} \frac{(x_i - m_i)^2}{m_i}$$

is distributed as X^2 with 2 degrees of freedom (Bishop, Fienberg, and Holland 1975). Alternatively,

$$G^2 = -2 \sum_{i=1}^{8} x_i \log_e \frac{m_i}{x_i}$$

has a similar distribution (Bishop, Fienberg, and Holland , 1975). Thus if X^2 or G^2 is larger than about seven, one may suspect that the particular pair of input maps shows some conditional dependence with respect to the occurrences.

WORKED EXAMPLE

In order to clarify the foregoing, the following artificial and rather trivial example may be useful to those wishing to work through an actual calculation (Table 1). For a real problem, the GIS (SPANS) can be used to produce a unique-conditions report showing the areas of each overlap class (Table 1A). Here we have appended a column showing the number of occurrences in each class, although using SPANS this can be produced as a separate report file and subsequently merged. We have written a FORTRAN program outside SPANS

Table 1. Test example of weights of evidence modeling. A, Unique-conditions table. B, Table of weights, variances, and contrasts. C, Table of posterior probabilities and variances. D, Conditional independence test, observed and (expected) number of unit cells.

A

Case	Map 1	Map 2	Occurrences	Area, km²
1	1	1	20	100
2	1	− 1	10	300
3	− 1	1	8	400
4	− 1	− 1	1	600
			$D = 39$	t = 1400

Unit cell u = 1 km²
Total area $T = t/u$ = 1400 unit cells
Area of binary pattern, $B_1 = b_1/u$ = 400 unit cells
Area of binary pattern, $B_2 = b_2/u$ = 500 unit cells
Intersection $B_1 \cap D$ = 30
Intersection $B_2 \cap D$ = 28
Prior probability, $P(D)$ = 39/1400 = 0.027857
Prior log odds, $O(D)$ = 0.028655
Variance of prior log odds = 0.02564

$$W_1^+ = log_e \left(\frac{30/39}{370/1361} \right) = 1.040$$

B

	W^+	$\sigma^2(W^+)$	W^-	$\sigma^2(W^-)$	C	$\sigma^2(C)$
Map 1	1.040	0.0360	− 1.149	0.1121	2.189	0.1482
Map 2	0.728	0.0378	− 0.840	0.0920	1.567	0.1299

For case 1 of unique conditions table:
Posterior log odds = 0.028655 + 1.040 + 0.728 = 0.796655
Posterior probability = 0.144
$\sigma^2(log_e O_{post})$ = 0.02564 + 0.0360 + 0.0378 = 0.09944
$\sigma^2(P_{post})$ = 0.002056

Table 1. (continued)

C

Case	Map 1	Map 2	P_{post}	σ_{wts}	σ_{miss}	σ_{tot}	$\dfrac{P_{post}}{1.96\,\sigma_{tot}}$
1	1	1	0.144*	0.0453	0	0.0453	1.617
2	1	− 1	0.038*	0.0133	0	0.0133	1.301
3	− 1	1	0.018	0.0077	0	0.0077	1.218
4	− 1	− 1	0.004	0.0019	0	0.0019	1.064
5⁺	1	0	0.075*	0.0186	0.0527	0.0559	0.685
6⁺	− 1	0	0.009	0.0033	0.0070	0.0077	0.594
7⁺	0	1	0.056*	0.0141	0.0566	0.0584	0.490
8⁺	0	− 1	0.012	0.0042	0.0135	0.0145	0.440

* cases with $P_{post} > P_{prior}$
⁺ cases not in Table 1 – A, but added here to show effect of missing data

For case of intersection $D \cap B_1 \cap B_2$, expected area $= 0.144 \times 100 = 14.4$

D

	$B_1 \cap B_2$	$B_1 \cap \bar{B_2}$	$\bar{B_1} \cap B_2$	$\bar{B_1} \cap \bar{B_2}$
D	20 (14.4)	10 (10.1)	8 (7.4)*	1 (2.3)*
$\hat{\bar{D}}$	80 (85.6)	290 (289.9)	392 (392.6)*	599 (597.7)

$X^2 = 3.4014$, $G^2 = 3.3801$, i.e., both test statistic values are less than tables χ^2 for 2 d.f., $\alpha = 0.99$, so hypothesis of conditional independence not rejected.

* notice that these values are slightly larger than those calculated from the P_{post} values in 1 – C, due to the small number of digits retained in 1 – C.

which calculates the weights and variances shown in the Table 1B. Notice that both maps have a positive value of contrast, C, indicating a positive correlation with occurrences, and that $C > 1.96\ \sigma$ (c) showing that for both maps the correlation is significantly greater than zero. It can be seen that map 1 is correlated more strongly than map 2, because B_1 occupies less area than B_2 yet more occurrences are located on B_1 than on B_2. Table 1C shows the posterior probability and associated standard deviations for all possible unique conditions. For illustrative purposes, we have added four unique conditions for one or other of the maps being missing, to illustrate the σ results. It is interesting to note that when binary pattern 1 is missing (Case 7) $P_{post} > P_{prior}$, but if pattern 1 is not present (Case 3) $P_{post} < P_{prior}$. However, $\sigma_{missing}$ for Case 7 is large, so that σ_{total} for Case 7 is 0.0584 as opposed to 0.0077 for Case 3. One would conclude that although in Case 7 the posterior probability is relatively favorable for an occurrence, it is uncertain because $P_{post} < 1.96\ \sigma$ (P_{post}); whereas in Case 3 the low posterior probability is more certain, $P_{post} > 1.96\ \sigma$ (P_{post}). As a guide to whether the posterior probability is significant, we have added a column to Table 1C showing $P_{post}\ /\ 1.96\ \sigma$ (P_{post}). Values less than 1.0 suggest that P_{post} is not significantly different from zero.

The test in Table 1D shows that there is no reason to reject the hypothesis that the two maps are conditionally independent with respect to the occurrences.

APPLICATION TO GOLD POTENTIAL IN NOVA SCOTIA

Preliminary results for this study are in Agterberg, Bonham-Carter, and Wright (1988), Bonham-Carter, Agterberg, and Wright (1988), and Wright (1988). The following description summarizes some of this previously published material, and describes new results on mapping the uncertainty of the probability estimates, and testing for conditional independence.

The Meguma Terrane of eastern mainland Nova Scotia (Fig. 2) mainly comprises lower Paleozoic turbidites (Goldenville and Halifax Formations), intruded by Devonian granites. Gold occurs in quartz veins, usually confined to the Goldenville Formation. Sixty-eight gold occurrences are recorded in the study area (McMullin, Richardson, and Goodwin, 1986), and about 30 of these have recorded production. No general consensus on the origin of the deposits has been achieved, as illustrated in Graves' and Zentilli's (1982) review.

The input maps are listed in Table 2, along with the weights, correlations, and variances. Two of the input maps are illustrated in Figure 3 (see Bonham-Carter, Agterberg, and Wright 1988). Notice that the geologic map is treated as a ternary map, and only the positive weights are employed, because the three classes are mutually exclusive. The anticlinal-axis neighborhoods (Fig. 3A) are optimized by determining the corridor width for which C is maximized. Similarly, the neighborhoods for Halifax— Goldenville contact, granite contact, and NW-trending structures are optimized by calculating C for a variety of neighborhood sizes, defined by corridor one-half-width. These calculations are discussed by Agterberg, Bonham-Carter, and Wright (1988). The lake-sediment geochemical signature was calculated by determining the linear combination of geochemical elements that best predicts lake-catchment basins containing one or more gold occurrences, and generating a regression-score map (Wright, Bonham-Carter, and Rogers, 1988). The map of regression scores was thresholded to binary form, finding the cutoff that optimizes C, (Fig.

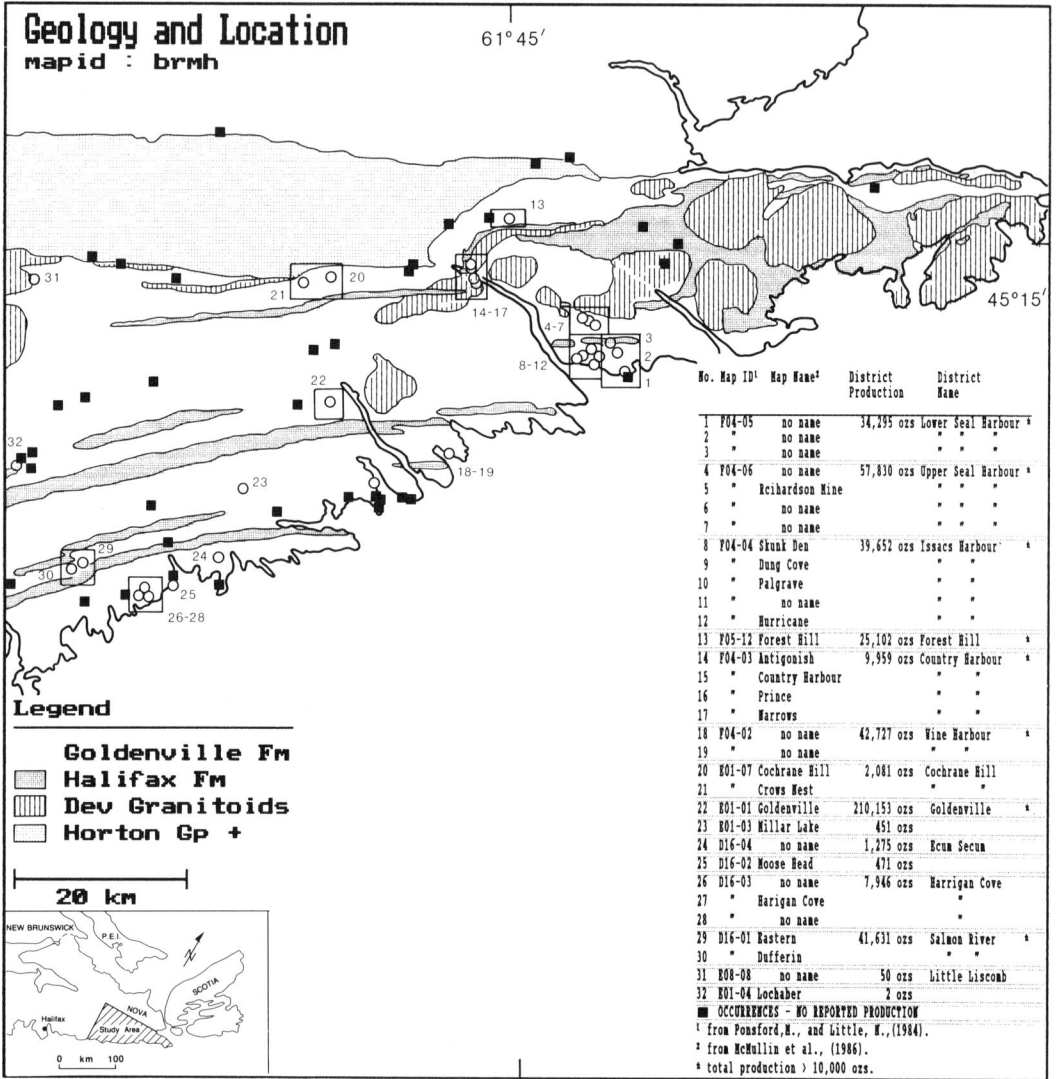

Geology and Location
mapid : brmh

Legend

- Goldenville Fm
- Halifax Fm
- Dev Granitoids
- Horton Gp +

20 km

No.	Map ID[1]	Map Name[2]	District Production	District Name
1	F04-05	no name	34,295 ozs	Lower Seal Harbour [a]
2	"	no name	"	" "
3	"	no name	"	" "
4	F04-06	no name	57,830 ozs	Upper Seal Harbour [a]
5	"	Reihardson Mine	"	" "
6	"	no name	"	" "
7	"	no name	"	" "
8	F04-04	Skunk Den	39,652 ozs	Issacs Harbour [a]
9	"	Dung Cove	"	" "
10	"	Palgrave	"	" "
11	"	no name	"	" "
12	"	Hurricane	"	" "
13	F05-12	Forest Hill	25,102 ozs	Forest Hill [a]
14	F04-03	Antigonish	9,959 ozs	Country Harbour [a]
15	"	Country Harbour	"	" "
16	"	Prince	"	" "
17	"	Narrows	"	" "
18	F04-02	no name	42,727 ozs	Wine Harbour [a]
19	"	no name	"	" "
20	E01-07	Cochrane Hill	2,081 ozs	Cochrane Hill
21	"	Crows Nest		
22	E01-01	Goldenville	210,153 ozs	Goldenville [a]
23	E01-03	Millar Lake	451 ozs	
24	D16-04	no name	1,275 ozs	Ecum Secum
25	D16-02	Moose Head	471 ozs	
26	D16-03	no name	7,946 ozs	Harrigan Cove
27	"	Harigan Cove		*
28	"	no name		*
29	D16-01	Eastern	41,631 ozs	Salmon River [a]
30	"	Dufferin		* *
31	E08-08	no name	50 ozs	Little Liscomb
32	E01-04	Lochaber	2 ozs	

■ OCCURRENCES - NO REPORTED PRODUCTION
[1] from Ponsford,N., and Little, N.,(1984).
[2] from McMullin et al., (1986).
[a] total production > 10,000 ozs.

Figure 2. Location map for gold-mineralization study.

3B). This avoids potential problems with the conditional independence assumption, because several of the geochemical elements are intercorrelated strongly. Although the geochemical regression scores represent an interesting map of gold potential, the rock type and structural information present in the other input maps cannot be captured adequately using the catchment-basin model. Furthermore, the catchment basins only cover part of the study area, and it is desirable to predict gold for the entire region.

Figure 4A shows the posterior-probability map, and $\sigma_{weights}$, $\sigma_{missing}$ and σ_{total} are in Figures 4B, 4C, and 4D, respectively. By comparing Figure 4A with Figure 4B it can be seen

Table 2. List of input maps for gold study, showing weights, variances, and contrasts.

Map	W^+	$\sigma(W^+)$	W^-	$\sigma(W^-)$	C	$\sigma(C)$
Geochemical Signature	1.0047	0.3263	−0.1037	0.1327	1.1084	0.3523
Anticline Axes	0.5452	0.1443	−0.7735	0.2370	1.3187	0.2775
NW Lineaments	−0.0185	0.2453	0.0062	0.1417	−0.0247	0.2833*
Granite Contact	0.3419	0.2932	−0.0562	0.1351	0.3981	0.3228*
Goldenville − Halifax Contact	0.3683	0.1744	−0.2685	0.1730	0.6368	0.2457
Halifax Fm.	−1.2406	0.5793	0.1204	0.1257	−1.4610	0.5928
Goldenville Fm.	0.3085	0.1280	−1.4689	0.4484	1.7774	0.4663
Devonian Granite	−1.7360	0.7086	0.1528	0.1248	−1.8888	0.7195

* H_0: $C = 0$ not rejected at $\alpha = .05$

that the uncertainty due to the weights, $\sigma_{weights}$, generally is correlated with posterior probability; areas of elevated probability are also areas with elevated uncertainty. In Figure 4C, the uncertainty due to the missing geochemistry, $\sigma_{missing}$, is zero in the lake-catchment basins, shown in white. Outside the basins, $\sigma_{missing}$, changes with the posterior probability, although the amount of variation is not large.

When the uncertainty components are combined as σ_{total} (Fig. 4D), we again notice that the total uncertainty varies with the probability. For example, the regions underlain by the Halifax Formation and Devonian granites are both low in probability and uncertainty. However, the ratio P_{post}/σ_{total} tells a different story, and shows that P_{post} in the Halifax and granite areas are not significantly different from zero. This can be seen in Figure 5 where the white areas are those with P_{post}/σ_{total} less than 1.5, superimposed as a mask on the P_{post} map. Here we see that most of the past producing gold mines are in the top two probability classes, as would be expected. However, a number of regions in the uppermost P_{post} class, indicating strong gold potential, are not associated with known gold mineralization.

Table 3 shows part of the unique-conditions report file, giving the posterior probability corresponding to the unique overlap combinations of the input maps. Using SPANS, an interactive screen query with the cursor can be used to investigate the conditions of any combination of maps in the database. As the cursor moves, a listing of the selected maps and their values appears on the monitor. This information is updated continuously as the cursor is moved. This is similar to manual checking the rows of the unique-conditions table (Table 3), but more powerful because of the spatial context involved in interactive query.

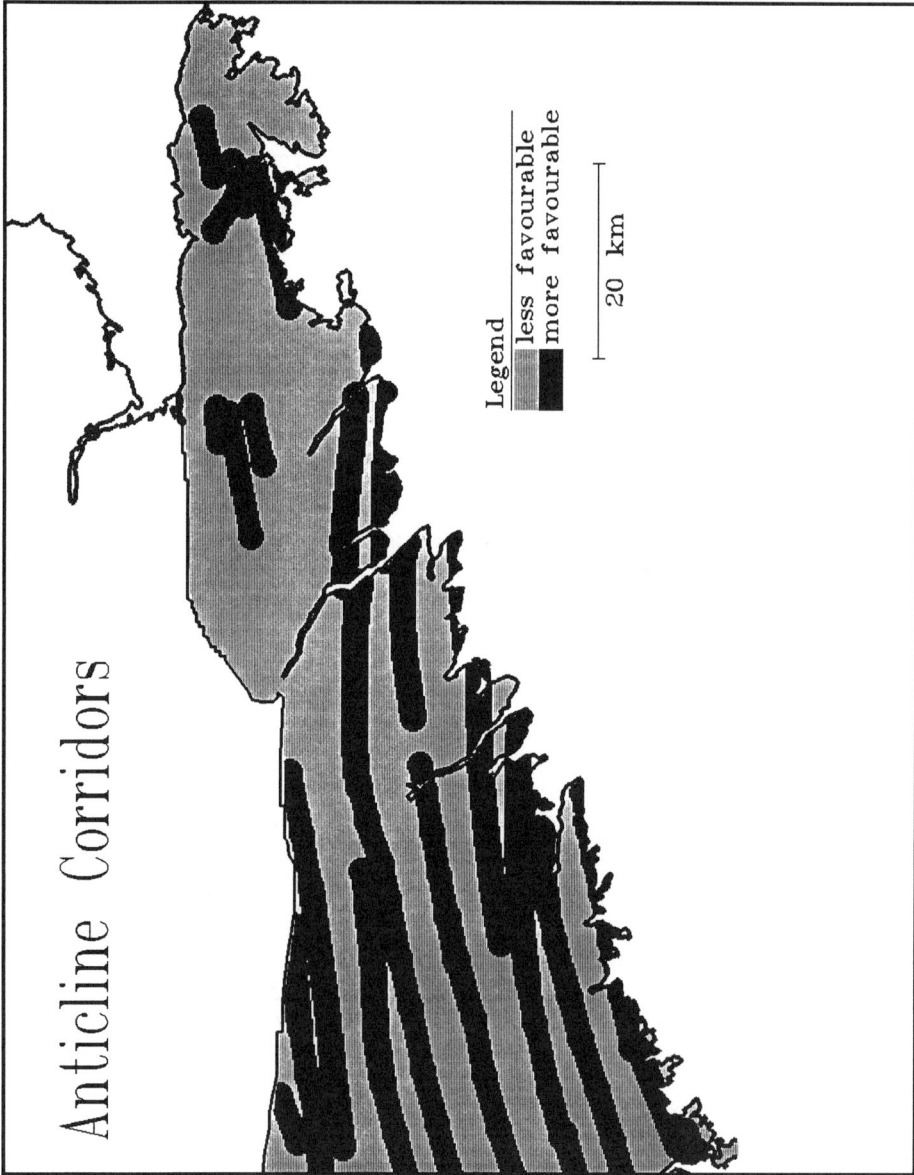

Figure 3. Two out of six input maps used in gold-mineralization study. A, Anticline corridors. B, Geochemical signature for gold, based on multielement lake-sediment survey.

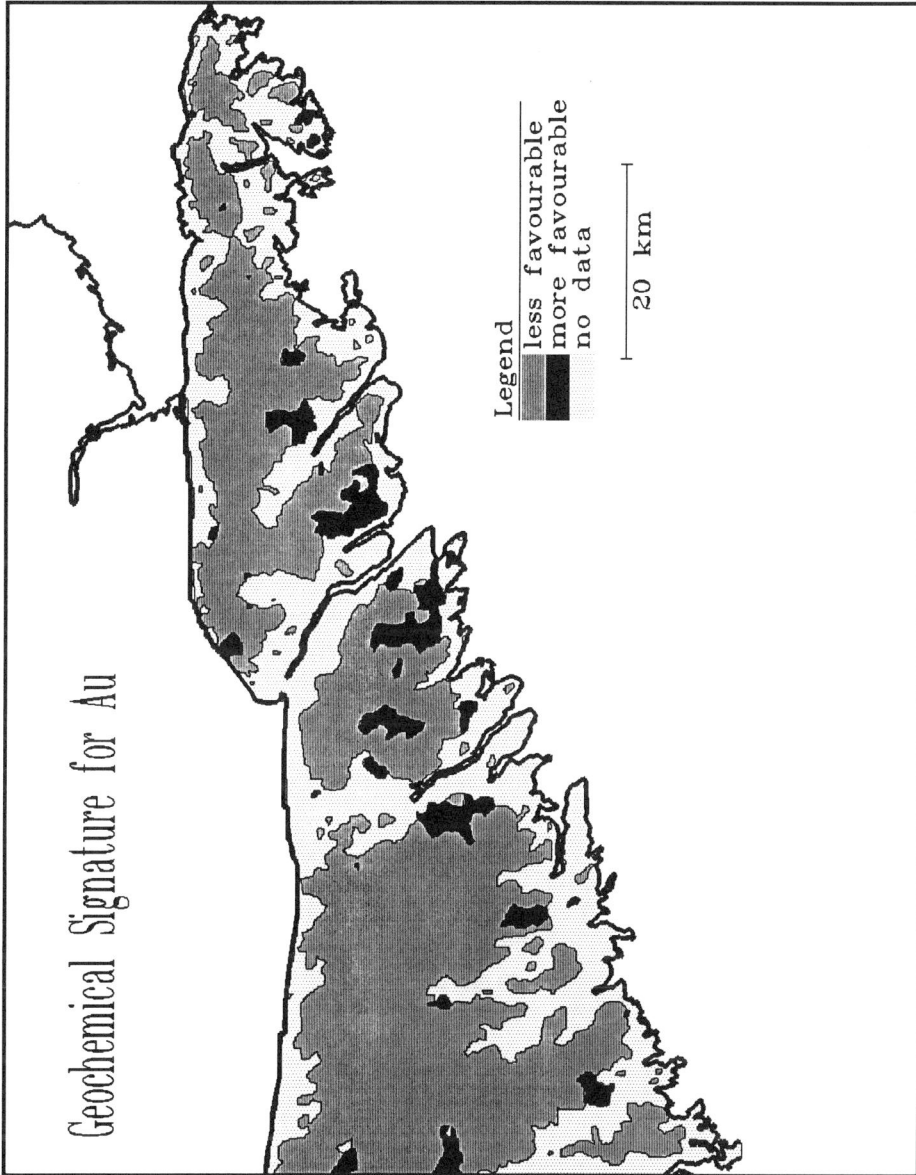

Figure 3. (continued)

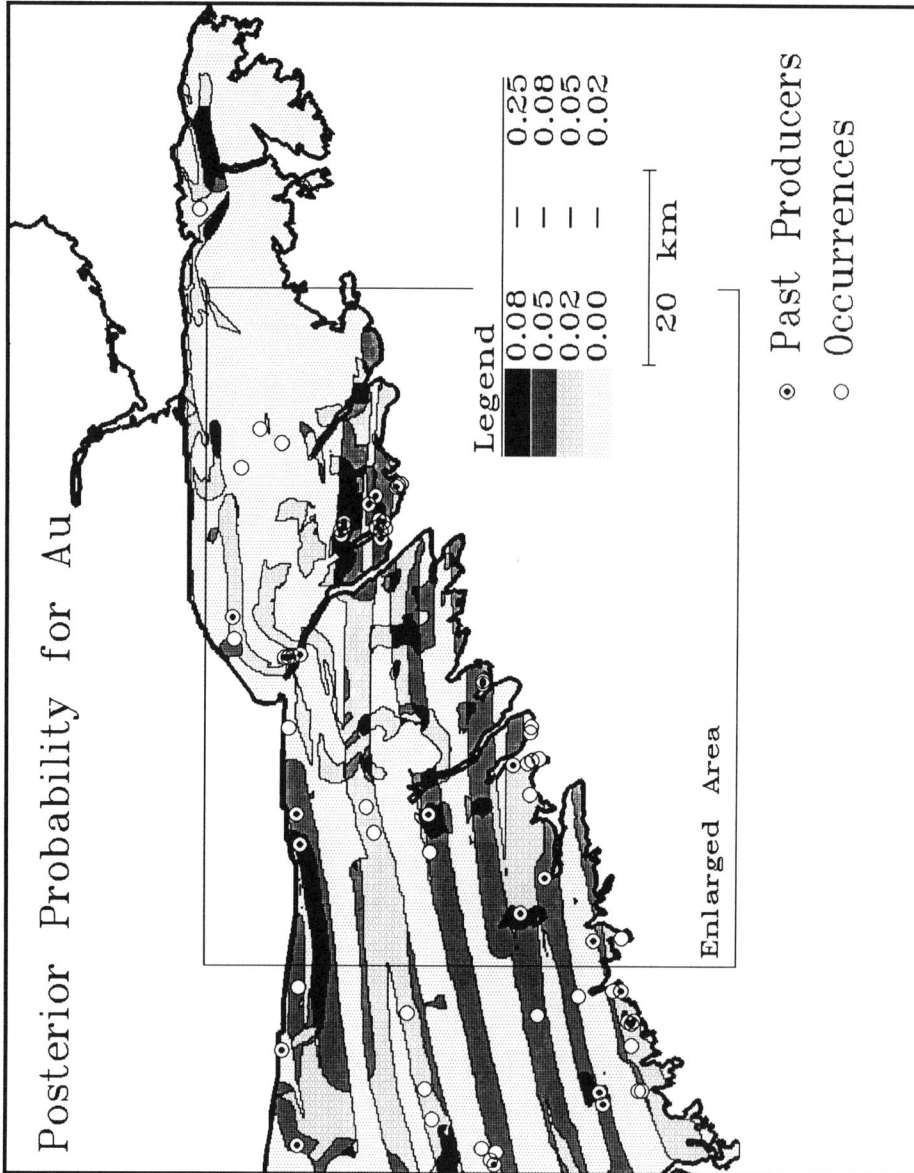

Figure 4. Output maps from gold-mineralization study. A, Posterior-probability map. P_{post}. B, Uncertainty because of weights map, $\sigma_{weights}$. C, Uncertainty because of missing geochemical-data map. $\sigma_{missing}$. D, Total uncertainty map. σ_{total}.

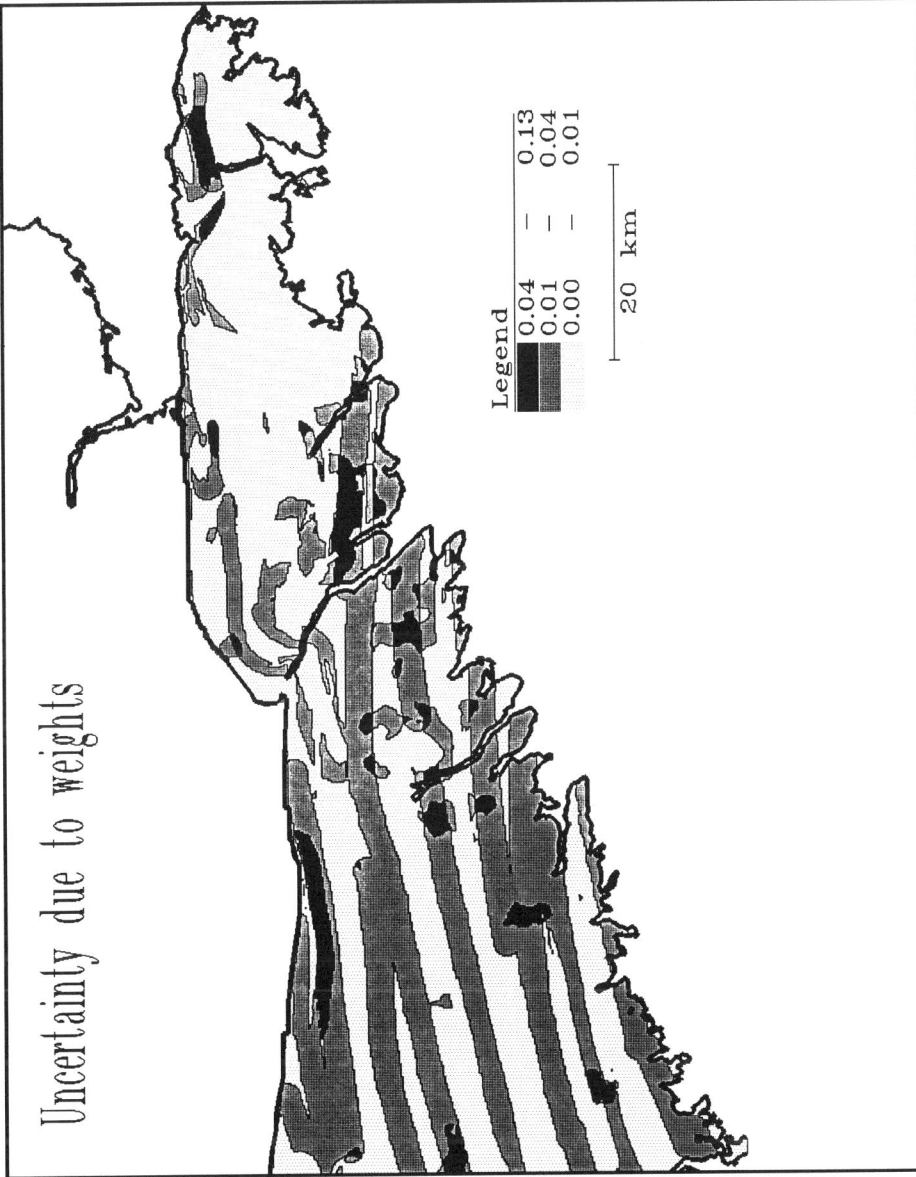

Figure 4. (continued)

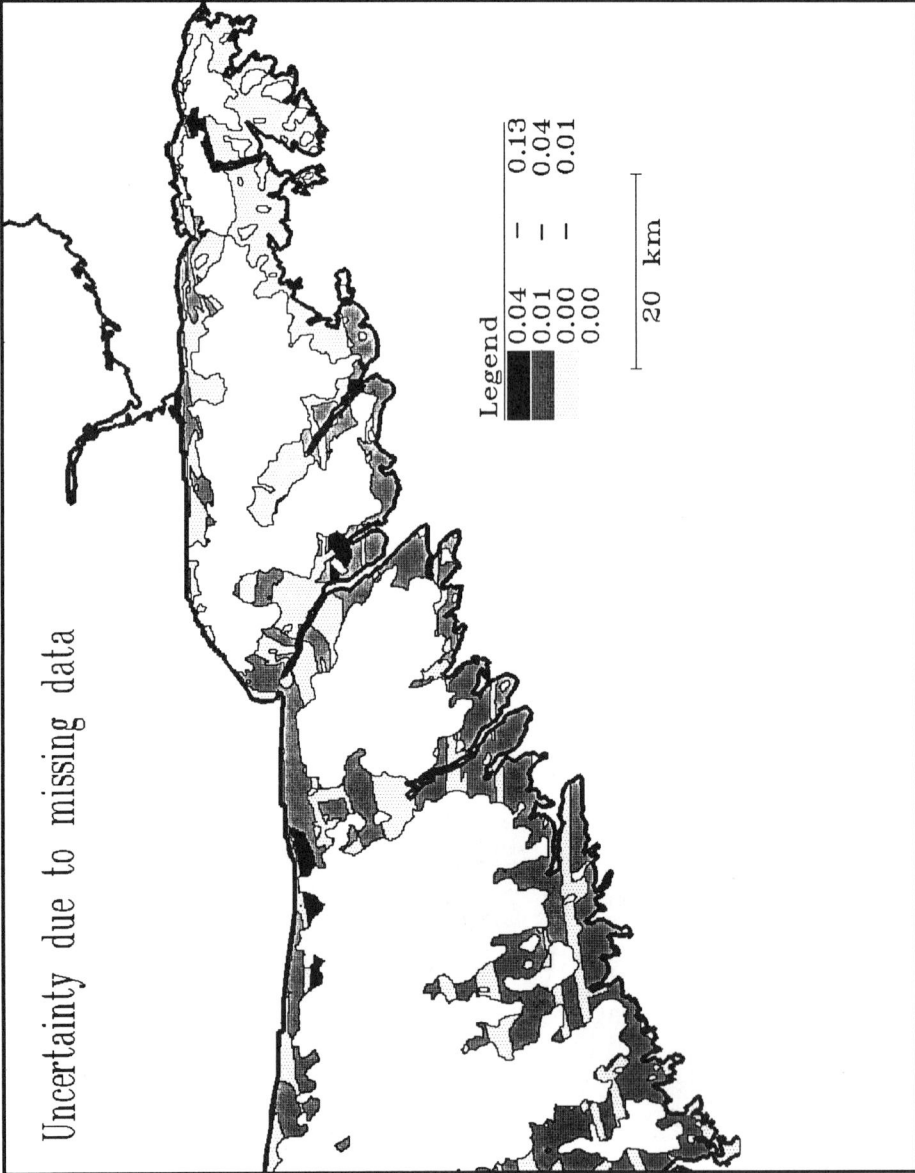

Uncertainty due to missing data

Legend

0.04	—	0.13
0.01	—	0.04
0.00	—	0.01
0.00		

20 km

Figure 4. (continued)

Total Uncertainty

Legend

	0.04	—	0.13
	0.01	—	0.04
	0.00	—	0.01

20 km

Figure 4. (continued)

GOODNESS-OF-FIT TEST

An overall comparison of goodness-of-fit between the observed distribution of occurrences, and expected distribution using the weights of evidence model, was described by Agterberg, Bonham-Carter, and Wright (1988) using a chi-square test. This showed a good fit, suggesting that the assumption of conditional independence was satisfied. One problem with the chi-square test is that the posterior-probability scale must be divided into classes, and the results can be sensitive to selection of cutoffs. The Kolmogorov–Smirnov test can be used as an alternative, and no "binning" is required, making it more suitable particularly where only a small number of occurrences are involved. We illustrate its use here.

For each occurrence, i, the unique conditions at that location are used to calculate a posterior probability from the model, P_i^*, $i = 1, 2, \ldots N_1$, where N_1 is the total number of occurrences. The P^* are sorted into ascending order. From the unique-conditions report file (Table 3), the area A_j, and posterior probability P_j for the $j = 1, 2, \ldots N_2$ unique-conditions polygons then are used. Note that these already are sorted into ascending order of P by SPANS. Then let

$$ S = \sum_{j=1}^{N_2} A_j P_j , $$

and the expected cumulative frequency distribution is

$$ F(P \leq P_i^*) = \frac{\sum_{j=1}^{N_2} A_j P_j}{S} \qquad \text{for } P_j \leq P_i^*, \ i = 1, 2, \ldots N_1 . $$

The observed cumulative distribution is

$$ F(P \leq P_i^*) = \frac{i}{N_1} \qquad \text{for } i = 1, 2, \ldots N_1 . $$

The Kolmogorov–Smirnov statistic is defined as

$$ d = \left| F_i - F_i \right|_{max} \qquad \text{over } i = 1, 2, \ldots N_1 $$

and the significance of an observed value of d is

$$ \text{Probability } (d > \text{observed}) = Q\left(\sqrt{\frac{N_1 N_2}{N_1 + N_2}}\, d \right) $$

where $Q(\lambda)$ is a function that can be calculated as described by Press and others (1986, p. 473).

For the gold data, Figure 6 shows the expected and observed cumulative distributions as a function of posterior probability. The maximum absolute difference, d, is 0.23369 giving a

Table 3. Extract from unique-conditions report for gold mineralization study.

Unique Conditions Report
Map nws2 – NW structures – Meguma (0.5 km)
     mgr8 – granite contact binary pattern
     br0m – geology – Meguma terrane
     dhg9 – Goldenville/halifax contact
     fbm2 – anticline binary pattern
     hmn4 – geochemical signature for Au
     prob – unique conditions

Case	nws2	mgr8	br0m	dhg9	fbm2	hmn4	prob	sq km	%	cum %	Post prob
75	2	1	3	2	2	1	21	36.40	.52	93.65	.0629
76	1	1	3	2	2	1	26	169.56	2.44	96.09	.0644
77	2	1	3	2	2	3	12	30.53	.44	96.53	.0693
78	1	1	3	2	2	3	11	138.39	1.99	98.52	.0709
79	2	2	3	2	1	2	76	0.07	0	98.52	.0731
80	1	2	3	2	1	2	75	1.87	.03	98.55	.0748
81	2	2	3	2	2	2	64	1.46	.02	98.57	.0887
82	1	2	3	2	2	1	60	15.34	.22	98.79	.0907
83	2	1	3	1	2	2	35	8.61	.12	98.91	.0971
84	2	2	3	2	2	3	62	3.25	.05	98.96	.0974
85	1	1	3	1	2	2	5	23.59	.34	99.30	.0993
86	1	2	3	2	2	3	63	10.23	.15	99.45	.0996
87	2	2	3	1	2	2	45	1.04	.01	99.46	.1349
88	1	2	3	1	2	2	48	1.34	.02	99.48	.1378
89	2	1	3	2	2	2	32	7.17	.10	99.59	.1689
90	1	1	3	2	2	2	27	26.07	.37	99.96	.1724
91	2	2	3	2	2	2	89	0.02	0	99.96	.2276
92	1	2	3	2	2	2	78	2.75	.04	100.00	.2320
TOTAL								6953.85	100.00		

2 = pattern present, 1 = pattern not present, 3 (br0m) = Goldenville, 3 (hmn4) = geochemistry missing

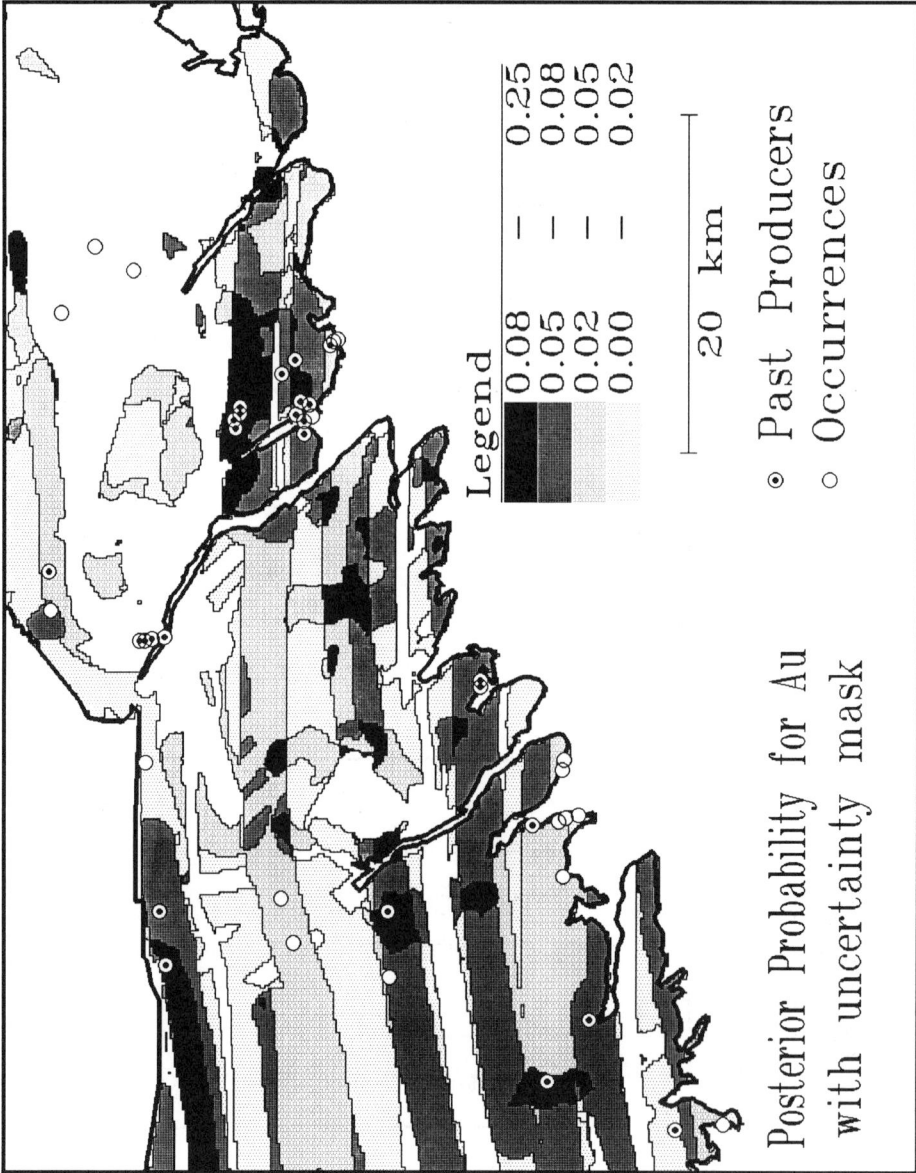

Figure 5. Enlargement of posterior-probability map in Goldenville region, with relatively uncertain areas masked out (white). Masked areas roughly correspond to Halifax Formation and Devonian granite.

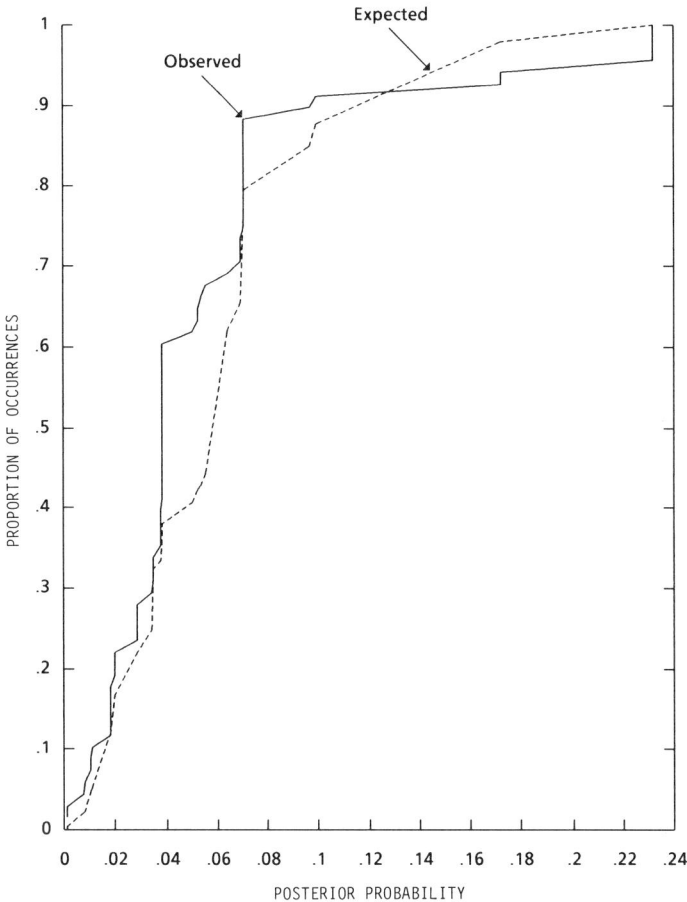

Figure 6. Plot of cumulative frequency distributions of occurrences (observed and expected) versus posterior probability.

probability (d > observed) of 0.032. Thus at a 95% confidence level, the hypothesis that the two distributions are the same would be rejected, but not at a 99% level. This result suggests that a weak violation of the assumption of conditional independence exists. The number of degrees of freedom to be used for the chi-square test is not clear, possibly explaining the different conclusion drawn from the Kolmogorov–Smirnov test.

To follow this up further, we carried out pairwise conditional independence tests, as described for the worked example, and the test statistic, G^2, values are shown in Table 4A. Most of the G^2 values are small indicating that the conditional independence assumption is not violated. However, for one map pair– anticline neighborhoods and Goldenville–Halifax contact neighborhoods– the hypothesis of conditional independence is rejected at the 95% level, but not at the 99% level. In Table 4B the observed and expected frequencies for all eight overlap classes are given. Notice, for example, that the model predicts 22.4 occurrences where both these patterns are present, whereas only 18 occurrences actually occur. Similar differences occur in the other classes. One solution to this problem would be to eliminate the map showing the

Table 4. Conditional independence tests between all possible pairs of maps. A, G^2 Values for pairwise tests of conditional independence. B, Comparison of observed and (expected) frequency of unit cells for Goldenville-Halifax contact (B_1) and Anticlines (B_2).

A

		1	2	3a	3b	3c	4	5
NW structures	1							
Granite contact	2	1.53						
Bedrock	3a	5.00	0.14					
	3b	0.39	0.24					
	3c	1.61	0.04					
GH contact	4	4.13	1.68	0.47	0.44	1.35		
Anticlines	5	0.99	0.14	0.82	2.53	1.79	7.29*	
Geochem. sign.	6	0.25	3.10	1.13	1.95	3.29	1.95	3.29

Note:

Tests between 1, 2, 4 and 5 were carried out using the full unique conditions table. Tests involving 3 or 6 were made using binary maps for each rock type separately, and eliminating unique conditions with missing geochemistry.

* Tabled χ^2, $\alpha = 0.05$, $df = 2$ is 5.99, suggesting that the hypothesis of independence be rejected, but not at $\alpha = 0.01$ when $\chi^2 = 9.21$.

B

	$B_1 \cap B_2$	$B_1 \cap \bar{B}_2$	$\bar{B}_1 \cap B_2$	$\bar{B}_1 \cap \bar{B}_2$
D	18 (22.4)	14 (9.3)	27 (22.4)	4 (8.8)
$\bar{D}$	423.2 (418.8)	575.3 (580.0)	811.9 (816.5)	1071.8 (1067.1)

Note:

Unit cell size of 1 km^2.

smallest correlation with deposits (Goldenville–Halifax contact), or combine the two maps into a single ternary map.

SUMMARY AND CONCLUSIONS

Geographic information systems can be powerful tools for spatial data integration of geoscience maps. They not only facilitate data capture and coregistration of maps, they also offer powerful computational tools such as unique-conditions mapping, that facilitate multimap modeling. Display capability and interactive query operations permit visualization and probing of data layers that enhance the understanding of results.

Weights of evidence modeling for estimating mineral potential is appealing because of its simplicity. The weights of evidence are estimated using locations of known mineral occurrences, but for underexplored areas, expert opinion could be used either instead of, or in addition to, known mineral occurrences. The ability to handle maps effectively showing proximity to linear features, and to determine the variance of probability estimates both due to uncertainties in the weights of evidence, and because of missing data, are useful aspects of this approach.

ACKNOWLEDGMENTS

We thank Danny Wright for his important contributions to this study and Bob Garrett for manuscript review. The work was supported partially by the Geological Survey of Canada under the Canada–Nova Scotia Mineral Development Agreement, 1984 - 1989.

REFERENCES

Agterberg, F.P., 1989, Systematic approach to dealing with uncertainty of geoscience information in mineral exploration: APCOM 89, Las Vegas, Nevada, in press.

Agterberg, F.P., Bonham-Carter, G.F., and Wright, D.F., 1988, Statistical pattern recognition for mineral exploration: Proc. COGEODATA Symposium on Computer Applications in Resource Exploration, Espoo, Finland, in press.

Agterberg, F.P., and Franklin, J.M., 1987, Estimation of the probability of occurrence of polymetallic massive sulfide deposits on the ocean floor, in Teleki, P. G. and others, eds., Marine Minerals: D. Riedel Publishing Co., Dardrecht, p. 467–483.

Bishop, M.M., Fienberg, S.E., and Holland, P.W., 1975, Discrete multivariate analysis: theory and practice: MIT Press, Cambridge, Massachusetts, 587 p.

Bonham-Carter,G.F., Agterberg, F.P., and Wright, D.F., 1988, Integration of geological datasets for gold exploration in Nova Scotia: Photogrammetry and Remote Sensing, v. 54, no. 11, p. 1585–1592.

Campbell, A.N., Hollister, V.F., and Duda, R.O., 1982, Recognition of a hidden mineral deposit by an artificial intelligence program: Science, v. 217, no. 4563, p. 927–929.

Graves, M.C., and Zentilli, M., 1982, A review of the geology of gold in Nova Scotia, *in* Geology of Canadian gold deposits: CIMM Special Paper, p. 233–242.

Harris, D.P., 1984, Mineral resources appraisal: Oxford Univ. Press, Oxford, 445 p.

McMullin, J., Richardson, G., and Goodwin, T., 1986, Gold compilation of the Meguma Terrane in Nova Scotia: Nova Scotia Department of Mines and Energy, Open Files 86-055, 056 (map with marginal notes).

Press, W.H., Flannery, B.P., Teukolsky, S.A., and Vetterling, W.T., 1986, Numerical recipes: The art of scientific computing: Cambridge Univ. Press, Cambridge, 818 p.

Samet, H., 1984, The quadtree and related hierarchical data structures: Computing Surveys, v. 16, no. 2a, p. 187–260.

Spiegelhalter, D.J., 1986, A statistical view of uncertainty in expert systems, *in* Gale, W. A., ed., Artificial intelligence and statistics: Addison-Wesley Publ. Co., Reading, Massachusetts, p. 17–55.

Spiegelhalter, D.J., and Knill-Jones, R.P., 1984, Statistical and knowledge-based approaches to clinical decision-support systems, with an application in gastroenterology: Jour. Royal Statist. Soc., A, v. 147, pt. 1, p. 35–77.

TYDAC, 1987, Spatial analysis system reference guide, version 3.6: TYDAC Technologies Inc., 1600 Carling Avenue, Ottawa, Canada, K1Z 8R7, 300 p.

Wright, D.F., 1988, Data integration and geochemical evaluation of Meguma Terrane, Nova Scotia, for gold mineralization: unpubl. masters thesis, Univ. of Ottawa, 82 p.

Wright, D.F., Bonham-Carter, G.F., and Rogers, P.J., 1988, Spatial data integration of lake-sediment geochemistry, geology and gold occurrences, Meguma Terrane, Eastern Nova Scotia: Proc. Can. Inst. Mining Metall., Prospecting in Areas of Glaciated Terrain–88, p. 501–515.

Stimulation Through Simulation: Geochemical Modeling

John C. Butler
University of Houston

ABSTRACT

A survey of classrooms and offices would detect microcomputers assisting with numerous tasks which include the gathering, interpretation, and synthesis of data and the processing of words, images, and ideas. For instructors or managers, there is a responsibility to ensure that the microcomputer is used as more than just a faster and neater way to accomplish tasks that were undertaken BmC (before microcomputers). The aspects of the microcomputer which make it suitable for the tasks noted here also can enhance understanding by encouraging experimentation with concepts and models.

Two broad categories of examples (chemical kinetics and geochemical cycles) are described to illustrate the ease with which dynamic models can be constructed.

INTRODUCTION

There is no doubt that the widespread distribution of microcomputers has had a major impact in the workplace and in academia. Several years of teaching with the aid of computing resources argues that the learning process is stimulated when the student (and the instructor!) has access to software that facilitates the simulation of a process. Almost every instructor (or manager) has engaged in the "what if" game to provoke a discussion which allows evaluation of the degree of understanding of a model. A particular chemical system is at equilibrium. What would happen if the amount of one component were increased? What would happen if the temperature were decreased or the pressure increased? How will the components of the system adjust to the disturbance? Will dynamic equilibrium be restored? What are the parameters of the new system and how long will restoration take? Will the change result in a transient or nonsteady state cycle? The longer it takes for those questioned to be in a position to respond, the smaller the benefits of the game. In both the classroom and the workplace, the ability to handle successfully the "what if" approach may provide an answer to a specific question as well as leading to a better understanding of the problem itself.

There is an abundance of information characterizing the chemical behavior of geologic materials and a knowledge of the principles of chemistry is central to many problems facing the geoscientist. The purpose of this paper is to illustrate how the principles of chemical kinetics and geochemical cycles can be modeled to solve a particular problem and to enhance the understanding of some of the important concepts. Experiments described in this contribution were conducted on Macintosh SE with STELLA as the modeling system. A brief discussion of modeling system follows.

STELLA

STELLA (Structural Thinking Experimental Learning Laboratory with Animation) is a dynamic modeling system for the Macintosh family of microcomputers (Richmond, Peterson, and Vescuso, 1987).   [STELLA can be obtained from High Performance Systems, 13 Dartmouth College Highway, Lyme, New Hampshire, 03768.] The system comes complete with its own language and a variety of procedures for displaying results from simulation studies; time series, x-y scatter diagrams, and as matrices. The software conforms to the standard Macintosh interface and it is easy to transport data matrices and graphics to other software packages.

STELLA is built around algorithms for solving finite difference equations which function as discrete approximations to differential equations. The modeling language is icon-based. When a STELLA worksheet is opened up, four tools (stock, flow, converter, and connector) and three tool manipulators (hand, ghost, and dynamite) appear on the left-hand side of the screen. The hand allows the user to select (a click and drag operation familiar to mouse users) and position the tools. The ghost allows reproduction of components of the model and the dynamite is used to remove portions of the model.

A model illustrating the concept of residence time is given in Figure 1. A dynamic system that is being modeled usually is divided into a number of storage units or reservoirs (Garrels, Mackenzie, and Hunt, 1975). The flow rate into or out of a reservoir is the flux of the substance. Residence time is defined as:

$$\frac{\text{Amount of material in the reservoir at a given time}}{\text{Instantaneous rate of addition (or subtraction) of material}}$$

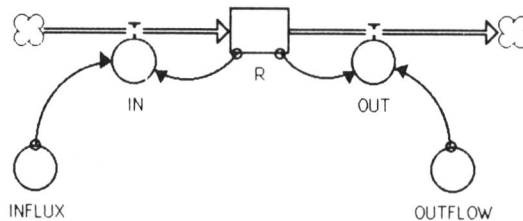

Figure 1. STELLA model for residence time.

The system modeled in Figure 1 is that of a constant reservoir (50 liters) with a constant flux in and out (6 liters per minute) and a single source and sink. Double clicking on an icon in the model opens a definition window which allows the model to be defined algebraically. For example, double clicking on R (Fig. 1) allows the initial value of the reservoir to be set at 50 liters. Stocks (R in Fig. 1) accumulate (or loose) something per unit time and are equivalent to reservoirs as defined. Flows (fluxes) into and out of the reservoir are controlled by the flow symbol (an arrow with a valved-container) with the arrow indicating the direction of flow. The cloud (the four-lobed shape at the end of the arrow) represents an infinite supply (source) or storage area (sink). Reservoir R is defined as a finite difference equation:

$$R = R + dt * (influx - outflow)$$

In more traditional notation, the finite difference equation could be written as:

$$R_t = R_{(t-\Delta t)} + \Delta t * (influx - outflow)$$

That is, the temperature at time t is equal to the temperature at one instant ago (t-Δt) plus the change in the contents of the reservoir that took place during that instant [(Δt*(influx-outflow)]. The flows influx and outflow are defined as 6 liters per minute. Converters (circles such as in Fig. 1) combine one or more inputs into a single output. Connectors (lines or curves with an arrow at one end) serve to link, stocks, flows, and converters to define the model. In Figure 1 the two converters (in and out) define the flux into and out of the reservoir as:

(1) in = reservoir/influx and

(2) out = reservoir/outflow,

respectively.

Once the model is defined completely, the user selects the time interval between steps (dT), the starting time, the total time the simulation will be performed and the simulation algorithm (Euler's method, 2nd-order, or 4th-order Runge-Kutta). Euler's method assumes that a flow is constant over the calculation interval (dT). The Runge-Kutta methods estimate the flow at either two or four points within the calculation interval. When the model is executed the initial conditions for the stocks, flows, and converters are calculated and the flows update the stocks. The new values for the stocks then are used to recalculate the flows and so on until the simulation time is exceeded. In the model given above the residence time is 50 liters/6 liters/minute = 8.33 minutes. The converters "store" the fluxes in and out which are computed at each time interval. Contents of stocks, flows, and converters can be displayed as time series or scatter diagrams.

CHEMICAL KINETICS

Studies of chemical equilibria and thermodynamics are concerned only with initial and final states and not with the mechanism(s) of the process or the rates of approach to equilibrium (Daniels and Alberty, 1963). Chemical kinetics, on the other hand, has as its domain an analysis of the steps that lead to the final products and the factors which determine the rates of such steps. Dynamic modeling is an ideal tool for examining chemical systems for which sufficient information exists for adequate characterization.

REVERSIBLE FIRST-ORDER REACTIONS

A STELLA model for a reversible first-order reaction is given in Figure 2 and can be represented as:

$$A \underset{k_2}{\overset{k_1}{=}} B \tag{1}$$

where k_1 and k_2 are the rate constants for the forward and reverse reactions, respectively.

As an example, k_1 and k_2 are fixed at 0.2 and 0.1 hour^{-1} respectively. Alternatively, these first-order rate constants could be characterized casting them as half-lives (t5) using the following relationship:

$$t_5 = 0.693/k. \tag{2}$$

Thus, the half-lives for the forward and reverse reaction rates are 3.47 hours and 6.93 hours, respectively.

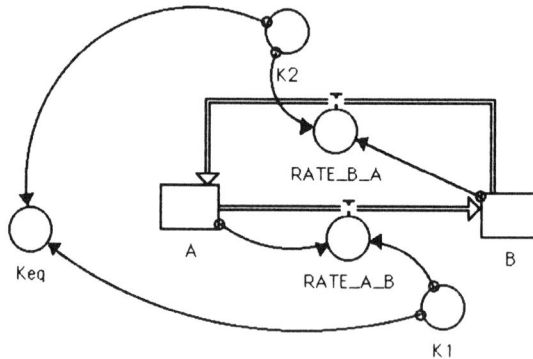

Figure 2. Reversible first-order reaction model.

The reaction rates are defined as:

$$\text{Rate A\_B} = k_1 A_0 \tag{3a}$$

$$\text{Rate B\_A} = k_2 B_0 \tag{3b}$$

where A_0 and B_0 are the initial amounts of components A and B, respectively. Time in hours versus the concentrations of A and B are plotted in Figure 3 using the rate constants as noted with A_0 and B_0 set at 1 and 0 units, respectively. At equilibrium $d(A)/dt = 0$ and the concentrations of A and B are represented by A_{eq} and B_{eq} and:

$$A_{eq}/B_{eq} = k_1/k_2 = K_{eq} \tag{4}$$

where K_{eq} is defined as the equilibrium constant; 2 for the reaction as stated.

If $A_0 = A + B$ (which implies that B_0 equals 0.0 at t=0) the equilibrium concentration of A can be formulated as:

$$A_{eq} = (k_2/(k_1+k_2))A_0 \tag{5}$$

The equilibrium concentration of component A is 0.333 concentration units so the equilibrium concentration of B must be 0.667 concentration units (Fig. 3). When chemical equilibrium is attained the forward and reverse rates are equal. At equilibrium the rates of the forward and reverse reactions are $k_1 A_{eq}$ and $k_2 B_{eq}$, respectively (0.0666 concentration units per hour). Of interest is how the two reaction rates approach the

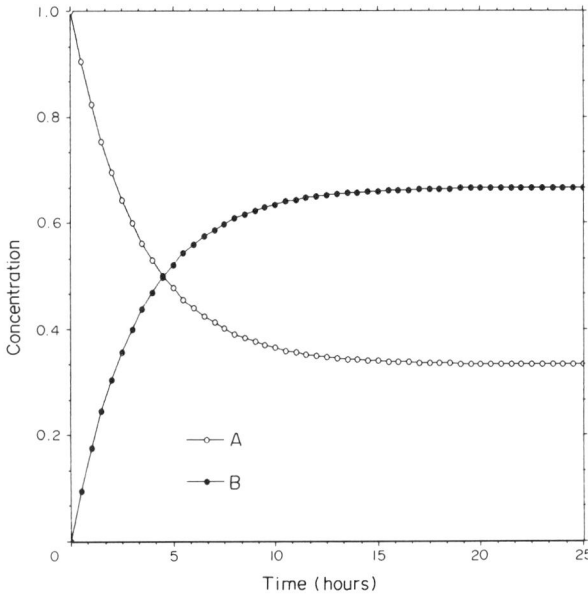

Figure 3. Concentration of A and B versus time for model in Figure 2 ($k_1 = 0.2$; $k_2 = 0.1$; $A_0 = 1$; $B_0 = 0$).

equilibrium value. The easiest way to explore their variation with time is through dynamic modeling. A time-series plot of Rate A_B and Rate B_A is given in Figure 4. This graphic illustration of equilibrium (Fig. 4) indicates more to most individuals than the formal verbal definition given. If both A and B are present at t = 0, Equation (5) does not hold. An easy way to determine the characteristics of such a system is through dynamic modeling. The model described was changed so that both A and B were present at one concentration unit each at the start of the reaction and a time-series plot for the concentrations of A and B is given in Figure 5. The conservation of material requires that $A + B = A_0 + B_0$. The rate constants are the same as the model depicted in Figure 2 and K_{eq} remains unchanged (2). This, coupled with the conservation of material as stated, requires that the equilibrium amounts of A and B are 0.667 and 1.333, respectively, as shown in Figure 5.

It is instructive to inquire as to what will happen to a system at dynamic equilibrium if a disturbance occurs. The model depicted in Figure 2 was modified as shown in Figure 6. Note that reservoir A is no longer connected to Rate A_B. Instead, a new converter AA is linked to reservoir A and defined as:

AA = IF TIME >25 THEN A+.333 ELSE A.   (6)

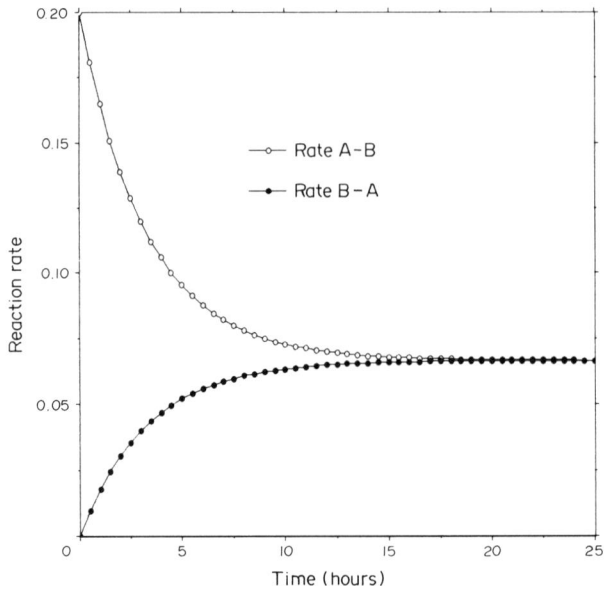

Figure 4. Time-series plot of reaction rates (A-B and B-A) for model in Figure 2.

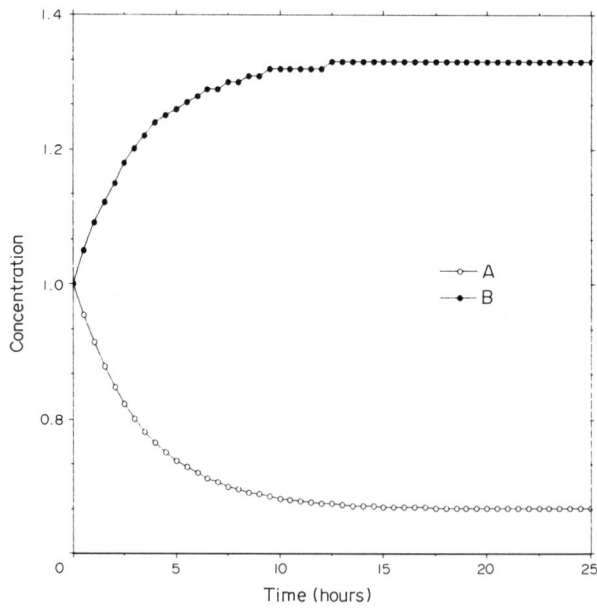

Figure 5. Concentration of A and B versus time for modification of model in Figure 2
($A_0 = 1$ and $B_0 = 1$).

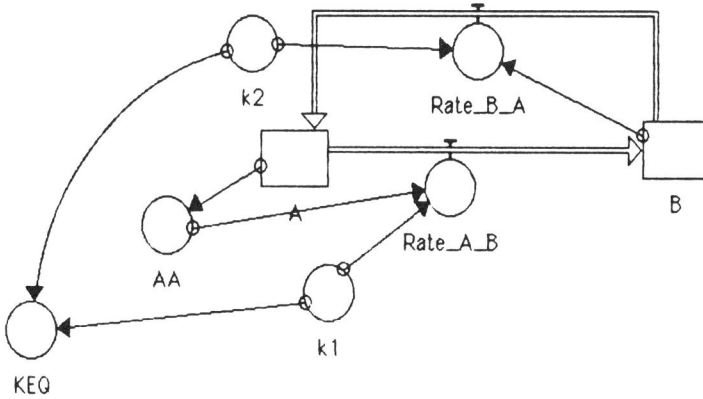

Figure 6. Modified first-order reversible reaction model.

That is, when TIME is less than 25 hours converter AA is equal to reservoir A. When TIME exceeds 25 hours AA is equal to the contents of reservoir A plus 0.333. This effectively doubles the concentration of component A after equilibrium is established between A and B. A general principle for assessing the effect of a disturbance on a system at dynamic equilibrium is that the system will react so as to undo the effect of the disturbance. Therefore, an increase in the concentration of A should favor the forward reaction with component B increasing until K_{eq} returns to a value of 2.0. A time-series plot of the concentrations of components A and B is given in Figure 7 and rates of the forward and reverse reactions are plotted versus time in Figure 8. The effect of Equation (6) is to make the concentrations of A and B equal at 25 hours. Component B increases at the expense of A until the equilibrium amounts of B and A equal 0.888 and 0.444, respectively. A new condition of dynamic equilibrium is established within approximately 10 hours after the disturbance (see Figs. 7 and 8).

The model described in Figure 2 was modified by increasing the rate constant for the forward reaction (k_1) from 0.2 to 0.4 hour^{-1} at 25 hours into the experiment. A time-series plot of the concentrations of A and B is given in Figure 9. When k_1 is increased at equilibrium, the forward rate is increased which results in an increased concentration of B. Prior to the disturbance, equilibrium concentrations of A and B are 0.333 and 0.667, respectively. Following the disturbance, the equilibrium constant (K_{eq}) is increased to a value of 4 and the equilibrium forward and reverse reaction rates are 0.080. The post-disturbance equilibrium concentrations of A and B are 0.200 and 0.800, respectively (Fig. 9). From Figure 9, the effect of the disturbance is eliminated after approximately 10 hours.

PARALLEL AND CONSECUTIVE FIRST-ORDER REACTIONS

Parallel and consecutive first-order reactions are represented by Equations (7) and (8), respectively.

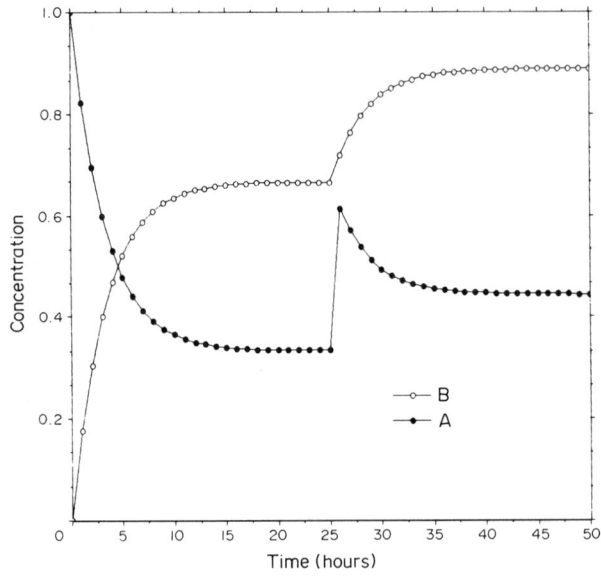

Figure 7. Effect of doubling concentration of A at 25 hours using model in Figure 6.

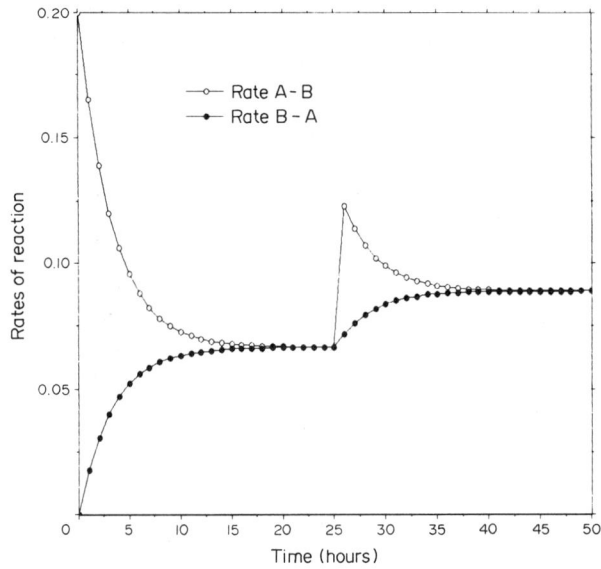

Figure 8. Time-series plot for reaction rates for model in Figure 6.

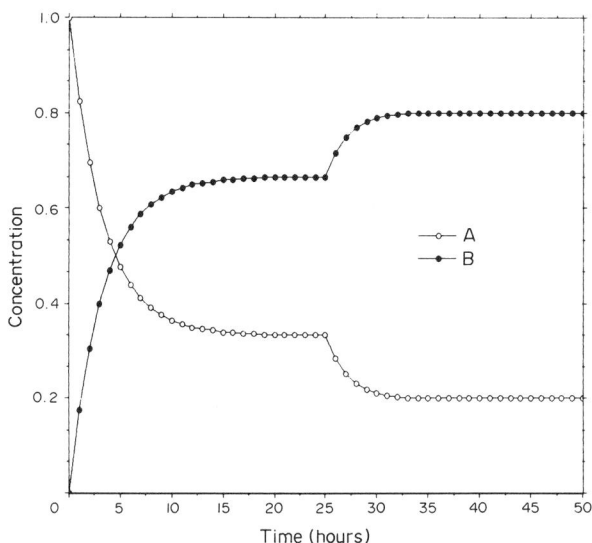

Figure 9. Concentration of A and B versus time for modification of model in Figure 6 where K_1 is doubled at 25 hours.

$$\begin{array}{c} k_1 \\ A \to B \end{array} \tag{7}$$

$$\begin{array}{cc} k_1 & k_2 \\ A \to B \to C \end{array} \tag{8}$$

Daniels and Alberty (1961) give the differential equations that must be solved simultaneously to obtain the solutions for these types of reactions. Although the mathematics are not complicated, many individuals might be tempted to accept the words and plots in texts rather than derive the relationships and evaluate the models. It is through hands-on experimentation, however, that understanding is assessed. The STELLA model for a consecutive first-order reaction is given in Figure 10. A time-series plot for the concentrations of A, B, and C is given in Figure 11 where $k_1 = 0.10$ hour^{-1}, $k_2 = 0.05$ hour^{-1}, $A_0 = 1$, $B_0 = 0$, and $C_0 = 0$. Note the presence of a slight induction period before the appearance of C. Daniels and Alberty (1961) note that the rate of production of C is complicated:

$$C_0 = A_0(1 + (1/(k_1-k_2))(k_2e^{-k_1t} - k_1e^{-k_2t}) \tag{9}$$

From Figure 11, however, the output from the model (Fig. 10) clearly displays the sequence of reactions taking place in the system.

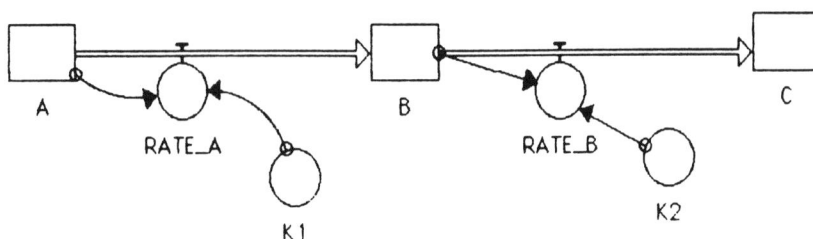

Figure 10. First-order consecutive reaction model.

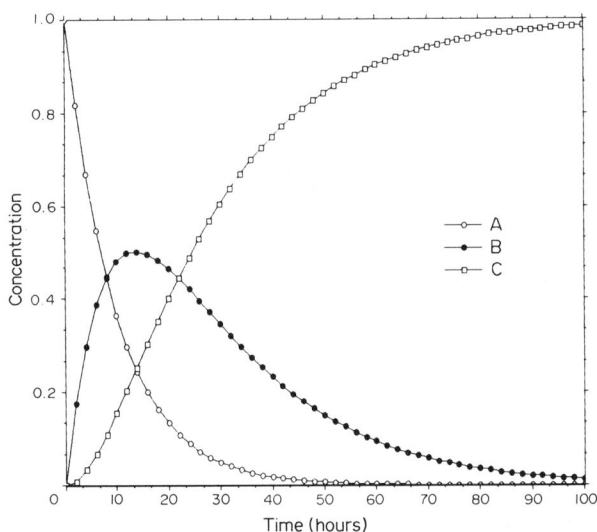

Figure 11. Concentration of A, B, and C versus time for model in Figure 10 (k_1 = 0.10; k_2 = 0.05; A_0 = 1; B_0 = 0; C_0 = 0)

As noted by Daniels and Alberty (1961), there is a relationship between the rates of the forward and reverse reaction rates and the equilibrium constant regardless of how complicated the chemical system is. The algebraic expressions required to describe more complicated first-order reactions increase in complexity as discussed by Alberty and Miller (1957).

The reactions modeled here are special classes of what might be termed a triple:

$$\begin{array}{c} A = B \\ \backslash\backslash \quad // \\ C \end{array}$$

A model for the triple reaction is given in Figure 12. Numerous special situations of the triple reaction are described in Alberty and Miller (1957). The rate constants (see Fig. 12) cannot all be independent according to the principle of microscopic reversibility (Alberty and Miller, 1957):

$$k_1 k_3 k_5 = k_2 k_4 k_6 \tag{10}$$

Alberty and Miller (1957) note that the general integrated equations for the triple are so complicated that they do not record them in their paper. However, it is a simple task to

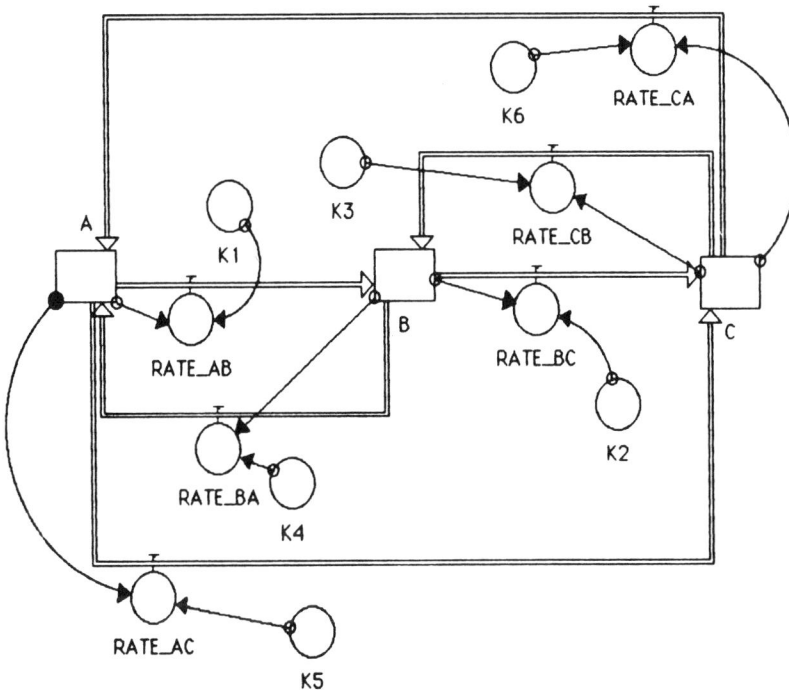

Figure 12. STELLA model for "triple" reaction involving components A, B, and C.

model the triple reaction. A time-series plot of the concentrations of A, B, and C is given in Figure 13 for $k_1 = .1$, $k_3 = .05$, $k_5 = .2$, $k_2 = .1$, $k_4 = .3$, $k_6 = .0333$, (all in hour^{-1}) $A_0 = 1$, $B_0 = 0$, and $C_0 = 0$.

Successful construction of such models confirms understanding, encourages experimentation, and facilitates incorporation of the underlying principles into independent projects. Experiences to date with students suggest that this type of modeling truly stimulates interest and enhances the learning process.

GEOCHEMICAL CYCLES

Many authors (for example, Garrels, Mackenzie, and Hunt, 1975; Broecker, 1971) have developed geochemical distribution models using a "box" approach. As with experiments with reaction kinetics, geochemical modeling allows the experimenter to assess the

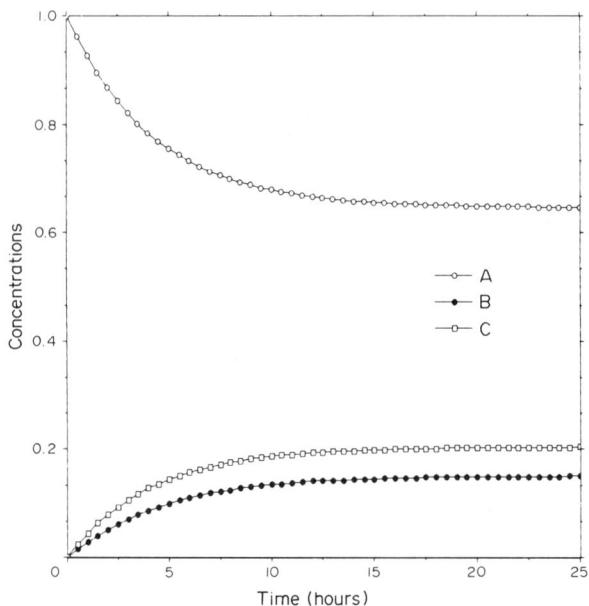

Figure 13. Time-series plot for model in Figure 12 using rate constants as defined in text.

"correctness" of an analysis of a particular problem. If a particular distribution pattern is thought to be represented by a steady-state or stationary cycle (Lerman, Mackenzie, and Garrels, 1975) the model should exhibit a constancy of fluxes and reservoir contents as a function of time. Failure to do so indicates that either the model is characterized incorrectly or that the distribution pattern may reflect a transient or nonsteady-state. If the geochemical cycle is steady-state, the user can experiment with various disturbances to assess the response of the system.

Garrels, Mackenzie, and Hunt (1972) describe several quantitative geochemical models. The distribution of lead between the bone, blood, and soft tissue reservoirs in humans was selected to illustrate the applicability of STELLA to geochemical cycle modeling (Fig. 14). Values of the fluxes (from Garrels, Mackenzie, and Hunt, 1972, p. 174) are given in Table 1. Note (from Figure 14 and Table 1) that there are two sources of lead — food and drink and the atmosphere. Lead leaves the body by excretion from the blood into urine and in feces, sweat, hair, and nails. The fluxes and initial reservoir contents were selected so that the system is in a steady-state. Residence times (reservoir content/sum of fluxes into the reservoir) for the bone, blood, and soft tissue reservoirs are 78 years, 120 days, and 35 days, respectively.

In Figure 15 the contents of the bone, blood, and soft tissue reservoirs as a function of time are invariant up to 50 days which reflects the steady-state behavior of the system. At day 50 the lead absorption is increased from 0.050 mg/day to 5 mg/day; perhaps the

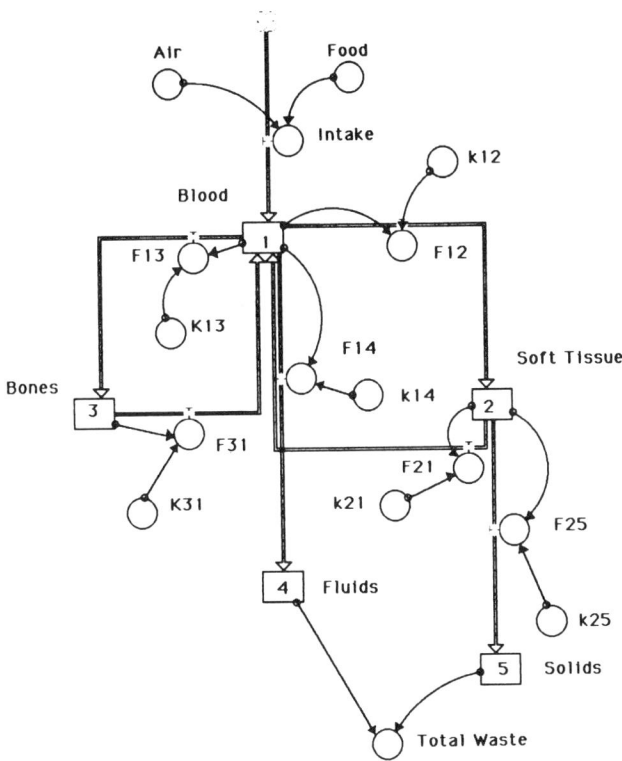

Figure 14. STELLA model of lead metabolism in humans (from Garrels,Mackenzie, and Hunt, 1975).

individual begins to drink orange juice from a ceramic jug. The contents of the bone, blood, and soft tissue reservoirs increase at a decreasing rate until the reservoir contents and fluxes are 100 times those recorded in Table 1. Given the relatively large concentration of lead in bones it is not surprising that after 500 days (Fig. 15), the blood and soft tissue reservoirs are within approximately 10% of the expected equilibrium values whereas the bone reservoir must increase by an additional 1,500,000 micrograms of lead.

SUMMARY

Models have become an integral part of the geosciences and those which are quantitative generally are amenable to simulation. Such an approach should lead to a better understanding of the principles underlying the model. As demonstrated by Lerman, Mackenzie, and Garrels, (1972) and Garrels, Mackenzie, and Hunt (1975), these models can be evaluated (ignoring complicated or "messy" mathematics) with pencil and paper. Experience suggests, however, that software packages such as STELLA can be a powerful stimulus for encouraging experimentation and understanding.

Table 1. Lead in humans

Reservoir	Initial Contents (micrograms of lead)
(1) Blood	1,800
(2) Soft Tissue	700
(3) Bones	200,000
(4) Fluids	0
(5) Solids	0

Fluxes	Rate (micrograms lead/day)	Rate Constants
(1) F12	8	0.0111
(2) F21	20	0.0114
(3) F13	7	0.00389
(4) F31	7	0.000035
(5) F14	38	0.0211
(6) F25	12	0.0171

(from Garrels, Mackenzie, and Hunt, 1975, p. 174)

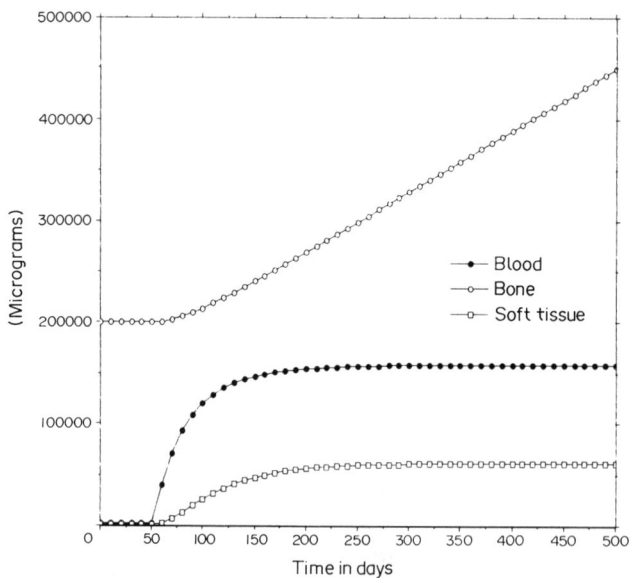

Figure 15. Time-series plot of contents of bone, blood, and soft tissue reservoirs (Fig. 12) after intake (atmosphere, food, and drink) is increased by factor of 100 at day 50.

REFERENCES

Alberty, R.A., and Miller, W.G., 1957, Integrated rate equations for isotopic exchange in simple reversible reactions: Jour. Chem. Phys., v. 26, no. 5, p. 1232-1237.

Broecker, W.S., 1971, A kinetic model for the chemical composition of seawater: Quaternary Res., v. 1, p. 188-207.

Daniels, F., and Alberty, R.A., 1961, Physical chemistry (2nd ed.): John Wiley & Sons, Inc., New York, 744 p.

Garrels, R.M., Mackenzie, F.T., and Hunt, C., 1975, Chemical cycles and the global environment: William Kaufmann, Inc., Los Altos, California, 206 p.

Lerman, A., Mackenzie, F.T., and Garrels, R.M., 1975, Modeling of geochemical cycles, *in* Quantitative studies in the geological sciences: Geol. Soc. America Mem. 142, p. 205-218.

Richmond, B., Peterson, S., and Vescuso, P., 1987, An academic user's guide to STELLA: High Performance Systems, Lyme, New Hampshire, 392 p.

The Evaluation of Pore-Geometry Networks in Clastic Reservoir Lithologies Using Microcomputer Technology

S. M. Habesch
Poroperm-Geochem Limited

ABSTRACT

The migration of fluids (hydrocarbon and aqueous) through reservoir lithologies is controlled mainly by the geometry of the pore networks. Geometric factors of most significance to fluid flow and reservoir quality, include pore size, pore network specific surface area (roughness), pore shape, pore connectivity, and pore tortuosity.

Direct three-dimensional measurements of these geometric properties are at present not possible and pore geometry or "porofacies" usually is inferred from volumetric porosity, permeability, and electroresistivity data. However, in most clastic reservoirs, the macropore geometry is well represented by resin-impregnated thin sections. Using back-scattered electron imaging of lithologic sections, large amount of two-dimensional pore geometry data can be acquired rapidly by image digitization, processing, and analysis software on a mainframe or microcomputer.

This article describes:

(a) The necessary stages for rapid, routine processing/analysis of back-scattered electron images of two-dimensional pore structures using a desktop microcomputer and commercially available software.

(b) The parameterization of pore structures by using a customized package termed POROS to generate measurements on the digitized pore images. These include the generation of both first-order (pore area, diameter, perimeter, orientation, and connectivity) and second-order (porosity, pore density, pore shape, and specific surface area) parameters.

(c) How large volumes of two-dimensional pore-geometry data are effective in modeling variations in reservoir potential of clastic lithologies. The pore-geometry data can be used to calculate reservoir parameters such as permeability and capillary pressure and also to generate lithologic classification schemes based on pore geometry "end-members." Several examples from the Brent lithologies of the North Sea Sector are provided to illustrate the techniques.

INTRODUCTION

At this time of generally low return prices for crude oil and increasing costs involved in core recovery in hydrocarbon field exploration and development programs, especially in offshore sectors such as the North Sea Basin in Europe, it especially is important to collect as much relevant data as possible from recovered core lithologies. Current reservoir modeling techniques employed to optimize production depend upon an in-depth knowledge of the nature and distribution of porosity within a reservoir system. One of the major controlling factors for hydrocarbon migration through reservoir lithologies is the geometric form of the pore structure. In this context of pore geometric systems, conventional techniques of porosity assessment in reservoir modeling (by conventional and special core analysis) are limited in the level of data provided, time consuming, may be sample destructive, and are becoming increasingly expensive at a time of oil price recession.

In established field development programs, there is an increasing demand among production reservoir engineers and petrophysicists for more detailed data on reservoir porosity networks— data which can be collected rapidly and accurately and which can be combined with other mineralogical and lithological information. One of the most effective methods of representing porosity and pore network structures is the use of resin impregnated two-dimensional sections with back-scattered electron (BSE) imaging (Fig. 1), in a scanning electron microscope. With present image-processing facilities on high-speed, large-memory microcomputers, these pore network images are digitized easily, isolated by grey-level (or structural) thresholding and selected measurements can be made on individual pores. The data-collection procedure is high-speed, user-friendly and can be automated with a minimum amount of user interaction. This procedure can be adopted for may of the commercial, low-cost image-processing software packages that are available currently for microcomputers and so could be introduced readily into a conventional sedimentological or petrophysics laboratory.

This article describes:
  (a) The necessary hardware requirement for this type of petrographic image analysis.
  (b) The analytical procedure (POROS) for the quantitative parameterization of pore networks observed in two-dimensional lithologies by image processing.
  (c) How large volumes of two-dimensional pore geometry data are effective in modeling variations in reservoir potential of clastic lithologies. The pore geometry data can be used to calculate reservoir parameters such as permeability and capillary pressure and also to generate lithologic classification schemes based on pore geometry "end-members."

IMAGING AND COMPUTER HARDWARE REQUIREMENTS

Figure 2 illustrates the general hardware requirement for a petrographic image-analysis system suitable for rapid pore geometry evaluation of reservoir lithologic samples. The samples should consist of polished stubs or sections, following vacuum or pressure impregnation with epoxy resin, araldite, or other suitable resin mediums.

Image Acquisition by Scanning Electron Microscopy (SEM)

The highest resolution images of pore networks in reservoir lithologies are obtained by a scanning electron microscope with a back-scattered electron detector system (Dilks, Parks,

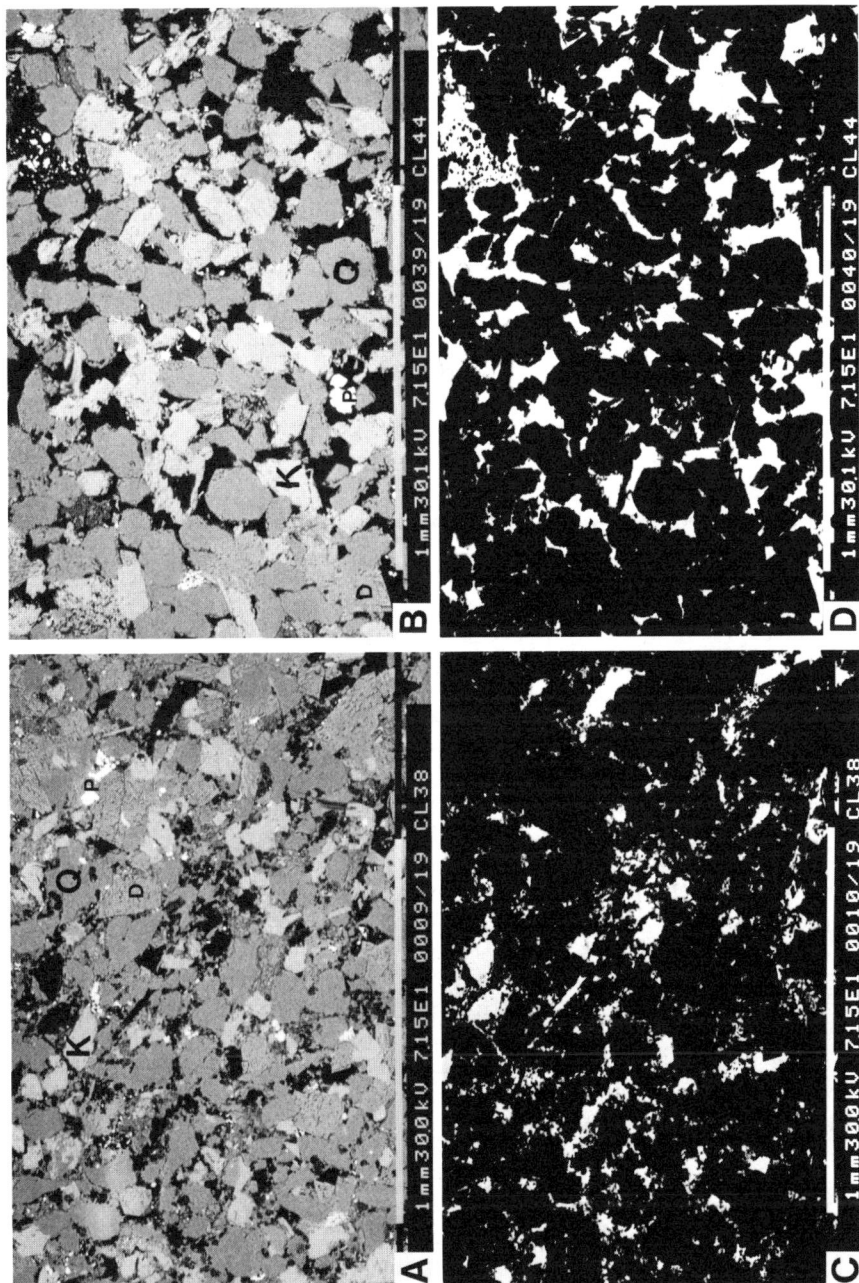

Figure 1. Two examples of Jurassic clastic lithologies from North Sea Basin, with back-scattered electron imaging and segmented pore structure. A, Low porosity, low permeability sample viewed by BSE imaging. Q - quartz, K - K-feldspar, D - dolomite cement, P - Pyrite. Porosity is black. B, Segmented pore structure of (A) after grey-level thresholding. C, High porosity, high permeability sample viewed by BSE imaging. Q - quartz, K -K-feldspar, D - dolomite cement, P - Pyrite. Porosity is black. D, Segmented pore structure of (C) after grey-level thresholding.

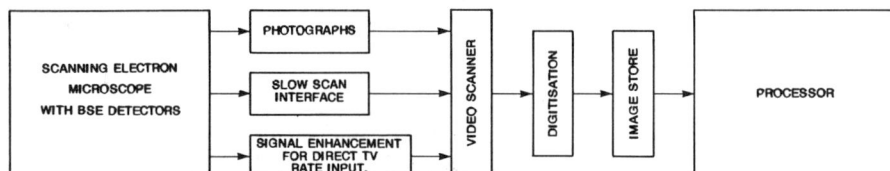

Figure 2.   Hardware requirements for a petrographic image-analysis system for pore-
geometry evaluation.

and Graham, 1984: Huggett, written comm., 1984; and Ruzyla, 1986). Back-scattered electron images are useful particularly in this context as BSE emission is proportional to the bulk atomic number of the analyzed material. As a result, in clastic lithologies (Fig. 1) heavier minerals such as barites, ferroan dolomite, and pyrite will generate a high BSE emission and therefore a relatively bright grey level. Lighter minerals such as quartz and K-feldspar will produce lower emission and a relatively lower grey level. The pore space (impregnated by resin) will produce little BSE emission and always will produce the lowest grey levels in BSE images (Fig. 1). The use of BSE images in pore network evaluation not only produces well-defined high-contrast images with the highest signal/noise ratios (reducing the amount of image "cleaning" and processing) but also extends the working magnification range, allowing the digitization of microporosity networks observed in clay-rich or chalk lithologies.

BSE images can be collected from many different scanning electron microscopes, using either solid state or scintillator BSE detectors. However, for the purpose of rapid, automated, and consistent image collection it is essential that the SEM working conditions (i.e., acceleration voltage, spot size and current, working distance, detector geometry, and gain/black level controls on the detector amplifier) are kept constant for the collection of different images from the range of samples.

Image Collection

The BSE images must be transferred next to the microcomputer. This can be achieved in three ways, two direct methods and one indirect technique.

(i)   Slow scan collection using a typical lines per screen value of 1000 and a line
        time of 32 ms. For most microcomputers a slow scan interface will be

necessary for this operation. The image is collected through a period of approximately 30 seconds and has a high signal/noise ratio.

(ii)   Fast TV rate collection using TV video rate collection conditions. The image is collected instantaneously by this method but the signal/noise ratio is low. This can be improved by an intermediate signal enhancement stage, involving the collection of many TV rate images of the same field before transfer to the image store.

(iii)   Video scanning of high-resolution BSE photographs is an alternative option where the microcomputer is not interfaced directly with an SEM as may be the situation in many laboratories.

Digitization, Image Storage, and Processing

Using POROS the collected BSE image is digitized, stored in random-access memory, and processed. Many commercially available hardware, software, or combined systems now are available at competitive prices. The microcomputer usually consists of a main processor, a math coprocessor, and an image-analysis coprocessor, one or preferably two color EGA text monitors, a hard-disk storage of at least 20 Mbytes, various hard-copy peripheral devices, and a library of general image-processing and statistical data-handling software routines. The data represented in this article were collected using a software based system (Minimagiscan) manufactured by Joyce Loebl, UK, and for reference, the hardware specification is provided in Table 1.

POROS — A SOFTWARE PACKAGE FOR PORE GEOMETRY EVALUATION

The general field of microscopic image processing and analysis is mature with the general principles, techniques, and applications in different disciplines now well-documented (Duda and Hart, 1973; Rosenfeld, 1984; Rosenfeld and Kak, 1976; Fabbri, 1980; Serra, 1982). Image-processing and analysis techniques have been applied to pore geometry assessment in reservoir lithologies previously (Rink, 1976; Rink and Schopper, 1978; Ehrlich and others, 1984; Dilks and Graham, 1985) with some small-scale studies based upon BSE imaging (Ruzyla, 1986; Doyen, 1988). POROS (Table 2) is an analytical procedure for the quantitative parameterization of pore networks, captured by BSE imaging, using a software library of image-processing techniques and data-handling programs. The procedure has been designed to process various clastic lithologic types and is used specifically for the batch processing of large numbers of samples under constant conditions.

Digitization

Digitization transforms the incoming analog video signal to a digital format. Spatial digitization generates a 512 x 512 pixel matrix, each pixel characterized by an intensity or grey level. Grey-level digitization places a numerical value on each detected grey level. In 6-bit systems the numerical range is 0-63 (black = 0, white = 63) and in 8-bit systems, 0-255 (black = 0, white = 255).

Table1. Image analysis microcomputer specification (Joyce Loebl) Minimagiscan.

IBM-compatible 3-processor image-analysis system (AT standard)		
Main processor	-	80286 16-bit
Coprocessor	-	80287
Memory	-	640 Kbytes
Hard-disk storage	-	40 Mbyte
Color text monitor (CGA)		
Color image monitor	-	14" High-resolution Analog RGB Analog O/P RS 334A/NTSC PAL/CCIR color
Single Diskette (File dump)		
Dot-matrix printer		
Microsoft mouse		
Images Analysis		
Coprocessor	-	High performance TMS 320C25 100ns pipeline DSP
Image memory	-	768 Kbyte
Spatial Resolution	-	512 X 512 pixels
Pixel Resolution	-	3 X 8 bit
Input look-up tables	-	3 X 256 X 88 bit RGB Tables Software programmable
Input Video signal	-	RS-343 standard, 8-bit ADC
Output Video signal	-	RS-343 standard, 75 ohm, RGB video signals (synchronized on green)
Camera system	-	NTSC and PAL formats are accommodated for CCD or tube types

Shade Correction

Shading causes the system to record different grey levels at different positions in the image although the viewed object is of uniform reflectivity. The major cause is the result of uneven illumination from unbalanced BSE detector geometry or unbalanced lighting when canning photographs. This effect is corrected by the generation of a standard image from a uniform reflector and using the standard to vary the grey-level information in the sample image multiplicatively.

Image Storage

The digitized grey-level image (with shade correction) then is stored in random-access memory prior to processing. Image storage usually is arranged separately from software and data-file memory so that the image-processing algorithms have fast access to the image store. As images are portrayed in two-dimensional pixel matrices, the memory is arranged into corresponding bit planes. Grey images are held in 4-, 6-, or 8-bit planes with

Table 2. Image-processing sequence for pore geometry parameterization by POROS.

IMAGE COLLECTION AND DIGITISATION	Monochrome image input from SEM with BSE mode facilities and digitisation to a 512 x 512 pixel matrix with grey level range of 0 - 255
SHADE CORRECTION ALGORITHM	Eliminates any variation in BSE signal caused by imperfect detector geometry set up.
NOISE REDUCTION ALGORITHM	Multiple image collection increasing the signal/noise ratio and elimiates noise artefacts at fast scan rates.
GREY LEVEL PROCESSING (a) CONTRAST ENHANCEMENT ALGORITHM (b) EDGE ENHANCEMENT ALGORITHM (c) NON LINEAR FILTERING ALGORITHM	 Increases the contrast range (grey levels) between pores and background Redefines the edges of individual pores Smooths out the overall grey level of the pores
SEGMENTATION	Selects and isolates the grey level range of the pores by grey level thresholding (Fig. 1)
BINARY PROCESSING (a) EROSION ALGORITHM (b) DILATION ALGORITHM	 Removes very small, unwanted artefacts which cannot be measured accurately.
CALIBRATION	Puts a size on each pixel (μm)
MEASUREMENTS	First Order Parameters (Table 3) are measured
DATA PRESENTATION	Statistics, histograms and correlations are provided
DATA REDUCTION	Pore geometry parameters converted to useful reservoir parameters.

one plane holding the most significant bit for each pixel, the next plane holding the next most significant and so forth.

Grey-level Processing

As high-quality, low-noise images are captured from BSE microscopy the amount of grey-level processing is minimal and the operation usually proceeds directly to the segmentation stage. However, in some lithologies, particularly clay-rich clastics, preliminary grey-level processing is required to enhance the resolution of the pore structure and a suitable option is provided.

On problematical stored images, grey-level processing has three functions: (i) increase the contrast between the pores and background, (ii) sharpen the edges of individual pores, and (iii) smooth out the overall grey level of the pores.

Contrast enhancement is carried out by applying a parabolic function (x^2) to the grey-level value for each pixel in the image. This has the effect of intensifying the larger grey levels, without affecting low grey levels, that is the pores. Edge enhancement involves the location of pore/mineral edges, regions of maximum grey-level gradient, and then increasing the gradient further using a nonlinear filter. The net grey level of the pores can be smoothed using a simple, linear mean filter based on a "neighborhood operation." A new grey level for each pixel is produced by combining the original level with those of its neighbors in a linear way, that is, each grey level in the neighborhood (a 3 x 3 pixel matrix) is summed and the mean value calculated which becomes the new grey level for each pixel in the neighborhood.

Segmentation

With the grey-level distribution of the pore space optimized, the next stage is to isolate the pores from the framework background. This is carried out by a process termed segmentation or grey-level thresholding, where a threshold is constructed which converts the digitized grey-level signal into a binary signal. A realistic hypothetical example is provided in Figure 3. The grey-level histogram represents the distribution of the low-intensity (0-60) grey levels in a processed digitized image. The pores are represented in grey levels 0-27 and a clear threshold exists between the porosity peak and the next peak, usually corresponding to quartz framework grains (Fig. 1A, 1C) or other low atomic number phases.

Segmentation places a window over the porosity grey levels (for example, 0-27 in Fig. 3) and loads the grey-level "look-up" table so that all levels up to and including a specific level result in a binary output of 0 and all greater levels give a 1 output. The resulting binary image, consists of only the pore structure (Fig. 1B, 1D). This operation probably is the most critical in the image processing of pore systems and, although automatic segmentation is possible within POROS, it is recommended that the manual operation option is used—the only stage of user interaction within the package. This is performed rapidly using a mouse-operated, sliding grey-level selector allowing the superimposition of the resultant binary image upon the digitized grey-level image.

Binary Image Processing

In spite of the high quality of the initial, grey-level processing operations, and correct segmentation, small amounts of unwanted noise persist in the majority of binary images.

There artifacts can be removed easily by binary processing operations, that is the use of erosion and dilation algorithms.

(i) Erosion involves stripping a layer (or layers) of pixels from the margin of an object. As a result the object will decrease in size and thin objects may be bisected.

(ii) Dilation is geometrically or morphologically the exact reverse of erosion where pixels are added to the periphery of objects.

These two algorithms are combined in an "opening" operation to remove small nice artifacts. The initial erosion will remove all areas which are thinner (i.e. noise) than the amount of erosion (a specified number of pixels) and these are removed irreversibly. In POROS a single, fine "opening" operation will remove the fine noise but will have only a negligible effect upon the larger pores.

Calibration

As the original BSE image is digitized to square pixels, size calibration is a relatively simple concept. One pixel width is equivalent to a real distance, microns in the situation of pore-geometry measurements. In practice a standard of known length or area is measured in pixels and a calibration factor is calculated. This factor then is applied automatically to any measurements made in pixels during the running operations of POROS, generating

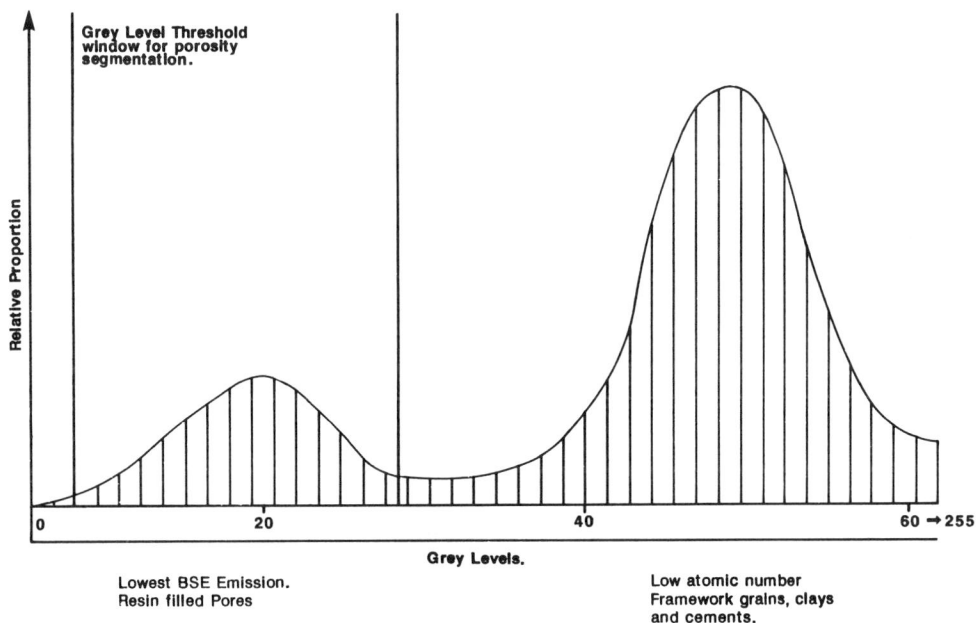

Figure 3. Segmentation of pore structure by grey-level thresholding in digitized back-scattered electron image with grey-level range of 0-255.

data in μm or μm^2. Several calibration factors have been calculated for many different combinations of operating conditions and in POROS the correct calibration is selected automatically.

Measurements And Data Presentation; Pore-geometry Data Generated By POROS

A typical output from POROS involves measurements of several parameters on up to 2000 pores for a single sample. The parameters are nested into three levels (Table 3). First-order parameters are the raw parameters measured on each isolated pore in the segmented image. Second-order parameters are calculated from the first-order parameters and apply to individual pores or characterize the entire pore network. Higher order (or reservoir) parameters involve calculations using combinations of second- and first-order parameters. As an example two different lithologic pore images with relevant first- and second-order data, are provided in Figure 4, showing frequency percentage histograms for pore area, pore specific surface area, and pore horizontal connectivity to illustrate the difference between these two samples. Model statistical parameters, that is mean, medium, and standard deviation values are selected from these histograms and can be used in correlation plots with other petrographic or petrophysical data.

DATA APPLICATION TO RESERVOIR POTENTIAL ASSESSMENT

Pore-geometry Parameters And Lithologic Permeability Correlations

Where a large number of lithologic samples have been processed by POROS, a good assessment of the influence of pore-geometry parameters on reservoir quality (permeability) can be made. Figure 5 shows a series of plots of permeability versus mean values of first- and second-order parameters for approximately 100 samples from a single clastic well study in the North Sea Basin. Good positive correlations are observed for pore area, diameter, perimeter, horizontal connectivity, and pore shape. A negative correlation is observed between permeability and specific surface area. These correlations illustrate that the reservoir quality of these clastic lithologies is controlled and can be modeled effectively by these parameters across a wide range of permeability (0.01 -10^4 mD).

Pore-Geometry Classification

The second-order (Field) parameters (porosity, pore density, specific surface area, and pore shape— Table 3) can be used in pore-network classification schemes for clastic reservoirs. An example is provided in Figure 6, consisting of approximately 130 data points collected from a single will study in the North Sea Basin and covering a permeability range of 0.01-4000 mD. The data spread in the classification diagrams suggest that considerable variation in pore geometry is observed between these samples and representative illustrative binary images are provided for potential pore structural "end-members."

These diagrams also can be contoured for other variables (Figure 7 A-D), for example; (a) air permeability, (b) capillary pressure, (c) facies type, and (d) authigenic cement content, to model the relationships between pore-geometry style and other important reservoir data. The purpose and advantages of these type of diagrams is not only to classify reservoir lithologies from a pore-network viewpoint, as opposed to the traditional grain framework or authigenic cement basis, but also to identify extreme end-member porotype lithologies which can be calibrated by engineering data (i.e. permeability, capillary pressure, etc.)

Table 3. Pore-geometry parameters

	Pore area (PA)	The number of detected pixels forming the pore's interior.
	Pore perimeter (PP)	For each point X(i), Y(i) on the pore's boundary, the distance to previous and succeeding points are calculated. The distance between the mid-point of these vectors is computed and the perimeter is the sum of the distances.
	Pore length (PL)	The center of gravity is calculated as a base point. A point on the boundary (X_i, Y_i) furthest from base is selected iteratively which will be the maximum chord length.
FIRST ORDER PARAMETERS	Pore breadth (PB)	Using the maximum chord length, the maximum normal distances of all boundary points is computed. Pore breadth is the sum of two maximum distances - maximum projection normal to length.
	Pore orientation (PO)	Pore orientation is the angle between maximum chord length and vertical (Y) axis.
	Pore horizontal connectivity (PW)	Pore width - the normal projection onto the horizontal X axis.
	Pore vertical connectivity (PH)	Pore height - the normal projection onto the vertical Y axis.
	Pore density (pores/mm^2)	The number of pores per unit area.
	Porosity (%)	Total detected pore area divided by analyzed field are.
SECOND ORDER PARAMETERS	Field specific surface area * μm^{-1}	Calculated by (PP/PA) * $(4/\pi)$ = Ss.
	Field shape factor *	Calculated by 1/3 * Ss/Km = ^{2}N/2PP (N = Number of pores).
	Pore aspect	Pore length/breadth ratio
	Vertical/Horizontal connectivity ratio	Pore width/height ratio (PW/PH)
	Permeability	Calculated from porosity and field specific
HIGHER ORDER RESERVOIR PARAMETERS	estimates (Kia)	surface parameters Kia = $f^3/5(1-f)^2 . Ss^2$
	Cumulative porosity/ pore area relationships +	Pore area frequency histograms are recalculated against measure % porosity (see text for examples)

* Shape factor and specific surface area also is calculated for individual pores
+ Capillary pressure calculations can be made from these relationships

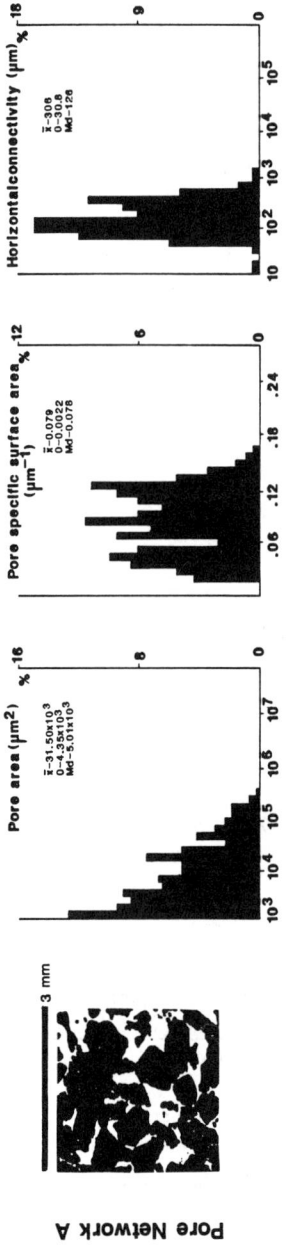

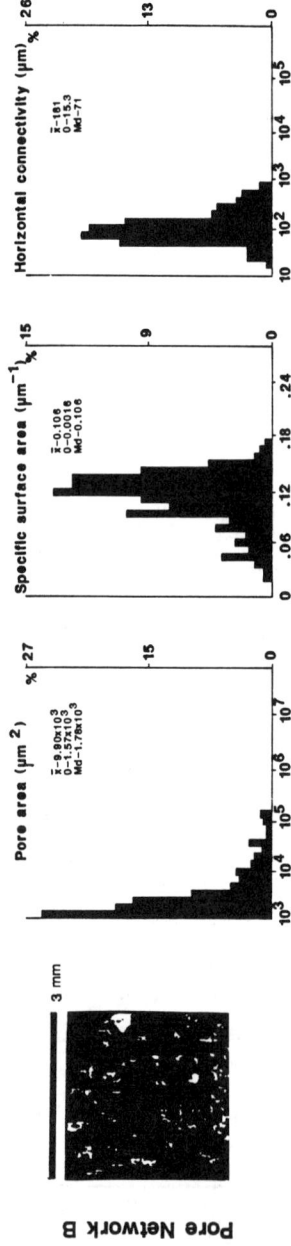

Figure 4.   First- and second-order parameter data for two extreme lithologic types. Percentage distribution histograms are provided for pore area, specific surface area, and horizontal connectivity. Binary images are processed, segmented images of effective pore structure, 3mm scale bar indicates size. Core analysis (helium porosity, air permeability) data, second-order field parameters and Kia estimates (see text) are provided. These two examples are clastic lithologies selected from North Sea well.

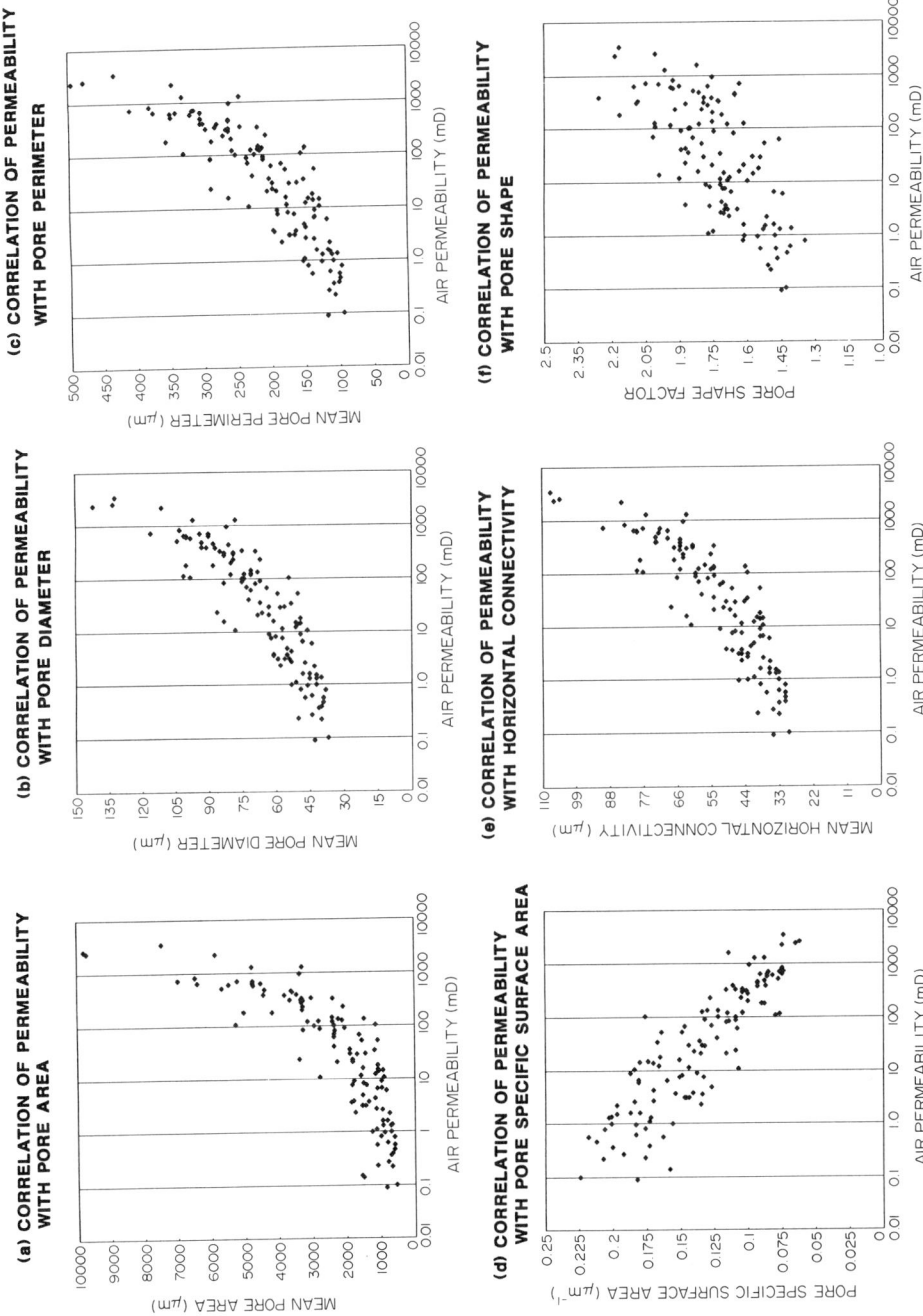

Figure 5. Correlation of air permeability (range of 0.01 to 104mD) against model statistics (mean values) of first- and second-order parameters. A, Pore area. B, Pore diameter. C, Pore perimeter. D, Pore specific surface area. E, Pore horizontal.connectivity. F, Pore shape. Sample data are compiled from single well study from North Sea Basin.

PORE GEOMETRY CLASSIFICATION

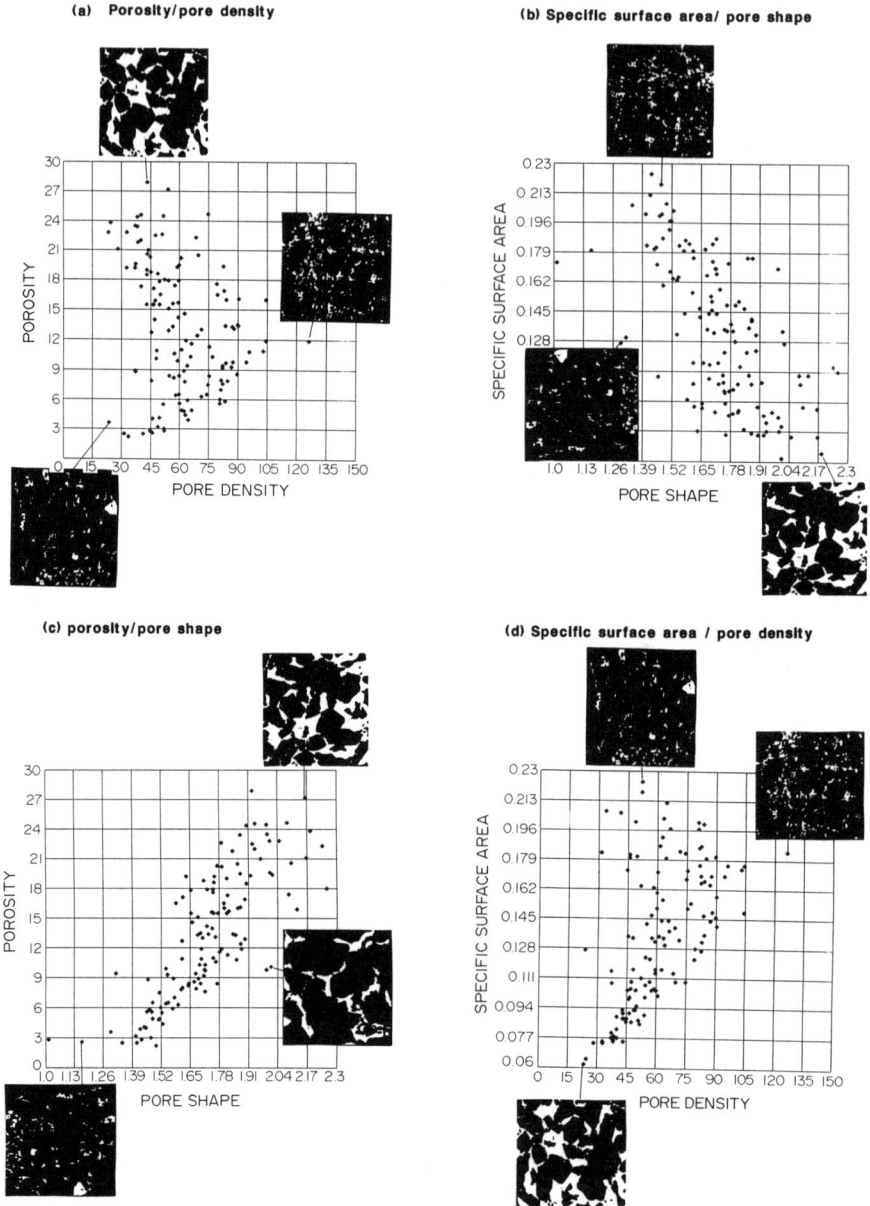

Figure 6.   Pore-geometry classification based on second-order field parameters.
Representative binary images of extreme, end-member pore networks also are
included. A, Porosity and pore density. B, Specific surface area and pore shape.
C, Porosity and pore shape. D, Specific surface area and pore density. Sample
data are compiled from single well study from North Sea Basin.

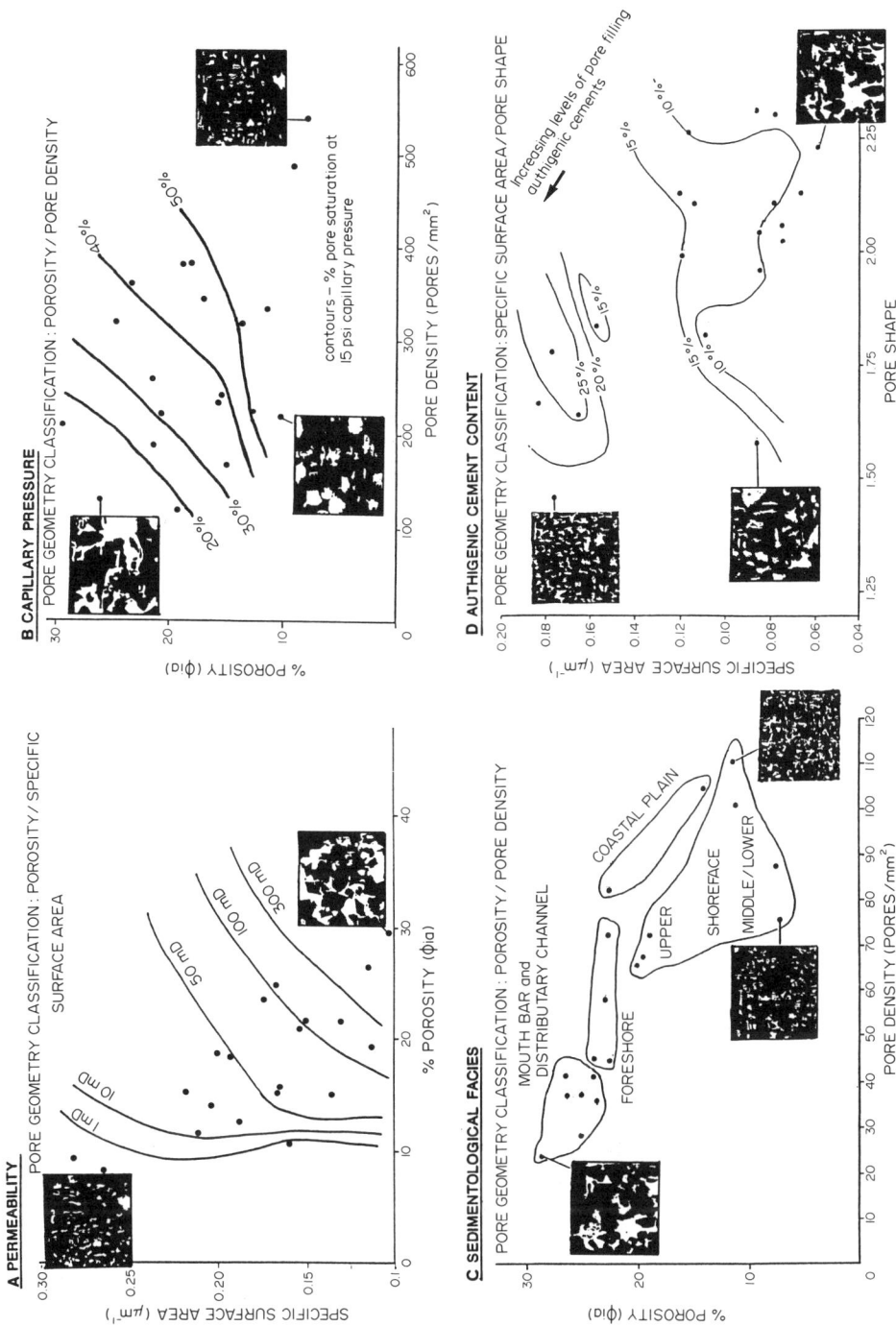

Figure 7.   Use of pore-geometry classification diagrams for modeling of other reservoir and petrologic data. A, Permeability.   B, Capillary pressure.   C, Sedimentologic facies.   D, Authigenic cement %.   Examples selected from Jurassic Brent lithologies of North Sea Basin.

Permeability Calculation Using Pore-geometry Data

The Carman-Kozeny equation (e.g. Collins, 1961; Archer and Wall, 1986) relates the permeability of a clastic lithology to the geometric structure of the pore space:

$$kck = \frac{\phi^3}{Ko \bullet (Le/L)^2 \bullet (1-\phi)^2 \bullet Ss^2}$$

where 0 = porosity fraction; Ss = specific surface area; Ko = Kozeny constant, and Le/L = tortuosity. Generally it is assumed that Ko • $(Le/L)^2$ = 5 (e.g. Ruzyla, 1986) and the equation may be rewritten as:

$$Kia = kck = \frac{\phi^3}{5(1-\phi)^2 \bullet Ss^2}$$

This model implies the porosity to be equivalent to a conduit, the cross section of which has a complex shape but an averaged constant area. Porosity (or pore fraction) and field specific surface area are determined clearly by POROS and can be substituted into the equation. Figure 8 illustrates the relationship between specific surface area, porosity, and Kia, with 100 data points determined from North Sea Basin wells and Figure 9 show how Kia determinations can be used to recreate effectively conventional permeability assessments on a well scale.

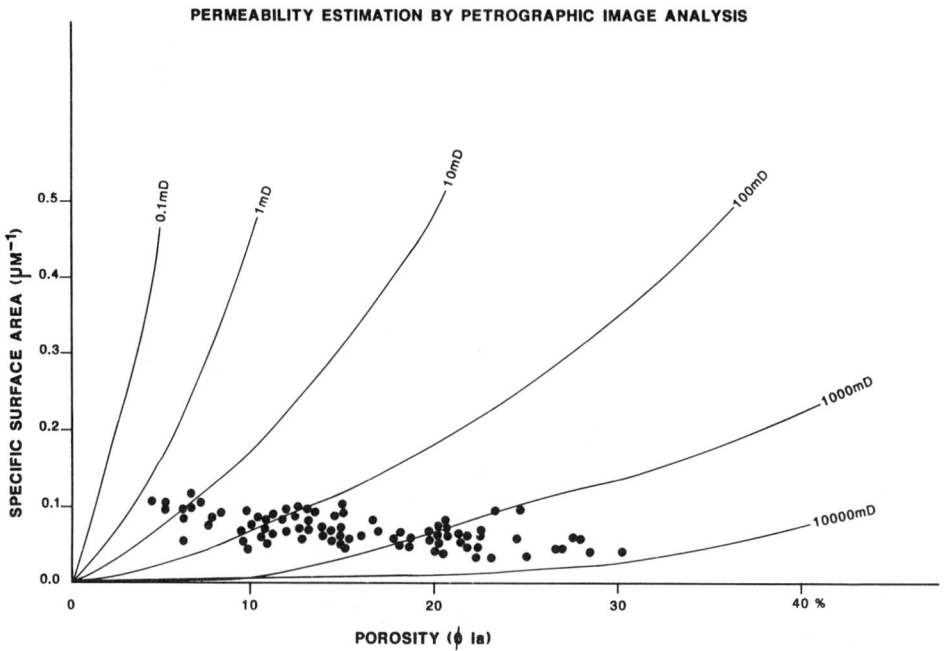

Figure 8.   Relationship between porosity, specific surface area (second-order field
            parameters determined by POROS), and permeability (Kia) calculated by
            Carman-Kozeny equation. Approximately 100 samples from North Sea well are
            used for representative illustration.

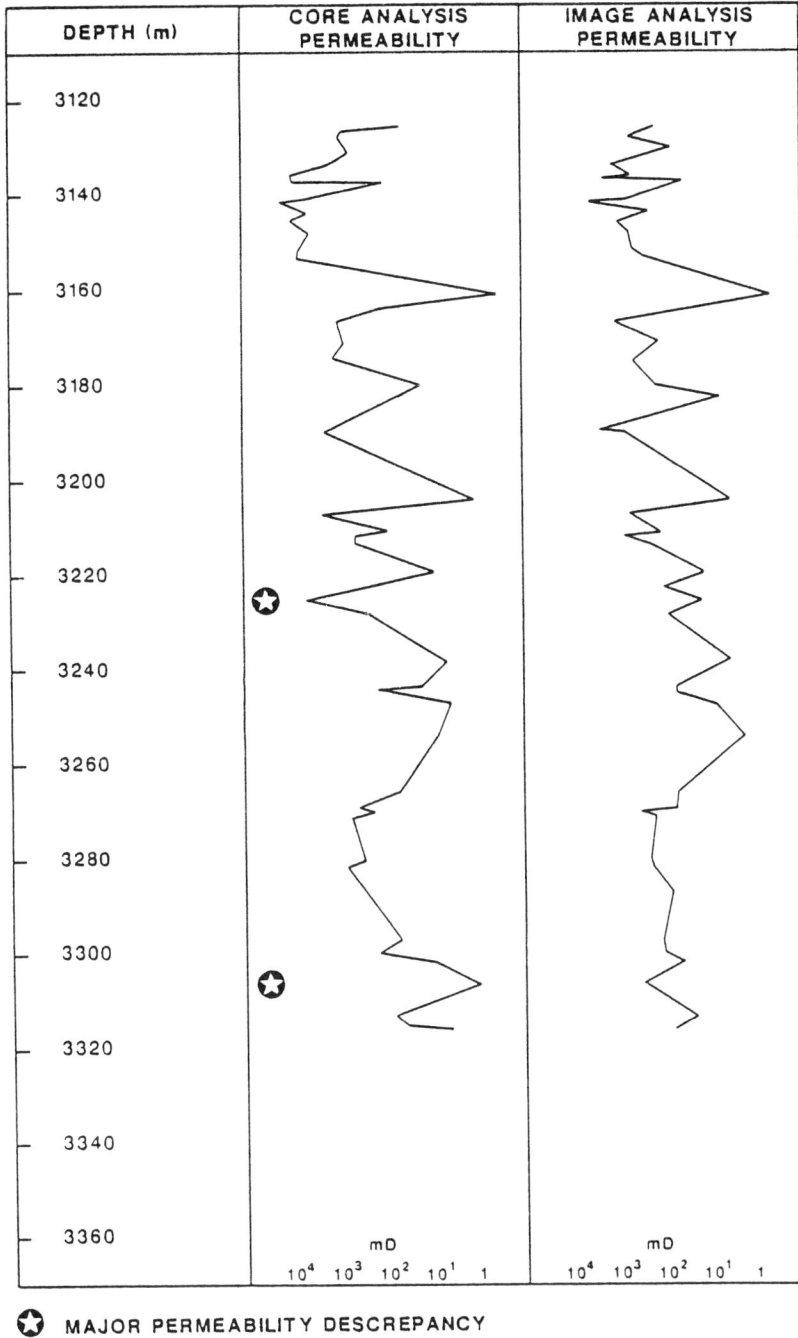

Figure 9. Comparison of measured and calculated permeability values from single North Sea Well (depth range– 200m). Note general coincidence of two data sets and only two anomalous points.

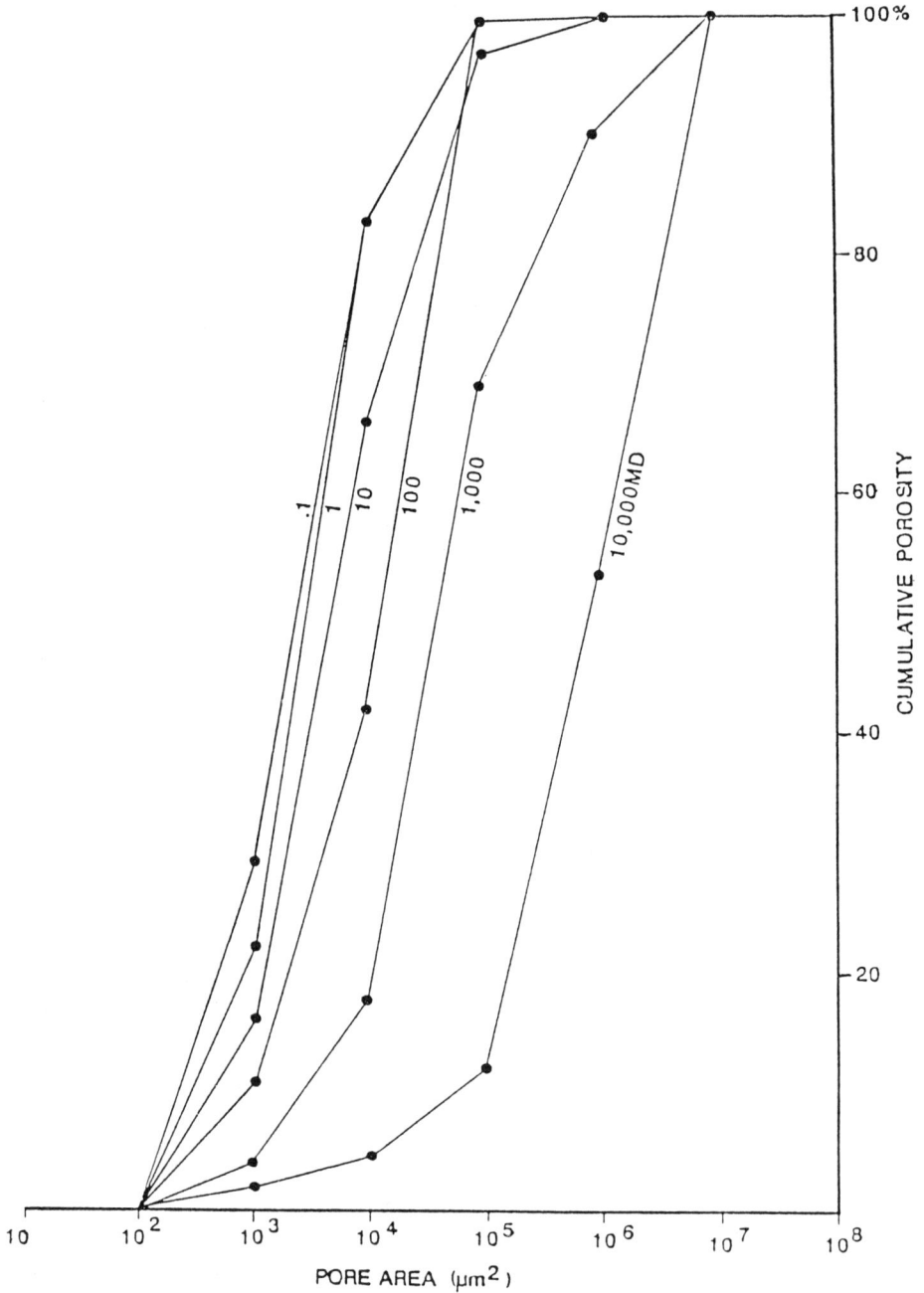

Figure 10. Pore area/cumulative porosity relationship curves compiled from pore area frequency histograms. Note how samples with different measured permeabilities (Kair) are isolated easily, highly permeable samples being deflected to the right—illustrating importance of pore-size distribution on reservoir quality and performance.

Cumulative Porosity/Pore Area Relationships

Pore area percentage histograms (Fig. 4) can be recalculated in terms of the proportions of effective porosity covered by different pore sizes. These data then are redisplayed as cumulative porosity plots (Fig. 10) through a large range of pore sizes $(1 - 10^7 \mu m^2)$. The shape and distribution of these curves is reflected in a variation in sample permeability with the higher permeability lithologies defected towards the right. This illustrates the strong influence pore-size distributions have upon effective permeability in reservoir sandstones. This form of data can be converted to a "pseudo" capillary pressure curve, with the calculation of capillary pressure (Pc) from the pore-size distribution data by use of the next equation (Archer and Wall, 1986):

$$Pc = \frac{2o \bullet Cos\,\theta}{r}$$

where o is the interfacial tension between wetting and nonwetting pore-fluid phases, is the angle between the wetting phase and the pore wall and is pore radii.

DISCUSSION

Using back-scattered electron imaging, POROS can provide a wealth of data on pore size, shape, and structure as represented in two-dimensional slices in reservoir lithologies. These data are an invaluable aid to engineers and petrophysicists concerned with modeling fluid flow in reservoirs.

(a) In reservoir studies, it is not possible always to take suitable core plugs for conventional core analysis measurements. However, porosity and permeability determinations can be made from pore geometry generated from smaller samples using image analysis and POROS.

(b) Volumetric porosity data are not sufficient always for the needs of reservoir engineers who are involved in modeling injection enhanced recovery techniques. Pore-geometry data will provide additional information on pore size, specific surface area (roughness), pore shape (coordination— number of throats per pore), and network connectivity.

(c) Pore-geometry parameterization will allow reservoir description, classification, and discrimination tn terms of pore structure rather than by the traditional approach of framework and authigenic mineralogy.

(d) Quantitative pore-geometry data can be correlated directly against: (i) Mineral data (e.g. % clay content, % blocky authigenic cements) so changes in pore shape and connectivity patterns can be modeled quantitatively in terms of authigenic events. (ii) Physical reservoir data (e.g. permeability, capillary pressure, and compressibility). Variations in this reservoir data can be modeled quantitatively by pore-geometry data.

REFERENCES

Archer, J.S., and Wall, C.G., 1986, Petroleum engineering; principles and practise: Graham and Trotman Ltd., London, 362 p.

Collins, R.E., 1961, Flow of fluids through porous materials: Petroleum Publishing Co., Tulsa, 274 p.

Dilks, A., Parks, D., and Graham, S.C., 1984, Characterisaton of sandstones and their component minerals by quantitative EPMA point counting in the SEM, in Romig, A.D., and Goldstein, J.I., eds., Microbeam analysis - 1984: San Francisco Press, p. 139-142.

Dilks, A., and Graham. S.C., 1985, Quantitative mineralogical characterisation of sandstones by backscattered electron image analysis: Jour. Sed. Pet., v. 55, no. 3, p. 347-355.

Doyen, P.M., 1988, Permeability, conductivity and pore geometry of sandstone: Jour. Geophys. Res., v. 93, no. B7, p. 7729-7740.

Duda, R., and Hart, P., 1973, Pattern classification and screen analysis: John Wiley & Sons, New York, 482 p.

Ehrlich, R., Kennedy, S.K., Crabtree, S.J., and Cannon, R.L., 1984, Petrographic image analysis. 1. Analysis of reservoir pore complexes: Jour. Sed. Pet., v. 54, no. 4, p. 1365-1378.

Fabbri, A.B., 1980, GIAPP. Geological Image Analysis Program Package for estimating geometrical probabilities: Computers & Geosciences, v. 6. no. 2, p. 153-161.

Rink, M., 1976, A computerised quantitative image analysis procedure for investigating featurers and an adopted image process: Jour. Microscopy, v. 107, p. 267-386.

Rink, M., and Schopper, J.R., 1978, On the application of image analyses to formation evaluation: Log Analyst, Jan-Feb. 1978, p. 12-22.

Rosenfeld, A., 1984, Picture processing 1983: Computer Vision, Graphics and Image Processing, v. 26, p. 347-393.

Rosenfeld, A., and Kak, A.C., 1976, Digital picture processing: Academic Press, New York, 457 p.

Ruzyla, L., 1986, Characteristics of pore space by quantitative image analysis: SPE Formation Evaluations, v. 1, no. 4, p. 389-398.

Serra, J., 1982, Image analysis and mathematical morphology: Academic Press, New York, 610 p.

The Israeli DTM (Digital Terrain Map) Project

John K. Hall
Geological Survey of Israel

Eric Schwartz
formerly of the
Environmental Protection Service, Ministry of the Interior

Richard L. W. Cleave
Historical Productions, Inc.

ABSTRACT

The Geological Survey of Israel, Historical Productions, Inc., and the Israel Mapping Center have undertaken the production of a new high-resolution digital terrain map (DTM) of Israel. The new DTM is based upon the topographic information contained in the 10 m contours of the 99 partial or complete 20-km square 1:50,000-scale topographic maps of the Israel Mapping Center. The grid used is 25 m, locked into the Cassini-Soldner projection of the standard Israel Grid. For each of the 1:50,000 topographic sheets, this grid can be visualized as an overlay of one-half millimeter squares, representing 641,601 elevation points per 400 km^2 map.

An inexpensive PC AT system with EGA graphics and high-resolution digitizer with 16 button cursor is used to edit a matrix of EGA screens of the maps which are scanned using commercial prepress scanners. FORTRAN programs with HALO graphics routines facilitate the editing and directly convert color-coded intervals between contours to a 25 m DTM using weighted means of points interpolated by bicubic splines from both N-S and E-W topographic profiles. Special modules allow inclusion of additional nonstandard contours, spot heights, and profile data, as well as steep discontinuities in the surface to be digitized.

The initial use of the DTM elevations will be to enable Historical Productions to calculate the heights of the individual 10 m pixels of a merged SPOT/LANDSAT image mosaic of all of Israel and the adjacent areas for the purpose of producing computer-generated oblique synoptic images of the satellite imagery and thus film or video movies of flight over the animated topography. The DTM and the computer methods used to produce and manipulate it constitute the first stage in a process which will allow representation of physical/geologic/geophysical data for the country in a far more visual and analytic manner. Examples are preparation of slope maps, solid 3-D models, shaded-relief displays, perspective images with quantitative color coding, stereopairs, or even the recontouring of the topography at different intervals or scales.

A byproduct of the project is the transfer of digital data from personal computers to Scitex color-separation equipment for inexpensive mass production of high quality color images. An immediate goal is the production of a similar DTM for the adjacent seafloor, prepared from various types of bathymetric observations. This DTM will be converted to an image by shading its slopes according to the sun angle present in the satellite image and coloring the DTM facets (pixels) according to the actual depth, and the image will be transferred to the Scitex equipment for merging with the detailed satellite imagery so as to print satellite images with the associated submarine topography shown as if it were located beneath water of unlimited clarity.

INTRODUCTION

The most basic ingredient of a digital Geographic Information System (GIS) is a digital representation of the terrain elevations, or a Digital Terrain Map (DTM - sometimes also referred to as a Digital Elevation Model -DEM, or a Digital Terrain Elevation Model - DTEM). A DTM consists of the heights of the terrain over a regularly spaced square or rectangular grid, which is represented in the computer as a two-dimensional array of numbers.

Now, in a project to produce computer-generated 3-D images using high-resolution merged SPOT/LANDSAT TM (STM) satellite imagery, Historical Productions, Inc. is joining with the Geological Survey of Israel and the Israel Mapping Center to produce a DTM on a 25 m grid. To our knowledge, this DTM will be one of the finest scale country-wide DTMs in existence, surpassing the limited coverage new generation of American 30 m DTMs which are "based on digitizing and gridding the original contour plate rather than scanning a stereomodel as a by-product of orthophotomap production" (Pike, Thelin, and Acevedo, 1987, p. 343). The new DTM elevations will be used to calculate the heights of all the SPOT/LANDSAT 10 m pixels, so that the images can be viewed obliquely and manipulated to produce films and oblique synoptic images, as well as 3-D physical models.

The DTM and the computer methods used to produce and manipulate it constitute the first stage in a process which will allow representation of physical/geologic/geophysical data for the country in a far more visual and analytic manner. Examples are preparation of slope maps, solid 3-D models shaded-relief displays, perspective images with quantitative color coding, stereopairs, and use of the heights to calculate the effects of topography in wind circulation, mass wasting, water runoff, etc. A byproduct of the project is the transfer of digital data from personal computers to Scitex color-separation equipment for inexpensive mass production of very quality color images.

The new DTM is based upon the topographic information contained in the 10 m contours of the 99 partial or complete 20 km square 1:50,000 topographic maps maintained by the Israel Mapping Center. The grid used is 25 m, locked into the Cassini-Soldner projection of the standard Israel Grid. For each of the hundred-odd 1:50,000 topographic sheets this grid can be visualized as a one-half-millimeter square overlay sheet, representing 641,601 elevation points per 400 km^2 map.

METHODS

The basic methodology of using topographic contours to produce a DTM is not new. However, the specific methods applied here, as far as we know, are new and innovative. The four stages and their innovative aspects are as follows:

(1) Scanning of the contour plate using standard prepress color-separation equipment, which for commercial purposes already is connected to the PC environment. Breaking down scanned images of any size into a matrix of adjacent PC screens for analysis. The amount of information on one PC screen represents almost the maximum amount of information that the PC has to be concerned with at a time.

(2) The use of inexpensive IBM-compatible personal computers with intermediate resolution EGA (640 x 350 pixel, 16 colors) graphics to carry out the editing and analysis of the scanned contour maps. Software is written in standard IBM Professional FORTRAN (Ryan-McFarland Corp.) with versatile HALO graphics routines (Media Cybernetics, Inc.).

(3) The use of rastor (rather than vector) editing of the contours; color fill of the intervals between contours to test contour continuity and uniquely define height bands and gradients across contours; analysis of the video screen to build up an immediate array of contour crossings; splining of N-S and E-W grid profiles for each screen, with continuity to adjacent screens, to produce separate DTMs for each screen and thus limit storage requirements; representation of the results as block diagrams at every stage to identify errors as they occur; addition of ancillary data from other sources to the scan or resulting contour crossing profile files via the digitizer.

(4) The use of a high-resolution (0.001") GTCO Corporation Digipad 5 digitizer with 16 button cursor to edit the screens and control the software. The present editing program has 10 different cursor indicated menus, each with up to 16 different functions or "degrees of freedom."

To digitize the 1:50,000 contour sheets, the original scribed contours on stable bromide paper are scanned at 12 lines/mm (304.8 dpi) on a Hell DC 300 B scanner. This scanning density is equivalent to scanning every 4.16 m on the ground. The digital scan file is stored on a Scitex Imager III image-processing system, sized to exactly 400 mm and with all the "pinhole" single black pixels removed. Then the scan file, typically 0.5-7 Mbytes, is transferred to the PC using the Scitex software "Handshake", a parallel GPIB (General Purpose Interface Bus) communications link, and software (Anonymous, 1987) for the PC from SHIRA Computers Ltd. On the PC, additional SHIRA software is used to pick the best image origin, and to break up the original 4801 x 4801 pixel image file into a matrix of 8 x 14 (or 112) EGA screens, which are stored in HALO's compacted .PIC format.

On the PC, each of the 112 EGA screens then is edited to remove dirt specks, make contours continuous, replace schematic cliffs and contour numbers by contours, color code each of the 10 height spans between 100 m index contours, and finally to determine the approximate height of the four corners of the screen. The screen then is analyzed to produce an intermediate file with the points of crossing of the contours with the 25 m grid (every 6th pixel on the screen). Once all the screens are edited, a process requiring several days of operator work, the intermediate contour crossing files are assembled to calculate a DTM for each screen. Points on adjacent screens are used to ensure continuity. On the map borders, adjacent map screens can be used to assure continuity between maps. The DTM is produced using a weighted average of the heights at each grid point determined from bicubic splines of the N-S and E-W profiles through that point. Straightline interpolation is used in areas with little relief.

The final DTM for a single 20 by 20 km 1:50,000 topographic sheet consists of 641,601 heights in a single file of 1,284,096 bytes, occupying about 6% more than one 1.2 Mbyte floppy disk. The heights are stored as two-byte integers giving the elevations in decimeters (between -3,276.7 and +3,276.7 meters). They are given as 801 sequential E-W rows of 801 points, beginning at the northwestern corner of the map and ending at the southeastern corner. The file is written in direct access binary with a 1024 byte record length, which allows any part of the file to be accessed simply and rapidly from a hard disk by the PC.

RESULTS

In the eighteen months since the DTM Project was conceived, including the period during which the software evolved from a Microsoft QuickBASIC prototype to the present multiprogram suite of more than 10,000 lines of FORTRAN code, fully 50% of the State of Israel has been digitized. The initial tests were carried out on the four sheets around Jerusalem. The resulting DTMs were used to test out the oblique image-generation phase of Historical Productions' program (see next section). Following the proof of the concept the area north to the Lebanese and Syrian borders and west to the Mediterranean was analyzed. Presently work is proceeding on the Negev Desert in the direction of Eilat on the Red Sea. The work force consists of three people working part-time on two work-stations belonging to the senior author and his 11 year old son. Computer usage, including extensive batch processing of the edited maps, averages 30 hours per day. As new types of morphological features and their cartographical expressions are encountered, or as repetitive editing tasks are identified and analyzed, additional software conveniences are added. The area analyzed to date includes an exotic choice of topography embracing pluvial and desert terrains, and ranging from absolutely flat valleys to wadis that are miniature Grand Canyons.

The time required to produce a DTM from a scanned map differs considerably. Rough topography which is represented on the maps by a schematic representation of a cliff must be reworked into contours (for slopes of up to 50° to 59° depending upon the contour orientation— or about 35-40 contours per cm), or else tagged as a radical discontinuity which must be bridged linearly. Editing time per screen can vary from 2-3 minutes for gently rolling foothills to 4-5 hours for the most rugged wadis. Areas with little topography present other problems. Areas of few or no contours, such as flat intermontane basins, playas, coastal flats, or lacustrine areas must be augmented by interpolated contours through isolated trig-points, additional 5 m contours from the 1:10,000 scale topo-cadastral (land registration) maps, or approximate formline contours to "flesh out" ravines, scarps, and watercourses with only a few meters of topography.

The accuracy of the finished DTMs is checked by several programs. One of these allows designation of any point on the map with the digitizer cursor, and displays the corresponding height, dip, and strike of the DTM interpolated for that point. The results on well-behaved contours are generally within 1-2 meters, while the values at isolated trig-point elevations are slightly more variable. In the process of determining the DTM at each mesh point from a weighted average of the N-S and E-W splined profiles, the root-mean-square (rms) differences are calculated routinely and determined to differ between 0.9 and 1.7 meters.

Another DTM checking program interpolates an array of 1418 x 1418 slopes from the original 801 x 801 height array by using the cross-product of the orthogonal N-S and E-W vectors intersecting the point to be interpolated to determine the vector normal to the facet at that point, and then determining the slope from that vector. This array, printed out in 330 colors on a Hewlett-Packard Paintjet printer at 90 dpi, produces an perfect slope overlay to the original map with different color tints every 0.273°. Color palettes can be varied to accentuate any particular slope or range of slopes. A similar scheme is used to color another interpolated array of 1418 x 1418 height points according to which 10 m interval (out of the 10 intervals between 100 m index contours) each height point occupies. The resulting plot of the repeated ten-color sequences can be used to assure that the DTM lies squarely on the original contours.

These tests show that the DTM is a good representation of the data in the original contours, especially because the methodology requires that the DTM equal a contour's height if that contour passes through a grid-point (unlike DTMs based on Triangular Irregular Networks - TINs, or surface fitting based on neighboring points, such as SURFACE II). However, minor but irritating artifacts crop up in areas with few contours.

Possible solutions include modification of the present splining technique, possibly using isolated trig-point elevations from the maps plus surrounding contour elevations to calculate an analytic surface using multiquadric equations (Hardy, 1971). These equations then can be used to produce intermediate control points on the profiles to be splined, or in the example of marine bathymetry, used to produce directly the DTM from spot soundings.

The DTM data are not limited to dry land. Where the 1:50,000 topo sheets encounter seas such as the Mediterranean, Red, and Dead Seas, or lakes such as Birkhat Ram and the Sea of Galilee, the appropriate marine data are being added to the screens via the digitizer and similarly analyzed. The architecture of the programs allows inclusion of spot soundings, profiles, and addition of contours other than the usual 10 m contours that are analyzed graphically.

FURTHER APPLICATIONS

Once a high-resolution DTM exists for the topography, the data can be used in a number of applications. Because the driving force for the project is the need to provide elevations for all the individual pixels within merged SPOT/LANDSAT TM images, so that these images can be manipulated on a computer to produce oblique synoptic images for video films, atlases, and posters, let us begin with actual uses to which the test data have been applied.

The first oblique scenes from these data were produced in early 1988 by Kevin J. Hussey at the Digital Image Analysis Laboratory (DIAL) at the Jet Propulsion Laboratory, California Institute of Technology, Pasadena, California. The satellite image (Historical Productions, 1988b) consisted of a small part of a cloud-free five-scene LANDSAT 5 thematic mapper (TM) sequence (Historical Productions, 1988a) taken on 18 January 1987, which covered the area between Jericho and Jerusalem. Three spectral bands of LANDSAT TM (Thematic Mapper) 28 m resolution pixels were reinterpolated to 10 m spacing according to the variations seen in a high-resolution 10 m pixel resolution panchromatic image from the French SPOT-1 satellite using proprietary algorithms developed by KRS (Kodak Remote Sensing) in Landover, Maryland. KRS has used the same technique to merge SPOT and LANDSAT data to produce complete 10 m color coverage for Historical Productions, Inc. for the area between Sidon-Damascus in the north and Eilat in the south.

The DTM data from the Jerusalem, Ramallah, Kallia, and Jericho sheets were reinterpolated to obtain the height for each of these 10 m resolution merged SPOT/LANDSAT pixels. The JPL-DIAL MicroVAX image-processing system then was used to produce several oblique images using the methodology described by Hussey,Hall, and Mortensen (1986). Four of these, at resolutions ranging from 1024 x 1024 to 2048 x 2048 pixels, are shown in Figure 1 (Color Plates 1 and 2). The vertical exaggeration is 250%. The figures illustrate the tremendous potential of the images for showing large areas with selective enhancement of the topography, and without the interference of air pollution, sun location, or Earth curvature. Once the database is in place, Historical Productions, Inc. intends to make motion pictures of simulated flights over the topography by having a computer produce 20-30 sequential frames for every second of film-flight. Because the 10 m pixel size limits such flights to elevations at which the closest topography will not appear at a scale of smaller than 1:25,000, it is possible to simulate flights at speeds of 2000-3000 km/hr, and thus show large areas in relative short times.

The system and software for generating oblique imagery is being developed in Jerusalem by Dr. Ehud Shimbursky of GISHA (Geographical Information Systems Ltd.). Depending upon the successful outcome of benchmark and acceptance tests with an Apollo DN4000 computer on a second area between Jerusalem and the Sea of Galilee (a continuation of that in Historical Productions, 1989), an advanced Apollo DN10000 graphic workstation with 60-100 MIPS (million instructions per second – a MicroVAX II is less than 1 MIPS) throughput will be used for this scene generation, and for generating higher resolution scenes or offset color printing via the Scitex equipment. In order to facilitate commercial production of oblique scenes, which require tremendous amounts of computing power, a second Apollo DN10000 unit of a more advanced design will be added at the end of 1990, giving a projected combined throughput of up to 400 MIPS (Apollo Computer, Inc., Chelmsford, MA, USA, pers. comm., 1989).

The slopes calculated from the test DTM files already are being used to produce slope maps of selected 1:50,000 maps at the Geological Survey of Israel. Presentation either can be as colored pixels denoting individual high-resolution slopes at each mesh point, as outlined, or as smooth contours separating areas of different slope distributions.

Actual 3-D models of some of the test case topography also have been made from the DTMs. Achiam Lifshitz at PAL Technologies Ltd. in Kiryat Shmona has used CNC (computer numerical control) techniques to drive a milling machine which shapes blocks of wax, plastic, and aluminum according to the original DTM. Possible applications include manufacture of 3-D molds so that thin plastic sheets printed with satellite imagery or other geographic information can be vacuum-formed into 3-D maps.

The DTM and the computer methods used to produce and manipulate it constitute the first stage in a process which will allow representation of physical/geologic/geophysical data for the country in a far more visual and analytic manner. In addition to the examples as mentioned are the following possible applications.

(1) The software offers a bridge between hardcopy data (maps to be scanned), ancillary data in digital or analog form (computer files of point data, interpreted airphotos, sketchmaps, etc.), and a computer image consisting of a large array of contiguous screens. This computer image can both be converted to a DTM (if applicable), or retransferred back up to the Scitex equipment for output as color separation films for mass production of high-resolution colored printed matter (maps etc.). Hence the software architecture would allow the construction of a final geologic map based upon a wide variety of source materials, including earlier additions of the map. Once editing was finished, the map would be ready for transfer to the Scitex machine for final editing and preparation of three/four color printing plates.

(2) The possibilities noted here allow further manipulation of the digital DTM data. Thus a DTM of the seafloor, prepared from various types of shipboard observations, can be transferred to the Scitex equipment as a detailed image. For merging with LANDSAT data it could be shaded according to the sun angle present in the satellite image, and colored according to the actual depth, so as to present the effect of looking into a pool – but a pool with underwater visibility of several kilometers. Alternatively, it could be presented as a standard single-color shaded-relief map, for printing via Scitex as background to all manner of navigational and scientific charts. Other possibilities are the merging of up to three data sets so that topography provides a shaded-relief map in black, the second data set is represented by the color spectrum, and the third data set provides the parallax needed to move pixels and produce a stereopair. Other manipulation possibilities lie in what is known in GIS as "multiprocessing" or mixing together different data sets in order to study and map their interactions. An example is mixing assay data with ore-layer thickness, overburden thickness, and excavation costs in order to determine profitibility in a mining operation.

(3) The generation and production of images by the described methods is an attempt to present efficiently considerably larger amounts of data to an observer than has been possible historically. By using the mind's perception of 3-D images, enhanced by artful use of color, it should be possible to use effectively Earth-science databases of ever-

increasing size, such as was accomplished by Haxby (1987) in using SEASAT satellite altitudes to produce a image showing the gravity field of the world's oceans. PC-based GIS systems already allow photo-DTM mergers which are modest in relation to those envisioned here.

(4) The DTM itself might become an important navigational tool once precision GPS navigational equipment becomes available. Results prove that a large area can be stored in a relatively small amount of memory and easily accessed to provide ground elevations as a function of location. Used in a remotely piloted airplane, the DTM might allow flying at a constant elevation above the ground, or unattended reconnaissance operations.

The FORTRAN programs developed for this project will be available for sale on diskette together with appropriate documentation at nominal cost in a manner similar to the open-file reports offered by the Geological Survey of Canada (Broome, 1986; 1987). The objective is to build a core of DTM users with experience and hardware installations capable of using DTM data and developing additional DTM-like databases of spatial data such as magnetic and gravity fields and their anomalies, seismic or drillhole-based subsurface horizons and layer isopachs, as well as seminumerical spatial properties such as geology, soils, and botany. An early version of the software, used to generate the topography seen in Figures 1A-D (Color Plates 1 and 2), has been contributed to the Computer Oriented Geological Society (COGS) public-domain software library as an accompaniment to this article. The latest version includes many improvements.

The two workstations used in the project consist of a 6 Mhz IBM PC AT with 80287 coprocessor and 20 Mbyte and 80 Mbyte disk drives (Norton CI = 5.9), and a Super Tech Taiwanese 100% PC-compatible 12.5 MHz half-wait state (Norton CI = 12.6) computer with 80287 coprocessor, 80 Mbyte Seagate ST-4096 disk, and a Cipher 1600/6250 bpi GCR CacheTape and Catamount controller which gives adequate streaming backup capability and future access to the world of Scitex and Apollo. A Hewlett-Packard Paintjet is used to output raster images with programming control over the 330 dithered colors. Both systems have 0.001" resolution GTCO digitizers with 16 button cursors. One digitizer is 11" x 17" whereas the other is 16" x 24". With the exception of the digitizers, the hardware requirements for using this software are minimal. Use of a PC XT computer should be possible, although the 360K diskettes would be a bothersome media for exchanging data. The software, which requires the coprocessor, was developed on the IBM system and is 100% AT compatible. It should run on less than 640K memory.

ACKNOWLEDGMENTS

We are grateful to Mr. Ephraim Arazi, founder and former CEO of Scitex Corporation Ltd. in Herzliya, for bringing authors Hall and Cleave together for this project. The continuing support of Drs. Ron Adler and Ya'akov Mimran, respectively, directors of the Israel Mapping Center (Ministry of Construction and Housing) and the Geological Survey of Israel (Ministry of Energy and Infrastructure), is gratefully acknowledged. The figures were produced by Kevin J. Hussey of the Digital Image Analysis Laboratory (DIAL) at The Jet Propulsion Laboratory, Pasadena, CA. Mr. Raymond Chazan and his staff at Spectra (Color Image Processing Ltd.) in Jerusalem scanned the 1:50,000 maps and provided the color separations for the figures. Software for the downloading of the scans from the Scitex was provided by Mr. Udi Zohar of SHIRA Computers in Tel Aviv. Ms. Bracha Mualem capably edited many of the later maps. We gratefully acknowledge these contributions to the success of the project.

This work was financed by Historical Productions, Inc., and initiated during the senior author's 1987-88 sabbatical in association with Scitex Corporation Ltd. in Herzlia, Israel.

REFERENCES

Anonymous, 1987, The Shira Interfacing System for image processing: BYTE Magazine, v. 12, no. 11, p. 64A-17.

Broome, J., 1986, MAGRAV2: An interactive magnetics and gravity modelling program for IBM-compatible microcomputers: Geol. Survey Canada, Open File 1334, 91 p.

Broome, J., 1987, Geophysical imaging software for IBM-compatible microcomputers: Geol. Survey Canada, Open File 1581, 47 p.

Hardy, R. L., 1971, Multiquadric equations of topography and other irregular surfaces: Jour. Geophysical Res., v. 76, no. 8, p. 1905-1915.

Haxby, W. F., 1987, Gravity field of the world's oceans: a portrayal of gridded geophysical data derived from SEASAT radar altimeter measurements of the shape of the ocean surface: Published for the Office of Naval Research by the National Geophysical Data Center, NOAA, Boulder, Colorado, NGDC Report MGG-3, colored image at approximate scale 1:40,000,000 Mercator projection.

Historical Productions, 1988a, Pictorial archive LANDSAT 5 satellite map: LANDSAT 5 TM imagery from January 18, 1987 between Eilat and southern Turkey, reproduced as 6 laminated posters at a scale of 1:150,000 in natural color .

Historical Productions, 1988b, SPOT/LANDSAT TM merge of Jerusalem and Jericho based upon the LANDSAT 5 imagery reinterpolated to 10 m pixels using SPOT-1 10 m panchromatic imagery: Single sheet laminated poster at 1:37,500 scale in natural color.

Historical Productions, 1989, SPOT/LANDSAT TM merge of eastern Galilee from south of the Sea of Galilee to Mount Hermon and southern Lebanon based upon the LANDSAT 5 imagery reinterpolated to 10 m pixels using SPOT-1 10 m panchromatic imagery: Single sheet laminated poster at 1:100,000 scale in natural color.

Hussey, K. J., Hall, J. R., and Mortensen, R. A., 1986, Image processing methods in two and three dimensions used to animate remotely sensed data, in Proceedings of IGARSS' 86 Symposium, Zurich, ESA SP-254, ESA Publications Division, p. 771-776.

Pike, R. J., Thelin, G. P., and Acevedo, W., 1987, A topographic base for GIS from automated TINs and image-processed DEMs, in Symposium proceedings of "GIS '87 - San Francisco", ASPRS-ACSM, p. 340-351.

Geostatistical Software For Evaluation of Line Survey Data Applied to Radio-Echo Soundings in Glaciology

Ute Christina Herzfeld
Scripps Institute of Oceanography

ABSTRACT

The program LSUNIMC is designed to evaluate regionally distributed geoscientific data, with specific options to adapt to line survey data. By assigning values to the nodes of a regular grid, a digital terrain model of the survey variable is set up that then can be contoured. The algorithms are based on the geostatistical method of universal kriging. LSUNIMC allows a wide range of kriging applications that can be directed by the user via input parameters. In order to meet the specific data distribution of surveys along track lines— as result from (aero) magnetic, gravimetric, bathymetric, and radio-echo sounding survey campaigns— search routines are open to adaptation to the individual survey device by the associated user-entered input parameters. Survey gaps between the tracks can be bridged efficiently.

Prior to the estimation of the digital terrain model, a structural analysis (variography) of the data has to be carried through, the result of which is entered into the kriging.

LSUNIMC is written in FORTRAN 77 and implemented on an IBM PC/XT/AT compatible.

To demonstrate the usage of the program, radio-echo sounding data from the Scharffenbergbotnen glacier, Dronning Maud Land, East Antarctica, are evaluated to investigate ice thickness and subglacial bed topography.

INTRODUCTION

Geophysical and geological surveys usually are carried out from vehicles that follow track lines: Aeromagnetics from planes, bathymetrics from survey vessels, submarine magnetics, gravimetrics from devices that are pulled by ships, radio-echo soundings from sledges driven over a glacier. The resulting data distribution in any situation is characterized by dense information along the survey track and gaps of information in between. The data may be recorded continuously (radio-echo soundings, Jonsson, Holmlund, and Grudd, 1988, Fig. 1, magnetic readings) or discretely along a line (single-beam bathymetric surveys), or on a stripe of varying width (multibeam deep sea sonar device, e.g. SEABEAM, see Fig. 2). The necessity of appropriate software to evaluate survey line data was outlined earlier (Briggs, 1974).

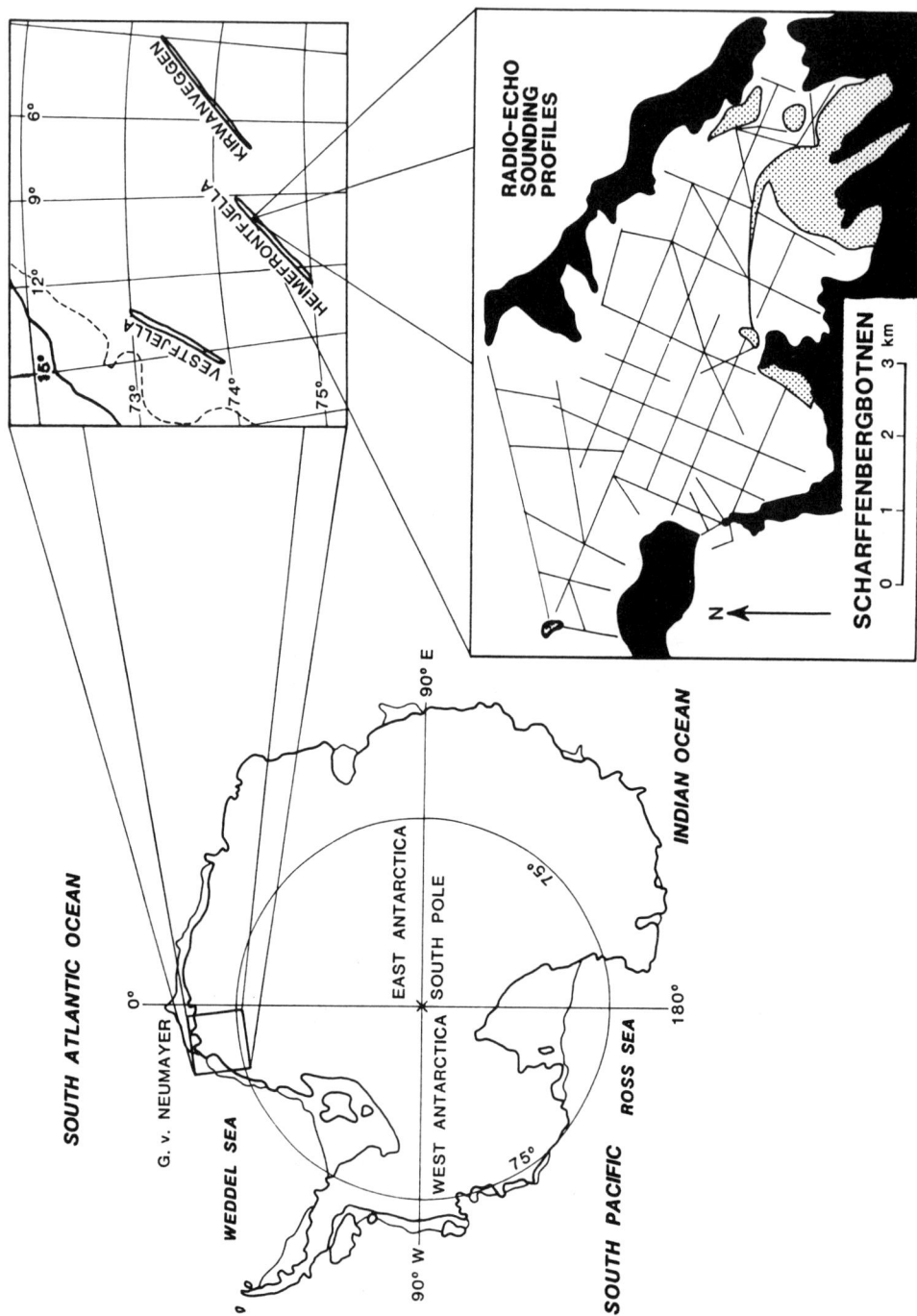

Figure 1. Location of Scharffenbergbotnen. Inserted maps show location of Heimefrontfjella and survey tracks of radio-echo soundings on Sczharfenberg-botnen glacier (from Herzfeld and Holmlund, 1989).

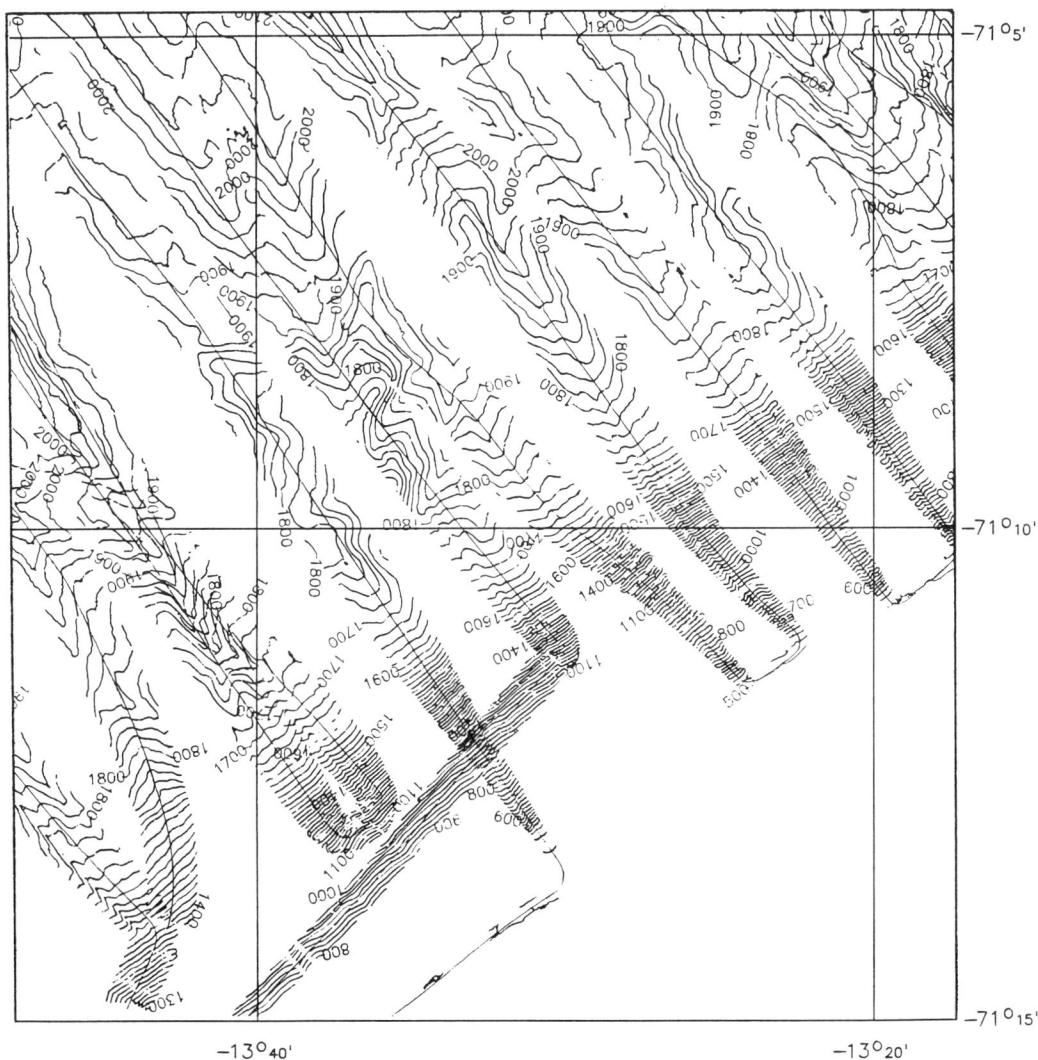

Figure 2. Survey pattern of SEABEAM multibeam deep-sea sonar system (General Instruments Corporation). Map shows part of Wegener Canyon, Weddel Sea, Antarctica, as surveyed during ANTARKTIS VI/3. Scale 1:100000 at latitude−70:55'. (Part of map produced by F. Niederjasper aboard RV POLARSTERN.)

The specific distribution of line survey data does not allow immediate interpolation by isolines. In order to achieve data on a regular grid, a digital terrain model is constructed that then is used as a basis for interpolation.

DEFINITION: A digital terrain model (DTM) is given by a regular grid and an algorithm assigning a value to each grid node.

Software packages generally are available that perform contouring of data on a regular grid. It is, however, of significant importance for the precision and reliability of the resultant map to select the method by which the DTM is set up carefully.

The approach realized in LSUNIMC uses the geostatistical method of universal kriging. Geostatistics— understood as the theory of regionalized variables in the sense of Matheron, (1963, 1971) and known best from mining applications— has been used for contour routines earlier (Davis,1986, p. 383-405). Kriging has optimal properties in as much as it yields minimal estimation errors, unbiased estimation, and an explicit description of the estimation error without additional computation effort. Prior to the estimation, a structural analysis (variogram analysis) of the data is necessary, that gives information on the spatial behavior of the investigated geologic variable (Fig. 3).

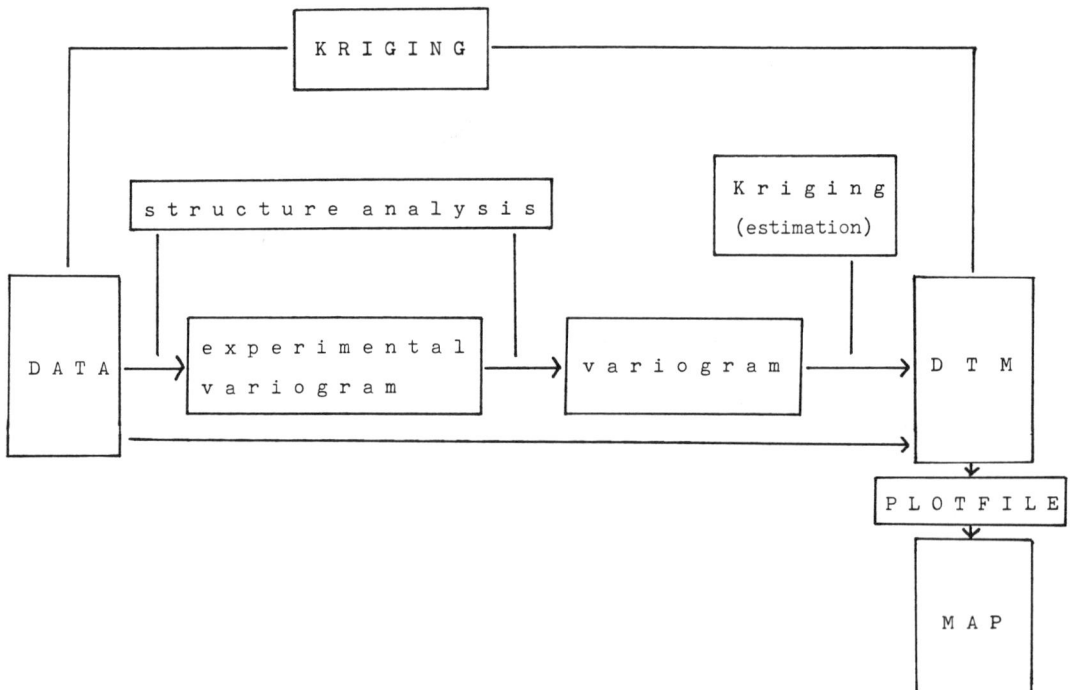

Figure 3. Diagram of kriging procedure.

But often kriging packages allow only restricted operations, in particular if they are available commercially or if they can be run on a microcomputer. The quality of the map depends on the following requirements:

- The variography has to be carried through carefully, and in particular be related to the geologic problem. The fitting of the variogram model must not be automated.

- The kriging program has to permit various variogram models.

- The selected kriging method must not assume stationarity, but must be able to account for drift components.

- The neighborhood search has to be adjustable to the data distribution (clustered or lined information) to avoid resemblance of survey pattern in the output map.

- The actual track directions of line surveys do not need to be known before program operation.

In this paper I present the microcomputer version LSUNIMC of a universal kriging program with special options designed for the evaluation of line survey data (see Note on Software). A short theoretical review of the geostatistical methods applied is followed by a software operation description and an example application from glacial radio-echo soundings. From this introduction, the reader is guided in the use of the program. In particular, possibilities to meet problem-oriented requirements to support specific geologic investigations are emphasized.

GEOSTATISTICAL BACKGROUND

Structural Analysis of the Data: Variography to Describe the Spatial Relationships

If measurements of a variable z are taken over an area D and split into a random part and a deterministic part dependant on a spatial structure, then the variable z may be termed a regionalized variable and as such be considered as a realization of a spatial random function Z in D (Matheron, 1963, 1971; Journel and Huijbregts, 1978). Ice thickness, subglacial bed elevation, seafloor depth, and gravity, for instance, are regionalized variables in this sense. The measurements have some irregular variation from one point to another, but show a high correlation between neighboring points. The latter determines the spatial continuity of the observed variable that is described in the variogram.

The experimental variogram is calculated as

$$\text{gam}^*(h) = \frac{1}{2n}\sum_{i=1}^{n}\left(z(x_i) - z(x_i + h)\right)^2 \tag{1}$$

where $z(x_i)$, $z(x_i+h)$ are samples taken at locations x_i, $x_i+h \in D$, respectively, and n is the number of pairs separated by h. It is assumed that the difference $(Z(x)-Z(x+h))$ for x, $x+h\in D$ is a second–order stationary variable (intrinsic hypothesis). This implies that the spatial correlation between two samples depends only on the vector h separating the sample locations. If the phenomenon additionally is isotropic, then the variogram depends only on the distance $r = |h|$. The experimental variogram is calculated in suitable distance classes. This then is fitted by a continuous variogram model (Figs. 4, 5).

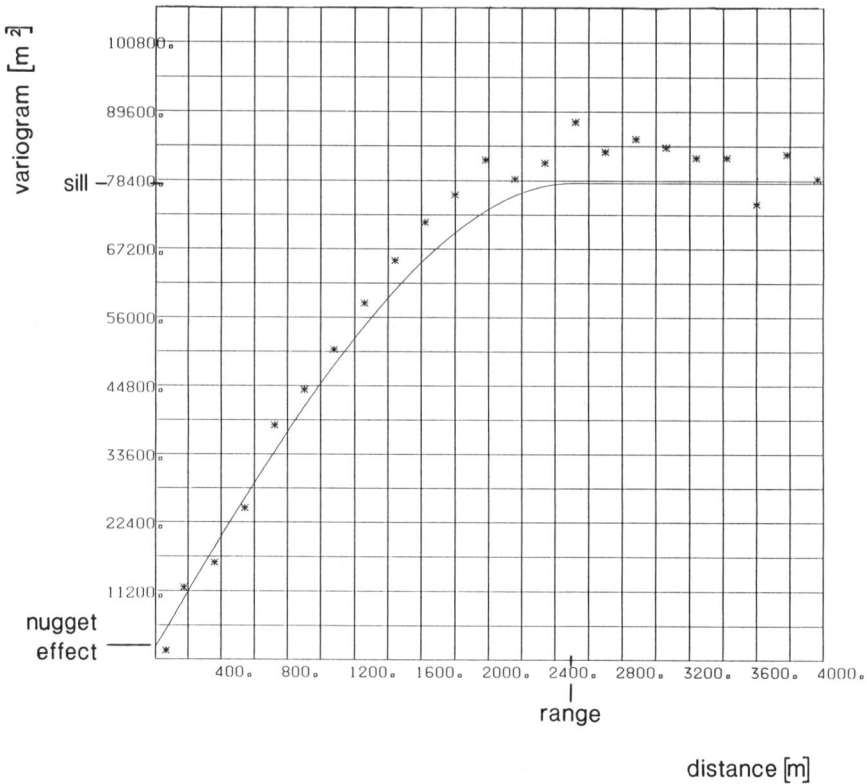

Figure 4. Global residual variogram of Scharffenbergbotnen ice thickness;
        *   - experimental variogram;
        --   - spherical variogram model (parameters as given in Table 1).

Geoscientific knowledge of the measured variable and experience in interpretation of spatial structures is necessary to adjust a variogram model to the experimental variogram. For methodological reasons, this step cannot be expressed by an algorithm. On the other hand, this necessity has the advantage of assigning weights relative to geologic meaning, whereas, in opposition, a numerical fitting routine would weigh all points by the same algorithm. This is the point that distinguishes geomathematical from purely mathematical methods.

The variogram model is determined by its type (that has to meet certain mathematical requirements), and the parameter's nugget effect Co, sill C (where Co, C are positive real numbers), and range (see Figs. 4, 5). A linear variogram with total sill Co+C and range a (a positive real number) is given by

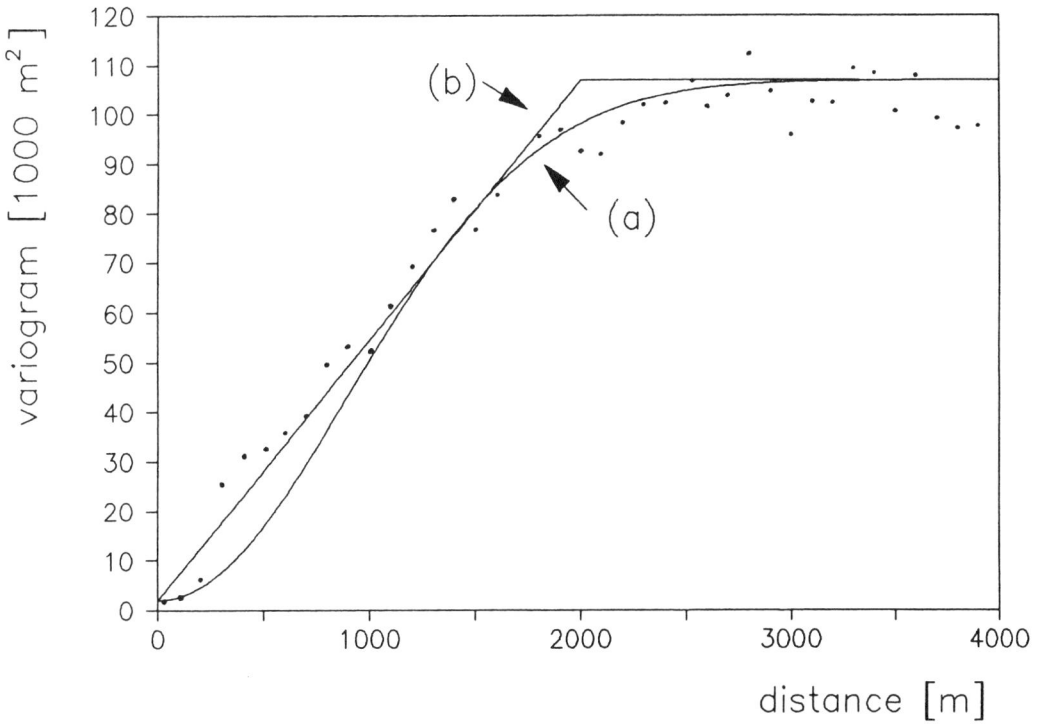

Figure 5. Global residual variogram of Scharffenbergbotnen subglacial bed topography;
  * - experimental variogram;
  (a) Gaussian variogram model;
  (b) linear variogram model; (parameters as given in Table 1).

$$\text{gam}(h) = \begin{cases} C_0 + \dfrac{C}{a}|h| & \text{if } |h| \leq a \\[2ex] C_0 + C & \text{if } |h| > a \end{cases} \qquad (2)$$

a spherical variogram by

$$
gam\,(h) \;=\; \begin{cases} C_0 + \left(\dfrac{3h}{2a} - \dfrac{h^3}{2a^3} \right) & \text{if}\,|h| \le a \\[2em] C_0 + C & \text{if}\,|h| > a \end{cases}
\tag{3}
$$

A Gaussian variogram defined by

$$
gam\,(h) \;=\; C_0 + C \left[1 - \exp\left(-\dfrac{h^2}{a^2} \right) \right]
\tag{4}
$$

has a practical (quasi) range of $a' = \sqrt{3}a$ with $gam(a') \approx C_0 + 0.95\,C$. The behavior near the origin reveals the spatial continuity of the regionalized variable. In our situation it is important for the smoothness of the resultant map and for its reliability. The range gives the maximal distance of points that can be used in the estimation neighborhood. Consequently, the range indicates the size of data gaps that can be bridged by kriging.

The geologic variables given as examples here, seafloor depth, ice thickness, etc., do not satisfy the hypothesis of stationarity in general. To detect a drift component m, I calculate

$$
m^*\,(h) \;=\; \frac{1}{n} \sum_{i=1}^{n} \, [\,z\,(x_i - z\,(x_i + h)\,]
\tag{5}
$$

and the residual variogram

$$
res^*\,(h) \;=\; gam^*\,(h) - \frac{1}{2}\,m^*\,(h)^2 \; .
\tag{6}
$$

Note that if the random function Z is stationary, then $m^*(h) = 0$, and $gam^*(h) = res^*(h)$ for all h.

Computing a Digital Terrain Model: Estimation of Grid Node
Values by Universal Kriging.

The method applied to estimate values on the regular grid is universal kriging. The unknown value $z(x_0)$ in a given grid node $x_0 \in D$, D the survey area, is estimated from measured values $z(x)$ in a neighborhood of x_0 by

$$
Z_0^* \;=\; \sum_{i=1}^{n} \alpha_i \, Z(x_i)
\tag{7}
$$

with weights $\alpha_i \in R$, $i = 1, ..., n$.

To take care of the fact that the geologic variable shows a trend, the random function $Z(x)$ is split into a deterministic drift $m(x)$ and a residual random function $Y(x)$:

$$
Z(x) \;=\; m(x) + Y(x)
\tag{8}
$$

Practically m(x) is taken as the local average of the samples whereas Y(x) is the random variation around m(x) and has zero expectation. It is assumed that the residuals are second-order stationary with variogram gam. (For reasons of numerical conditioning, this assumption is made in kriging software usually.) The form of the drift is hypothesized to be known as

$$m(x) = \sum_{l=1}^{k} a_l \, f_l \, (x) \tag{9}$$

with unknown coefficients $a_l \in R$ and some $k \in N$. Usually low-degree polynomials are taken.

In order to get an unbiased estimator, the conditions

$$\sum_{i=1}^{n} \alpha_i f_l(x_i) = f_l(x_0) \qquad \text{for } l = 1, \ldots, k \tag{10}$$

are sufficient. According to the number $k \in N$ the estimation also is termed unbiased kriging of order k. The coefficients i are determined such that the estimation variance $E[Z^*(x_0) - Z(x)]$ is minimal with respect to the unbiasedness conditions (Eq. 10). Here the basic assumptions on the spatial model are used: if covariances are substituted by variogram values for the residuals according to the definition

$$gam_{x,y} = \frac{1}{2} E\left[Y(x) - Y(y) \right]^2 \qquad \text{for } x, y \in D \tag{11}$$

then I get n conditions that together with the k unbiasedness conditions form a system of linear equations, the universal kriging system. For the general variogram models, it can be shown that the kriging system has a unique solution. The coefficients are substituted in Equation (7) to give the estimated value and additionally yield the estimation standard deviation

$$s^2 = \sum_{i=1}^{n} \alpha_i \, gam(x_i, x_0) + \sum_{l=1}^{k} \lambda_l f_l(x_0) \tag{12}$$

where λ_l, $l = 1, \ldots, k$ are Lagrange multipliers.

> REMARK: Theoretically there are problems to determine the proper variogram belonging to the universal estimator. In practice the residual variogram [(variogram of the residuals (Eq. 6)] can be taken (Matheron, 1971, p.188-195). However, it is not possible to derive both residual variogram and drift from a single realization, so *a priori* information on the variogram is necessary (Armstrong, 1984). In the glaciology investigation presented below the residual variograms could be estimated because of good data supply and by repeated variogram analysis.

DESCRIPTION OF DATA PROCESSING USING LSUNIMC

Referring to Figures 3 and 6 there basically are three steps to get from the survey data to the contour map:

(1) Variogram analysis
(2) Estimation of the DTM by kriging
(3) Contouring and plotting

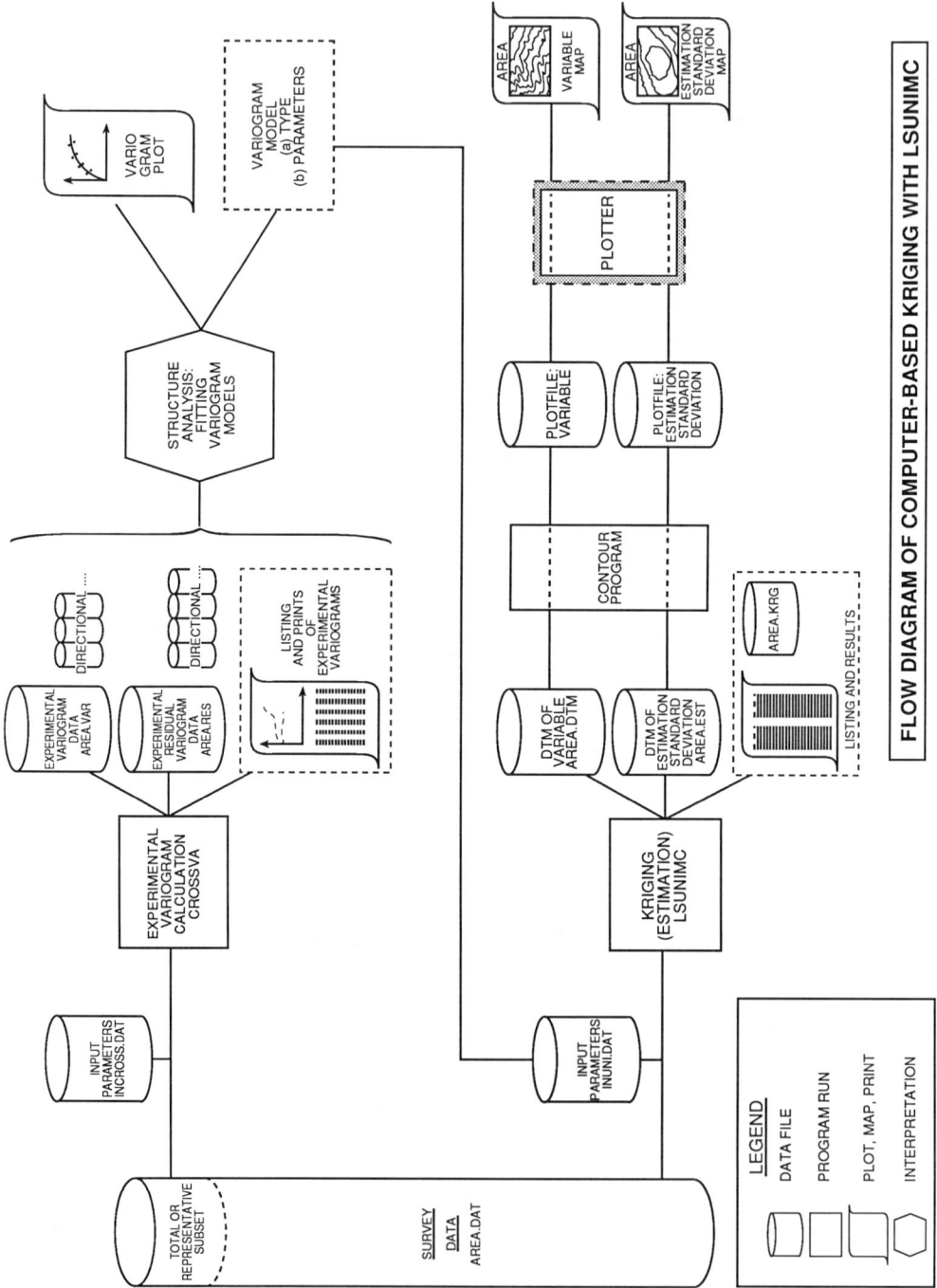

Figure 6. Flowdiagram of computer-based kriging with LSUNIMC.

As the program LSUNIMC performs step 2, steps 1 and 3 are treated just as necessary in the context. I assume that the survey data are given in coordinate form (x,y,z) with x being the East coordinate, y being the North coordinate, z being the measured value of the geologic variable at location (x,y).

Variogram Analysis

A program calculating experimental variograms and experimental residual variograms has to be used (for instance, CROSSVA of V. Pavlowski, cf. Note on Software). The program outputs data files containing the experimental variograms and experimental residual variograms in several user-defined directions, and line-printer output provides graphical display of the variograms.

The next step, which is the fitting of the variogram model to the experimental variogram, should not be automated. Of course, interactive graphics software can be employed to visualize the approximation of the fit until a convincing solution is formalized.

The following information is necessary to run the kriging program: the function-type of the variogram model (e.g. spherical, linear, Gaussian), and the characteristic parameters (range, sill, nugget effect).

Universal Kriging With LSUNIMC

LSUNIMC is written in FORTRAN 77 in modular form, compiled by a standard FORTRAN compiler, and implemented on an IBM PC/XT/AT compatible. Several options can be set by the user, these enter the program via a parameter file named INUNI.DAT. The input parameters are described in Table 1 and an example is given in Table 2.

LSUNIMC sets up the digital terrain model as follows: Grid node locations are either entered via the parameter file or calculated automatically, if the size of the grid spacing is given. At each grid node a neighborhood search is performed in order to select the closest sample points from the sample data set. The size of the radius (or edges) of the search circle (or rectangle) can be entered (parameter RAY). Points that are too close to the node might be left out on purpose for reasons of numerical conditioning (parameter EPSILO).

The search routine can be adjusted to the distribution of line survey data, as the algorithm depends on the location of the grid node relative to the survey tracks: if the grid node happens to lie inside or close to survey coverage the data supplied near that node is good, so no complicated search is necessary. This is tested by checking whether in a small distance (parameter SHRAY) of the grid node some data points are determined already (parameter ISHCNT), and the grid node value is derived from these ISHCNT nearest measurements. If the grid node lies in the gap between two coverages or at the edge of the survey area, then the estimation has to be based on measurements that are further away. In this situation the number of input data points needs to be higher and the radius larger. The decision between the two situations is made automatically.

The radius, RAY, should be set at 1.5 times the range approximately.

The program LSUNIMC produces the following output files:

- AREA.KRG contains information on the program run, plus a list of data associated to each grid node: coordinates, number of samples used for estimation, average distance, standard deviation of the samples, grid node value, and its estimation standard deviation. A printout of AREA.KRG proves useful for better understanding of the estimation process.

- AREA.DTM contains the DTM for further processing by a contour program.

Table 1. Input parameters for LSUNIMC.

Line	Parameter	Description
line 1:	title	(1-80 characters)
line 2:	NX	number of grid nodes along x (east)
	NY	number of grid nodes along y (north)
	NZ	number of grid nodes along z (NZ=0 for maps), if NX,NY,NZ are set 0,0,0, then the appropriate number of grid nodes to cover the map area is calculated from the other parameters
	XC,YC,ZC	coordinates of the SW-corner, the first node has the coordinates (XC+PASX/2, YX+PASY/2)
	PASX, PASY, PASZ	step size, distance between grid nodes along x, y, z
	NUMDSK	set 1
line 3:	RAY	maximal distance of sample (from grid node) used for estimation at that node; if RAY is set 0 then RAY=SQRT((2*PASX)**2+(2*PASY)**2+(2*PASZ)**2))
	MAXNOT	maximal number of samples used to krige a grid value; if MAXNOT is set 0 then MAXNOT=24
	EPSILO	minimal distance of sample (from grid node) used for estimation at that node; set >0
	ISUCH	set 1 for quadrant search (default) set 2 for distance search
	DELTX, DELTY, DELTZ	maximal distance of sample along x, y, z, respectively, (from grid node) used for estimation at that node; if set 0,0,0 then DELTX=XMAX-XMIN, DELTY=YMAX-YMIN, etc
	SHRAY	radius of search circle around grid node in areas of dense sampling; if there are ISHCNT samples closer to the grid node than SHRAY, then the kriging is based on ISHCOUNT points (to save time); if SHRAY is set 0 then SHRAY=MIN(PASX/3,PASY/3)
	ISHCNT	number of samples for kriging in SHRAY; if ISHCNT is set 0 then ISHCNT=MIN(2+3*NDRIFT,4)
line 4:	IP	set 1
	NVAR	number of variables to krige
	NDRIFT	degree of drift of (0,1,2)
	IVALI	set 1 unless IVALI ist set 0, at sample location values are kriged for validation of the estimation model
	NDAVAL	number of sample points to be kriged for validation (if set 0, all sample points)
	IVOUT	output channel for validation
line 5:	INBND	set 0
	UGRENZ	set 0
	OGRENZ	set 0
	FORMAT	set '*'
line 6:	INDATA	file name of input data file (eg area.DAT)
line 7:	FMT	format of data cards (set '*')
line 8:	OUTFILE	name of output file containing kriging results (eg area.KRG), uses UNIT 32
line 9:	DATFILE	name of output file containing digital terrain model (eg. area.DTM), uses UNIT 34
line 10:	ESTFILE	name of output file containing estimation standard deviation on DTM (eg area.EST), uses UNIT 35
line 11:	DNOM	name of variable (character*8, in ' ')
	LOGTRN	parameters for logarithmical data transformation; if LOGTRN is set 1, basis is e; BETA is additive constant
	BETA	
	IVM, VALMIS	missing-value-option to be used in case there are sample locations without sample values, otherwise set (0,0); if IVM set 1 then missing values >= VALMIS if IVM set -1 then missing values < VALMIS
	IGAM	index of the variogram function, options: 1 spherical model, 1-dim, (x,y); 2 spherical model, 2-dim; 3 spherical model, 3-dim; 4 bispherical model, (x,y); 5 bispherical/spherical model; 6 Gaussian model (x,y); 7 Gaussian-spherical/linear model; 8 spherical-linear or linear model (x,y); 9 Gaussian-linear model (x,y)
line 12:	GAMPAR	variogram parameters
following lines:		for following variables, repeat lines 11,12

Table 2. Example of LSUNIMC input parameter file

```
SCHARFFENBERGBOTNEN BED TOPOGRAPHY
0,0,1,0,0,0,100.,100.,0,1
3000,12,0.1,2,0,0,0,500,4
1,1,1,0,0,33
0,0,0,'*'
MORPH.DAT
'*'
MORPH_P100GAU.KRG
MORPH_P100GAU.DTM
MORPH_P100GAU.EST
'BED TOPOG',0,0,0,0,6
2.,103.,2200.,0,0,0,0,0,0,0,0,0,0,0,0,0
```

- AREA.EST contains the estimation standard deviation associated to each grid node value, destined as input for a contour program.

Contouring and Plotting

In this last step contour maps are produced from the results of the kriging. Here any software can be used that interpolates the values given on a regular rectangular grid. Of course, the output is not restricted to contour maps only, for instance three-dimensional figures or cross sections can be constructed as well, once the DTM is set up.

AN EXAMPLE OF APPLICATION IN GLACIOLOGY: CARTOGRAPHIC REPRESENTATION OF ICE THICKNESS AND SUBGLACIAL BED TOPOGRAPHY

Setting of the Study

In order to give a demonstration of the use of LSUNIMC, radio-echo data are evaluated to map ice thickness and subglacial bed topography. Data were sampled on Scharffenbergbotnen, Dronning Maud Land, East Antarctica, in the glaciology part of the Swedish Antarctic Research Program (SWEDARP) during the German Antarctic Expedition VI/3 1987/88 (Jonsson, Holmlund, and Grudd, 1988).

Scharffenbergbotnen is a glaciated valley in the Heimefrontfjella, Dronning Maud Land, East Antarctica (S 74:35'/ W 11:40', see Fig. 1). The elevation of the ice surface is about 2500 m asl on the upstream side and about 1300 m asl on the downstream side of the glacier. The innermost part of the valley is situated about 100 m lower than the outlet, thus the ice flux is opposite to the flow direction of the surrounding inland ice. The glacier is approximately 7 km long. It has an inflow of ice from the inland ice.

Measurements were taken with radio-echo equipment based on a low-frequency sounder constructed and produced by M. Sverrisson and H. Bjornsson at the University of Reykjavik (Sverrisson, Johannesson, and Bornsson,1980). The transmitter produces a pulsed signal of 8.1 MHz with a wavelength of about 26 m. The equipment was operated in both continuous profiling mode and spot mode. The signal was received on an oscilloscope and recorded by a 35mm camera fixed to the screen of the oscilloscope. The transmitter and receiver were transported on sledges by a snow mobile (Jonsson, Holmlund, and Grudd, 1988).

The sampled radio-echo data typically are distributed along profiles of differing orientations (Fig. 1), with continuous measurements along the profiles and gaps of information in between. The ice thickness data set contains about 1000 sample points that were digitized from the profiles. The data for subglacial topography are determined by subtracting ice thickness data from surface data. The surface data are taken from a manuscript map, scale 1:25000, provided by the Institute for Applied Geodesy (IFAG), Frankfurt, West Germany.

Geostatistical Data Evaluation

Variograms and residual variograms are calculated from the sets of ice thickness and subglacial bed elevation data. A comparison of variograms and residual variograms displays the divergence that is the result of nonstationarity of the regionalized variables. In this situation the method of universal kriging can be applied. In a test of the drift polynomials, models with a linear drift gave best results.

As a result of the structure analysis, the variogram models given in Figures 4 and 5 and in Table 1 were fitted. Ice thickness as a regionalized variable is continuous in the mean square sense, whereas bed topography is differentiable (the Gaussian model provides a better fit in the origin than the linear model), in other words, it is smoother than ice thickness. The small size of the nugget effect of ice thickness compared with the sill indicates that radio-echo measurements contain only relatively small errors. For a discussion of the geologic implications of the variography the reader is referred to Herzfeld and Holmlund (1989).

Besides the mentioned Gaussian variogram model and drift degree, the following input parameters were set to krige the subglacial topography, all other parameters were taken as defaults (notation as used previously in Table 1):

        PASX   = PASY = 100 [m]
        RAY   = 3000 [m]
        MAXNOT = 12
        EPSILO = 0.1 [m]
        SHRAY = 500 [m]
        ISHCNT = 4

By contouring the resultant DTM, the map of bed topography given in Figure 7 is obtained. Ice thickness is estimated on the same grid net with the spherical variogram given in Table 3, the map is presented in Figure 8.

Although the grid spacing is finer than the average profiling, survey patterns cannot be recovered from the contoured evaluation. For both the maps of ice thickness and subglacial topography, absolute estimation standard deviation is less than 5 m in the whole area. Thus (under the assumption of a Gaussian data distribution) with a likelihood of 0.95, the map values have a deviation of less than 10 m.

Glaciology Conclusions

The maps (Figs. 7, 8) show that Scharffenbergbotnen is a deep U-shaped valley following the northwest-southeastern direction of the mountain range (Spaeth and Fielitz, 1987). A shallow ridge in the central part of the valley can be observed on the bed topography map

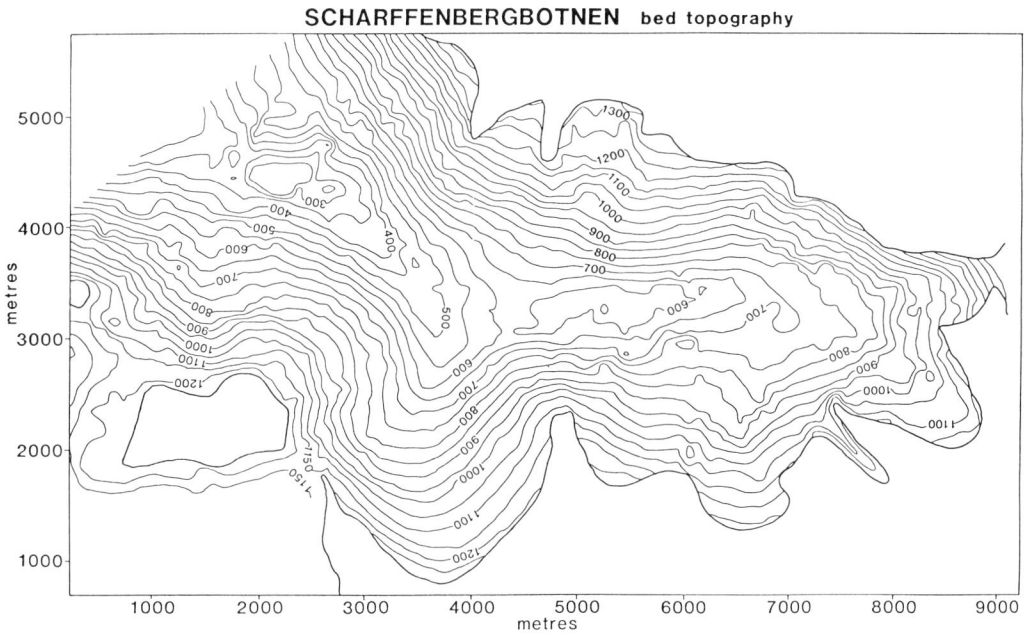

Figure 7. Subglacial bed topography of Scharffenbergbotnen basin, contoured from DTM with 100 m grid width, 4794 grid nodes, based on 857 data points. DTM was calculated using universal kriging with gaussian variogram given in Table 3 and parameter file INUI.DAT of Table 2 (from Herzfeld and Holmlund, 1989).

Table 3. Global residual variogram models of Scharffenbergbotnen ice thickness and subglacial bed topography

VARIABLE	VARIOGRAM TYPE	VARIOGRAM PARAMETERS		
		RANGE [M]	NUGGET EFFECT [M^2]	SILL [M^2]
ICE THICKNESS	SPHERICAL	2500	2000	78000
SUBGLACIAL BED TOPOGRAPHY	GAUSSIAN	2200	2000	105000
	LINEAR	2000	2000	105000

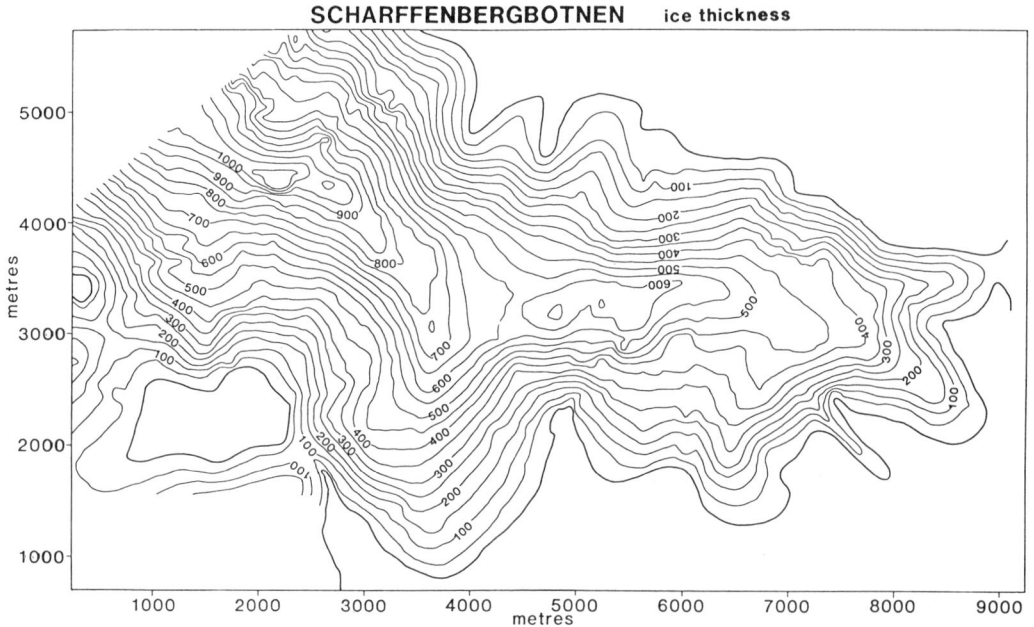

Figure 8.   Ice thickness in Scharffenbergbotnen basin, contoured from DTM with 100 m
            grid width, 4794 grid nodes, based on 857 data points. DTM was calculated using
            universal kriging with spherical variogram given in Table 3 (from Herzfeld and
            Holmlund, 1989).

(Fig. 7). The ice flow follows the main northeast-southwestern direction of the valley only
partially, whereas another part branches off southwards just before the shallow ridge.

Both the existence of the ridge and the branching of the ice flow were not detected by
nonmathematical interpretation of the field data. Consequently, the geostatistical
evaluation of radio-echo data brought some new information to the glaciologists. As the
maps have a high precision and reliability, they could be used as a basis for investigations
on ice flow, glacier dynamics, and mass balance. With the help of the geostatistical
structure analysis, connections between ice flow directions, bedrock morphology, and
tectonic lineaments of the mountain range were studied (Herzfeld and Holmlund, 1989).

CONCLUSIONS

By use of the universal kriging program LSUNIMC, a high-precision cartographical
evaluation of geologic and geophysical data is possible, even if the survey pattern shows
large gaps of information as is the situation in widely spaced line surveys. The survey
pattern does not reappear in the cartographical representation, and the map of
estimation standard deviation shows that even in survey gaps the estimated values are
reliable.

Because of the flexibility of the program LSUNIMC, special geologic requests can be met
corresponding to different geoscientific questions (for an example, see Herzfeld, 1989).

Furthermore, the method applied has an impact on survey optimization: If the sizes of the theoretical range of the variogram (from variogram parameters), of the effectively used maximal distance of sample points to a grid node (from AREA.KRG), and of the line survey gaps in average are viewed synoptically, it is possible to determine the largest size of gaps that can be bridged by the estimation method. An optimal sampling strategy is achieved, if a presurvey is carried through to obtain this information, such that in the main survey the profile distances can be planned in a way to save survey time. Because LSUNIMC is available on microcomputers, the necessary calculations can be performed in the field.

ACKNOWLEDGMENTS

The major part of the software development was carried through during the German expedition ANTARKTIS VI/3 1987/88, Alfred-Wegener-Institute for Polar and Marine Research, Bremerhaven, West Germany. The radio-echo data were sampled by Dr. Per Holmlund and Hakan Grudd in the field surveys on Scharffenbergbotnen, Dronning Maud Land, East Antarctica as part of the glaciology project (Dr. Stig Johnsson) in the Swedish Antarctic Program (SWEDARP) during ANTARKTIS VI/3. My thanks to Prof. Dr. W. Skala, Dr. Heinz Burger, and Dr. Reinhard Schoele, Free University of Berlin, for permission to use the initial UNIKRG program, to Holger Klindt for helpful discussions on the system development aboard RV POLARSTERN, to Per Holmlund in particular for the friendly cooperation during my stay at the Tarfala research station (University of Stockholm) in Northern Sweden, to Prof. Dr. Dan Merriam for encouragement to submit this contribution, to Dr. Tim Herbert (Scripps Institution of Oceanography) for reading the manuscript, and to Evelyn Hegemier of Scripps for drawing Figure 6.

NOTE ON SOFTWARE

LSUNIMC is the microcomputer version of a universal kriging program (UNIKRG, SEABEAM-version) developed and implemented on the VAX 11/750 computer onboard the RV POLARSTERN as part of the SEABEAM-post-processing during the expedition ANTARKTIS VI/3 1987/88 at the Alfred-Wegener-Institute for Polar and Marine Research, Bremerhaven, West Germany (Herzfeld, 1988). The basic program (UNIKRG) was developed at the Free University of Berlin Institute for Geology/Mathematical Geology by Dr. R. Schoele for resource assessment and updated by Dr. H. Burger. For calculation of the experimental variograms, a slightly altered version of the FORTRAN program CROSSVA, written by Dr. V. Pavlowski at the FU Institute for Geology/Mathematical Geology was used. The interpolation of the DTM was realized by the commercial software package UNIRAS/GEOPAK.

REFERENCES

Armstrong, M., 1984, Problems with universal kriging: Jour. Math. Geology, v. 16, no. 1, p. 101-108.

Briggs, J. C., 1974, Machine contouring using minimum curvature: Geophysics, v. 39, no. 1, p. 39-48.

Davis, J. C., 1986, Statistics and data analysis in geology (2nd ed.): John Wiley & Sons, New York, 646 p.

Herzfeld, U. C., 1988, Geostatistische Verfahren im SEABEAM-post-processing, *in* Fuetterer, D. K., ed., Die Expedition ANTARKTIS VI mit FS "POLARSTERN" 1987/88, Ber. Polarforschung (Rep. Polar Res.) 58, Bremerhaven, p. 133-139.

Herzfeld, U. C., 1989, Geostatistical methods for evaluation of SEABEAM bathymetric
   surveys: Case studies of the Wegener Canyon, Antarctica, Marine Geology.

Herzfeld, U. C., and Holmlund, P., 1989, Geostatistical analyses of radio-echo data from
   Scharffenbergbotnen, Dronning Maud Land, East Antarctica: submitted to Z.
   Gletscherkunde Glaz.

Johnsson, S., Holmlund, P., and Grudd, H., 1988, Glaciological and geomorphological
   studies of Scharffenbergbotnen, *in* Fuetterer, D. K., ed., Die Expedition
   ANTARKTIS VI mit FS "POLARSTERN" 1987/88, Ber. Polarforschung (Rep. Polar
   Res.) 58, Bremerhaven, p. 186-193, 195.

Journel, A., and Huijbregts, C., 1978, Mining geostatistics: Academic Press, London,
   600 p.

Matheron, G., 1963, Principles of geostatistics: Econ. Geology, v. 58, no. 8, p. 1246-1266.

Matheron, G., 1971, The theory of regionalized variables and its applications: Les Cahiers
   du Centre de Morphologie Mathematique de Fontainebleau, 211 p.

Spaeth, G., and Fielitz, W., 1987, Structural investigations in the Precambrian of Western
   Neuschwabenland, Antarctica: Polarforschung, v. 57, no. 1/2, p. 71-92.

Sverisson, M., Johannesson, A. E., and Bjornsson, H., 1980, Radio-echo equipment for
   depth sounding of temperate glaciers: Jour. Glaciology, v. 25, no. 93, p. 477-486.

Regional Geophysical Data on a Compact Disk

A. M. Hittelman and H. Meyers
National Geophysical Data Center, NOAA

ABSTRACT

The "Geophysics of North America" compact disk project was an outgrowth of research and compilation of regional studies data. For more than a decade, scientists, their professional societies (i.e., SEG and GSA), and their governments (i.e., U.S., Canada, and Mexico) have been working toward a consolidated collection of land and marine geophysical data for North America. Much of this important compilation was done under the auspices of the Geological Society of America's Decade of North American Geology - including topography, magnetics, gravity, earthquake seismology, crustal stress, and thermal aspect data.

Applications for this data compilation include the analysis of regional trends, such as the removal of long wavelength fields from a more detailed set of data, or the analysis of interrelationships between gravity and magnetics in comparison with topography or earthquake seismicity.

The compact disk format was selected because it represents a state-of-the-art medium that is capable of storing a large quantity of data, whereas providing easy and cost-effective retrieval in a workstation environment.

To encourage innovative use of the data, accession software was developed that simplifies data selection. The software also allows the user to obtain "quick-look" retrievals of the data— complete with geographic references, color raster images, data contour overlays, and data profiles. The software employs modern user interfaces such as pull-down windows and on-line documentation. Innovative access utilities, including geographic partitioning and indexing of data, increase the speed of the access.

PROJECT BACKGROUND

During the 1980's, the Earth-science community placed great emphasis on the compilation of data sets that crossed national borders. North America was one region that received high attention; and, several geophysical data sets were developed with the cooperative efforts of hundreds of scientists and the endorsement of numerous governments and their agencies, professional societies, academic institutions, and private sector organizations.

In developing the North America compilations, scientists selected the map as the preferred presentation media. At the same time, they recognized the long-term benefit in the development of the digital database which was the foundation for this display. Following compilation, the data sets were placed in national and international data centers for distribution.

An important function of data centers is the packaging of data in an informative and useful manner. This task may include quality control, as well as ensuring high documentation standards. A new emphasis is on developing simplified data-access facilities for personal computers. In response to user needs, data have been made available on floppy diskettes (occasionally with companion accession software). For voluminous data sets, compact disk technology has provided a PC-based solution. Usually referred to as CD-ROMs (Compact Disk - Read Only Memory), these data platters provide a compact and rugged format in which sizeable quantities of stored information can be made available economically (Fig. 1).

The "Geophysics of North America" compact disk project includes the disciplines of topography, magnetics, gravity, earthquake seismicity, crustal stress, and thermal aspect data. Accession software, which simplifies data selection and provides the user with graphical browse capabilities, employs modern user interfaces such as pull-down windows and on-line documentation.

ENTREPRENEURIAL STYLE OF PROJECT MANAGEMENT

Most software development projects are managed using classical life-cycle approaches which dissect development efforts into phases— such as requirement definition, analysis and design, programming, testing, acceptance, and installation/operation. Some of these phases may overlap in time, or be dissected into finer components. Testing, for example, can occur during all of these phases. In general, these phases provide an important communication framework in which developer and user can track progress. The exact form of the approach typically conforms to the culture of the enterprise. Each organization usually supports its project management techniques with documents such as project plans, design specifications, and acceptance criteria and milestone charts— all of which add to the communication and control efforts.

In using many of these techniques, criticisms keep arising concerning: (a) the enormous overhead associated with documenting the life-cycle effort, and (b) the difficulty in defining exactly what is going to be done. Software developers are frustrated continuously with the lack of clear direction and the continuous changes (a.k.a., enhancements) which are being requested. Many organizations have instigated change control procedures to handle these issues, because every change delays project completion and adds to the development costs. Change is necessary, however, for no software developer wants the users to say, "You did give me what I asked for, but it's not what I really need."

An unorthodox approach to project management is to define the project when it is completed. With new software tools that support prototyping, this is possible. It allows

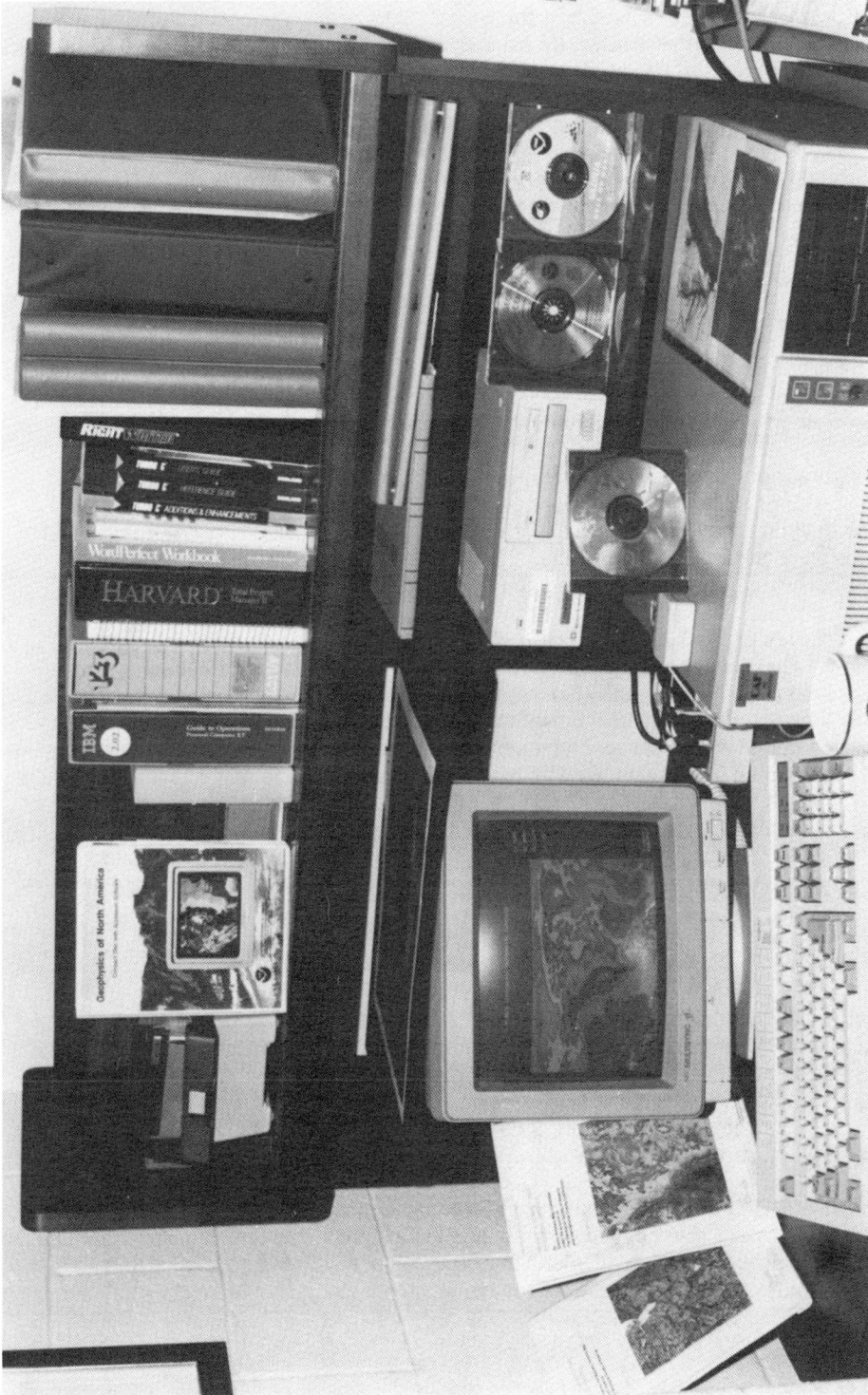

Figure 1. CD-ROM readers can be added easily to personal computers, significantly expanding amount of information available on one's desktop.

one to define a product by iterative experimentation, creating a closer communication environment between the developers and the users. This entrepreneurial approach to software development operates under the premise that it is more important to develop the right product than to develop the product right. Once the product reaches a stage in the prototype cycle where it is "robust enough", a release of the product can be made— even if it is marketed as a preliminary (usually referred to as beta-test) release.

Such a prototype methodology was used in developing the "Geophysics of North America" CD-ROM. The project was structured around a number of development groups, all of which had the same project leader. The groups included:

- Software Design Group - Defined requirements and acted as a user entity, reacting to growing versions of the product.

- Data Preparation Group - Developed the data files and performed quality control.

- Programmers - Produced application and user interface code.

In classical approaches to software development, analysts precisely define the user interface for the application programmers. Interestingly, in this project the user interface was developed by programmers other than the application programmers and their activities were performed in parallel. The software design group reacted to prototypes of both the user interface and the applications.

Appropriate tools are critical to the success of this approach. The computer language selected was "C", the user interface was written with a "windows" application development package (which actually generated "C" code), and the graphics were focused towards EGA/VGA standards. Another software product that proved helpful in the prototype-environment was a demo developer, which was used for creating slideshow-style poster sessions and a tutorial for the final product.

Quality control was an integral part of the project— from data preparation through product release. Because much of the accession software supported a graphic color representation of the data, data errors became apparent rapidly (Fig. 2 – Color Plate 3).

CD-ROM WORKSTATION ENVIRONMENT

CD-ROM technology was introduced in the early 1980's. Its growing popularity partly is because of agreement on standards and cost breakthroughs associated with inexpensive platter production and CD-ROM readers. Although the audio industry has pioneered much of this technology, CD-ROM drives designed for computers differ from those designed for music. The primary differences are the omission of digital-to-analog circuitry and the addition of error-handling circuitry.

The power of the CD-ROM lies in its ability to supply sizeable quantities of stored information in a compact and rugged format. A compact disk can store approximately 600 megabytes of data, which is equivalent to 300,000 pages of written text or four standard 6250-bpi tapes. Cost of readers for personal computers is falling rapidly some are available for under $750 (in 1988).

The "Geophysics of North America" CD-ROM was designed to operate with an IBM PC/AT-compatible machine with color graphics capabilities. The typical cost of such a system, which includes a CD-ROM reader and the CD-ROM with its accession software, is less than $3,600 (in 1988). Those who already have a suitable computer simply can upgrade for about $1,000.

In using the CD-ROM product, the data and software can be shared over a local area network. This works well if the product is used infrequently by more than a single user at any one time. One of the benefits of a CD-ROM, however, is that data can be brought to a user in such a way as to relieve contention on the overall network. Nobody likes to wait for their data; and, when their data become part of their dedicated machine, they do not have to wait.

DATA STRUCTURE PHILOSOPHY

All of the original (archival) data were presented on the CD-ROM in ASCII code. These archive data were maintained in their original data structure as much as possible. For example, some of the potential field data were sorted by columns originally , having the advantage of making maps on plotters more convenient. Displaying data on the screen, however, is easier when the data are sorted by rows in a common grid structure. Consequently, it was decided to store the data on the CD-ROM in a display format as well. The display formats were designed to facilitate quick-look capabilities and were recorded as binary code. Display data included raster images, computed contours, and condensed point-data information.

All display images were computed in a latitude-longitude spatial domain with grid sizes being multiples of 2.5 minutes. The grid spacing was selected to support the original resolution of the primary data sets.

Many of the data sets have been contoured and stored on the CD-ROM in vector format. Utilities used to produce the contours are less sophisticated than those available to much of the scientific community. The contours were meant to support only quick-look reviews of the data, not high-resolution studies.

Point-data also are presented in display formats. The display formats used for earthquake seismicity, crustal stress, and thermal aspects all contain fewer fields than their archival counterparts, including primarily information that would be used in the quick-look maps.

DATA CONTENTS

The information contained on the CD-ROM include data from a wide range of geophysical disciplines throughout the North American region. Both land and marine environments encompassing nearly one-half of the Northern Hemisphere are covered by this compilation.

Topography includes 5-minute gridded elevation and bathymetry values. A file containing 30-second gridded data is present for the conterminous United States and its coastal waters.

Gravity data include free-air and Bouguer anomalies on 6-km and 2.5-minute grids. Isostatic, free-air, and Bouguer gravity anomalies are available for the mainland United States on both 4-km and 2.5-minute grids.

Magnetic-anomaly data are presented in 2-km and 2.5-minute grids. MAGSAT satellite-derived anomalies are present on a 2-degree and 10-minute grids (up to 50 degrees North latitude).

Satellite imagery on a 10-minute grid covers several popular AVHRR wavelengths, as well as the vegetation index, during both winter and summer months.

Point data includes earthquake events (1534– 1985), crustal stress, and thermal aspects.

Boundary data include geopolitical information such as coastlines, country, state and province borders, and U.S. county outlines. In addition, the software allows latitude and longitude grids to be displayed (Fig. 3 – Color Plate 3).

ACCESSION SOFTWARE

Capabilities

Perhaps the most enjoyable element of the project was developing the software to access and browse the data. Originally, it was believed that the primary use of the product simply would be the storage and the retrieval of the data for use in studies by an investigator. During the development of the software, it rapidly became apparent that the quick-look features of the software was fun to use. Furthermore, many visitors who saw early versions of the product kept asking for enhancements— wishing to expand the basic capabilities into a robust Geographical Information System (GIS). While it was tempting to do this, the resources were simply not allocated for this activity.

Although there are about two dozen major data sets on the CD-ROM, there are about 500 indexed files. Heavy indexing may be desirable to speed retrievals. Data were segregated based on data type and geographic distribution. Data that were maintained in grid structures were separated logically into areas with "reasonable" limits. Data that had a random distribution— point data (e.g., earthquake events), contour data, or boundary data— were regionalized based on a computerized scheme. The regionalization technique, developed by Ray E. Habermann, partitioned data geographically into compartments with an equal number of records. Similar to "balanced tree indexing", a common Database Management System (DBMS) technique, regionalization reduces the access time for searching the database— typically by an order of magnitude.

It may be stated that "data are not sexy." Indeed, it is difficult to become excited about tables of numbers, reports, or even monochromatic graphic representations of data. When color display capabilities are added to the product— producing images with overlays of point data, contours, and geopolitical boundaries— one begins to get enthusiastic about the information. This is especially true when you get rapid response with easy-to-use features that support considerable flexibility (such as palette painting features that allow you to customize color presentations) (Fig. 4 – Color Plate 3).

The applications within the accession software were designed and coded (in "C") independently of the user interface. Basic requirements included the ability to select data based on discipline, area of interest, and parameters unique to the data itself (such as magnitude range of an earthquake). If data sets are stored on the CD-ROM in multiple forms (e.g., different grid intervals), the user is given data set retrieval options. This occurs with respect to archival formats, which were stored in Mercator-projected kilometer grid spacings, and display formats, which were stored in latitude-longitude grid spacing. Selection criteria are repeated (for example, an investigator may be interested only in Canada or California); therefore, the software has options to store selection criteria.

The browse facility opens graphic views of the data, with many options to mix-and-match. Most of the gridded data can be viewed as color-map images, cross-section profiles, or contours (with optional interval spacings). Point data can be presented with various symbols. The color-map images can be painted using standard palettes defined for each data type, new palettes interactively designed by the user, or previously defined palettes customized by the user. Color cross-section profiles are available for all gridded data sets; by specifying the end-points desired, the profile is drawn identifying the first and last values, the zero datum line, and the high and low values. Contour presentation options provide several selections of pre-computed lines with a user-specified color selection.

One interesting application of the browse facility is to view how topography might seem if the ice sheets melted—resulting in a 25-foot rise in sealevel. By retaining oceanic colors (i.e., blues) in the topography image through 25 feet above sealevel and overlaying coastline boundary data as a reference, one can plan readily where to buy new beach front property in this scenario. Other applications for the topographic data include: aviation problems, hazardous-waste plume dispersion, weather modeling, surface run-off analysis, and base-map referencing (Fig. 5 – Color Plate 3).

Some of the data support temporal studies. Earthquakes, for example, can be viewed as a series of time-lapse images. To facilitate this capability, a series of images have been stored on the CD-ROM and can be viewed sequentially—each image representing events in a specified region during different time intervals. This type of analysis could be useful in tectonic or natural hazard studies.

One of the limitations of CD-ROMs is the inability to add new information or correct erroneous data that may have slipped into the product. When errors are located, some are correctable within the accession software that accompanies the compact disk. The user may not be aware of the data error if he or she is using the accession software (as opposed to browsing the compact disk with one's own software). New data can be added to the product on floppy diskettes or on one's hard disk. To accomplish this one must use the same data format as the compact disk and modify search path statements within the set-up procedure.

The User Interface

An application development package was used to create screens with pop-up menus. The user-interface product actually produced "C" code which interfaced smoothly with the application code. Included within the user-interface package was support for "help" facilities—a useful form of on-line documentation.

The selection-criteria prompts always provide the user with information as to the maximum and minimum values, as well as the units of measure of each parameter being browsed. As examples: (1) if one was selecting an area of interest, the latitude and longitude limits of the entire data set would appear on the selection screen as defaults, (2) if one only wished to view elevations above sealevel, in selecting the desired parameter range for elevation, the user would simply change the minimum value from the default to zero.

The user interface permits the saving of selection criteria. Thus, if you were studying a certain region, such as the San Andreas Fault Zone, you could save your geographic search window as well as your range of earthquake magnitudes of interest. In addition, you can save a specialized color palette which you may wish to use again.

To help individuals learn how to run the product, a "demo generator" program was used to create tutorials and on-line demonstrations. The on-line demonstrations produce a screen "slide show", which is useful in poster session presentations at technical meetings.

Additional "help" facilities are available in the printed manual that accompanies the CD-ROM, the "read me" files stored on both the CD-ROM and the accession software's floppy diskettes, and the on-line help facilities that are available through a function-key while running the accession software.

THE FUTURE OF CD-ROM TECHNOLOGY

Other Projects

Many government agencies are involved in projects to provide data in CD-ROM formats. The U. S. Geological Survey, Defense Mapping Agency, National Aeronautics & Space Administration, Bureau of the Census, U. S. Postal Service, National Geophysical Data Center, and National Climatic Data Center— to name just a few— all have optical disk data-management efforts underway. Many of the projects contain spatial data, supporting mapping and scientific applications.

The private sector also is involved in massive publication efforts. Public or corporate libraries without a CD-ROM reader may soon be a thing of the past. Reference works, catalogs, and journals are all popular items for publishing in CD-ROM formats.

Global Science

Much of the emphasis in environmental science today is in the recognition that global systems need to be understood in greater depth. With this new consciousness prevalent, it becomes increasingly more important to bridge the scientific gap between developed and developing nations. CD-ROM technology is one of the newest breakthroughs which may facilitate this process. Although many countries cannot afford large mainframe computers or high-technology graphic workstations, most can afford personal computers for their scientists. Adding CD-ROMs to their computers is inexpensive relatively. Some international organizations have suggested already that they may fund the hardware and distribution costs to supply applicable data sets to the appropriate countries.

By involving all countries in global science, greater international cooperation results with respect to both data exchange and the resolution of global environmental issues.

STAFF CREDITS

The motivational force for this project was Herbert Meyers, Chief of the Solid Earth Geophysics Division. Mr. Meyers has spent many years on several of the national and international committees facilitating contributions of data and talent. He also provided the vision to publish these data in a compact disk format.

Numerous individuals within the National Geophysical Data Center participated in the project. Allen M. Hittelman served as project leader and, as such, participated in all phases of the development. Helping with data preparation and documentation were: W. Minor Davis, Ray E. Habermann, John O. Kinsfather, Allen M. Hittelman, David A. Hastings, John J. Kineman, Carl C. Abston, Susan E. Godeaux, Ronald W. Buhmann, David T. Dater, and Peter W. Sloss. Support was provided by Richard Hansen and Cemal Erdemir of the Cooperative Institute for Geoscience Data Management and Applications (CIGMA) at Colorado School of Mines. Accession software and publication support were provided by: John O. Kinsfather, Ray E. Habermann, Allen M. Hittelman, Gerald H. Orita, Joy A. Ikelman, Carl C. Abston, and Chris Wells.

In any project there are those who accomplish tremendous feats that are above-and-beyond the call of duty. Ray E. Habermann and John O. Kinsfather were two such stars—developing code and innovative procedures far in excess of what normally is expected of an employee. All those involved in this project are especially indebted to them.

Dissecting Variograms

Michael Edward Hohn
West Virginia Geological Survey

Maxine V. Fontana
Northern Virginia Community College

ABSTRACT

Advanced geostatistical methods demand easy-to-use programs for examining observed data before the resource assessments. An interactive graphics program implementing Journel's h-scattergram is useful for identifying statistical outliers, detecting spatial discontinuities and subpopulations, examining alternative measurements of a regionalized variable, and for modeling a variogram. Written in Pascal for an IBM PC AT, the program also can be used to teach how a variogram relates to spatial autocorrelation, and how outliers and other problems affect the variogram, and consequently, estimation.

INTRODUCTION

Geostatistical methods are simple, but tedious, methods made easy by computers. Computer programs are available for obtaining variograms, linear estimates, and nonlinear estimates such as those provided by disjunctive kriging and nonparametric kriging. Geostatisticians can fit models easily to observed variograms through interactive graphics, spreadsheets, or curve-fitting algorithms using generalized least squares or maximum likelihood. Given a set of data, the practitioner can determine it almost too easy to perform a complete analysis without looking in depth at the data. A resource assessment can be carried out, maps drawn, and the report written up before someone questions the source of a "bullseye" on the map, or realizes that the kriged estimates reflect too much smoothing because some statistical outliers inflate the nugget effect. The sophistication of many geostatistical packages insulates the geologist from the data.

Journel (1983) took a step back to the data by showing the relationship between a simple display he terms an h-scattergram and the variogram. Given an observed value of a regionalized variable $z(x)$, and a second value $z(x + h)$ situated h units from $z(x)$, the h-scattergram is simply a plot of $z(x)$ versus $z(x + h)$ for all n pairs of values (Fig. 1). An h-scattergram can be drawn for each discrete value of h, or when the spatial distribution is

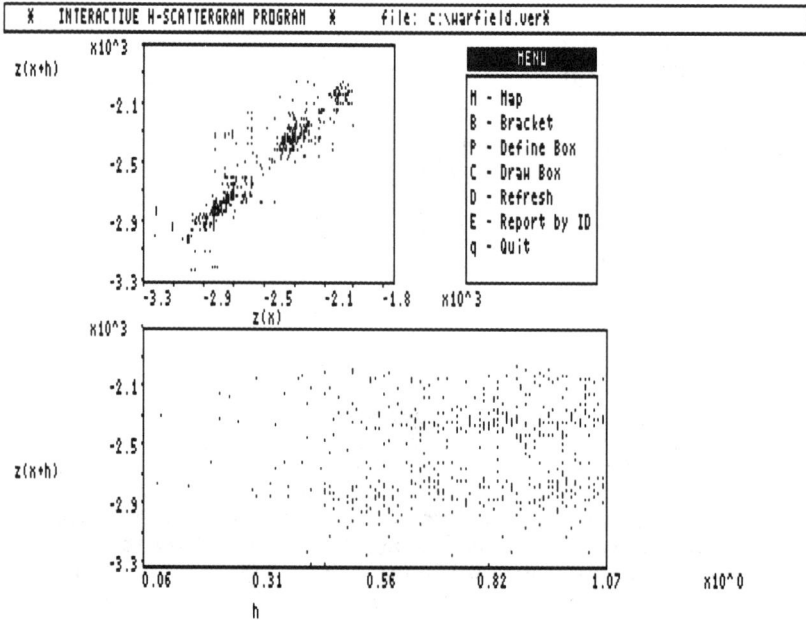

Figure 1. Main screen of program SCATPKMP described in text.

irregular, for each value of h ± d, where d is a tolerance. Journel shows that the moment of inertia about a line of unit slope on the h-scattergram:

$$\frac{1}{2n}\sum_{i=1}^{n} (z(x_i) - z(x_{i+h}))$$

equals the variogram. Thus, desirable properties of the h-scattergram include its direct view of a set of data, and its simple relationship to an important statistic in resource assessment.

Without going into details, Journel (1983) suggested that an h-scattergram could be used for obtaining robust semivariogram models. Indeed, a precursor of the computer program described in this paper was shown useful in detecting outliers and data-entry errors (Hohn, 1988). This paper describes an interactive computer program, SCATPKMP, that displays an h-scattergram on a microcomputer. The capabilities of the program have been improved since the original description of a similar program in Hohn (1988). It now includes an additional graphics display and methods for identifying individual points on the h-scattergram.

This paper consists of two parts, one outlining the computer program and its features, the second presenting a number of case studies that illustrate how the program is used.

THE PROGRAM

SCATPKMP was written with Borland's Turbo Pascal on an IBM PC AT, using the *Turbo Graphix Toolbox* available from Borland for defining graphics windows, and a Tektronix 4696 color-ink jet printer for obtaining hard-copy output. In writing the program, we were most concerned with a rapid display, and a menu that was easy to use. All commands available to the user are contained in one menu (Fig. 1), which is displayed at all times except when a map of data locations is on the screen.

Once a user specifies the name of an input file and labels for three axes, SCATPKMP plots scattergrams in two windows, and the menu in a third. The basic h-scattergram has three parameters, an observed value $z(x)$ of a regionalized variable, at location x, an observed value $z(x + h)$ of a regionalized variable at location $(x + h)$, and the distance h that separates those observations. The observed values can be measured from cores, grab samples, weather stations, or in this situation, gas wells. Furthermore, as we shall see, the "observations" may have been either measured or estimated by the leave-one-out procedure used in validation, in which instance h may be a generalized "distance" such as estimation variance (for an example, see Hohn, 1988, chap. 3). Taking the simplest example of the h-scattergram described by Journel, the upper left window in Figure 1 represents measured tops of a formation in southwestern West Virginia; values range from about 2,000 to 3,000 feet below sealevel (i.e. -2,000 to -3,000 feet above sealevel). Although each pair of wells, say z_i and z_j, could be plotted twice on the scattergram, storage and computational requirements are minimized by plotting only one point, usually (z_i, z_j), where $i < j$. In addition, the number of pairs needing to be plotted are limited by judicious selection of a maximum value for h.

The lower window shows the same points with distance h plotted on the abscissa. Together, the two windows are an attempt to display a three-dimensional scattergram and to provide a visual link between the h-scattergram and the variogram. As stated in the introduction, the degree of scatter about a line with unit slope is proportional to the variogram. For instance, assume a set of data exists with a small nugget effect in the variogram. For small values of h, $z(x)$ nearly equals x $(x + h)$, and the pairs $(z(x), z(x + h))$ should lie along a line. As h increases, the difference between pairs of values $z(x)$ and $z(x + h)$ also tends to increase, and scatter about the line becomes pronounced. Beyond the range, the scattergram shows no statistical correlation between $z(x)$ and $z(x + h)$. Keeping in mind that the correlogram is the inverse of the variogram, one can use h-scattergrams to demonstrate that the variogram indeed does measure spatial continuity in a sensible way.

The Menu

The Menu commands allow the user to subset and identify points on the scattergram in a number of ways.

B - Bracket: The user can define a range of values for h, shown as a bracket on the lower window. As the bracket is moved along the abscissa, only those pairs in the h-scattergram lying within the range specified by the bracket are plotted in the upper window (Fig. 2). One can see the affect of h on the appearance of the scattergram.

P - Define Box: The user is provided with a cursor for defining the lower left and upper right corner of a rectangle in the upper window.

C - Draw Box: Once the extent of a rectangle is defined, this command draws the rectangle and reports the identification numbers for each pair of observations (Fig. 3).

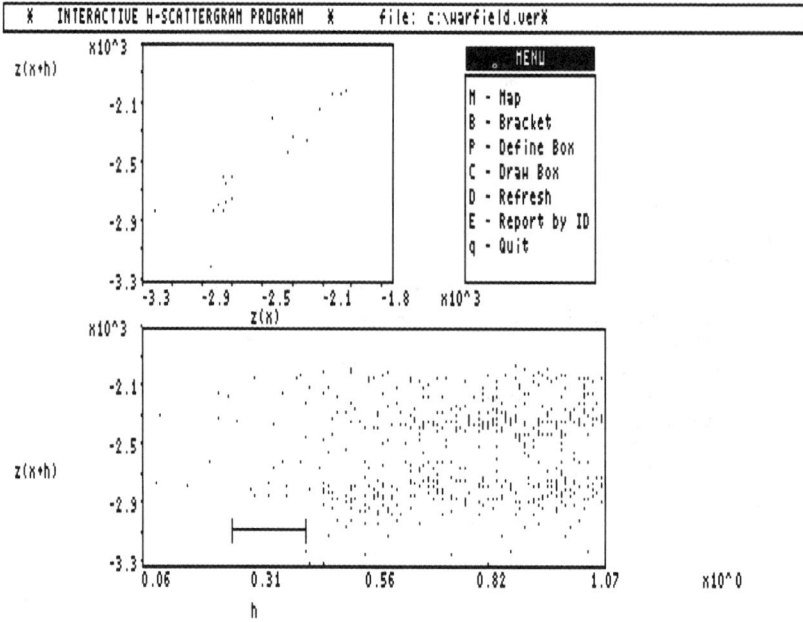

Figure 2. User has specified a range for h, shown as bracket in lower graph. As user moves bracket with right and left arrow keys, program displays corresponding h-scattergram in upper graph.

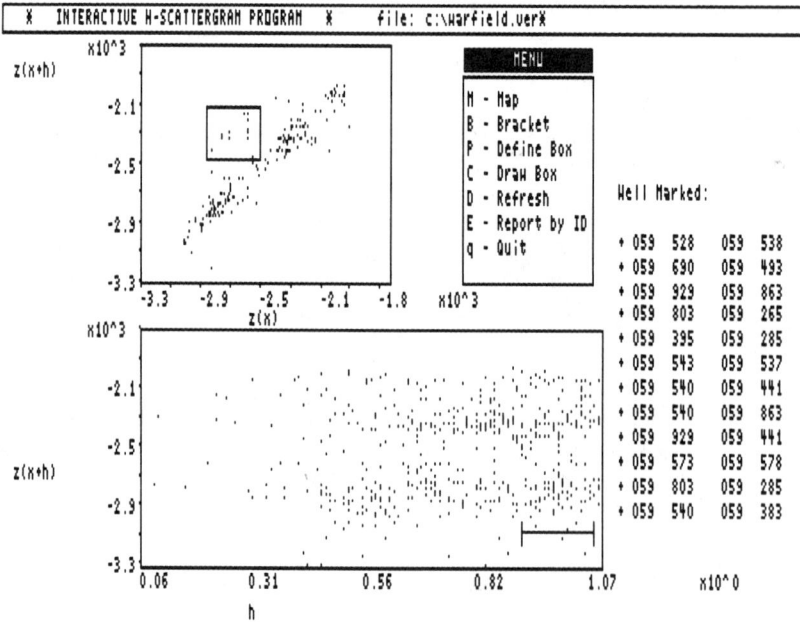

Figure 3. User has defined range of values for z(x) and z(x + h), shown as rectangle on upper graph. Right are county-permit identifiers for pairs of wells lying within box.

E - Report by ID: This command reports the value of the regionalized variable for a
                     location specified by the identifier. In addition, SCATPKMP highlights
                     points for which the location specified by identifier forms one member
                     of the pair (Fig. 4).

D - Refresh: The user can "clean-up" the screen of rectangles and data after the box
                     commands.

M - Map: A second screen (Fig. 5) shows the geographic location of each
                     observation as a dot; the larger symbols represent observations that
                     occur in pairs falling within the region of the h-scattergram defined by
                     the box as described. Using the box and map commands, one can
                     identify unusual observations and determine their geographic
                     locations relative to the rest of the data.

Data

Each input record comprises data for two observations and the distance separating them.
Data for each observation consist of an identifier, the value of the regionalized variable,
an easting value, and a northing value.

EXAMPLES

Structure

The Upper Devonian Huron Member of the Ohio Shale is a black, organic-rich unit that
overlies gray shales, and is mappable across southwestern West Virginia (Neal and Price,
1986). The h-scattergram of the base of the Huron (Fig. 3) within distances of 1 km shows
little scatter because well control is good and the horizon is recognized easily on both
gamma-ray and driller's logs.

One group of points stands out from the others in the scattergram of Figure 3; highlighting
these points on a map of well control shows that they correspond to wells lying along an
east-west trend (Fig. 4). This example was selected because a fault zone— the so-called
Warfield Fault— roughly bisects the area. This fault zone is expressed on a computer-
generated contour map (Fig. 6) where east-to-west contour lines are pinched together and
there occur several closed contours. This area lies over the southern limb of a large
anticline with an east-west orientation.

These data reflect a discontinuity on the h-scattergram and ultimately on a variogram.
With better well control in the area of the fault, more points would diverge from the main
trend in the scattergram, resulting in a poorer correlation and increased values in the
variogram with perhaps a larger nugget effect. The nugget effect plays an important role
in determining the degree of smoothing imposed during estimation. Note that the graphs
and map illustrate clearly that two populations of wells have been sampled, and that they
are separated geographically by what turns out to be a fault zone. This observation
suggests that the northern and southern one-half of the area should be mapped separately
or computer-mapped using programs that take discontinuities into account.

Thickness of Devonian Shale

The Devonian shales include gray shales, black shales, and siltstones, for a total
thickness of about 2,500 feet in northwestern West Virginia. Over the western part of the
State, shales increase in thickness from about 1,000 feet in the west to 3,000 feet in
central West Virginia. The thickening appears uniform in state-wide maps, but local

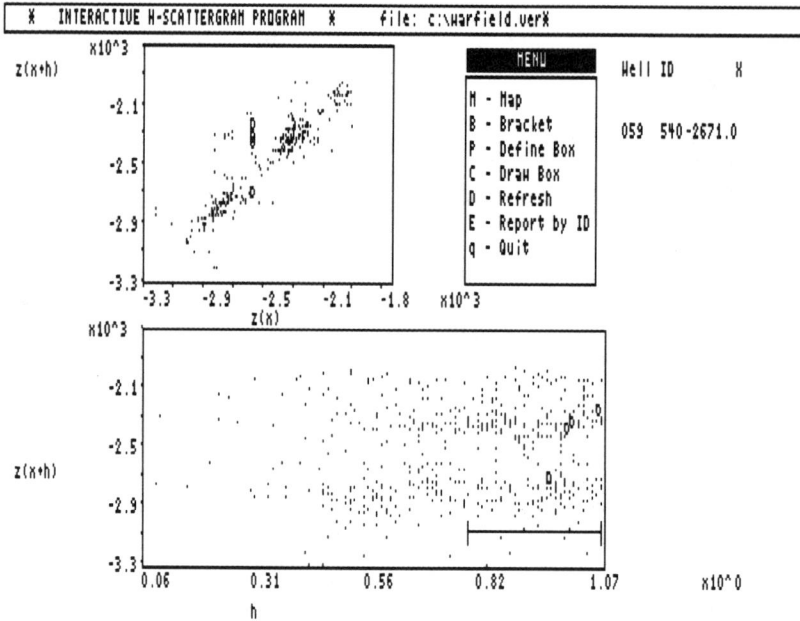

Figure 4. User has requested program to report value of subsea depth (-2671) for one well. Pairs in which this well is member are shown as open circles on both graphs.

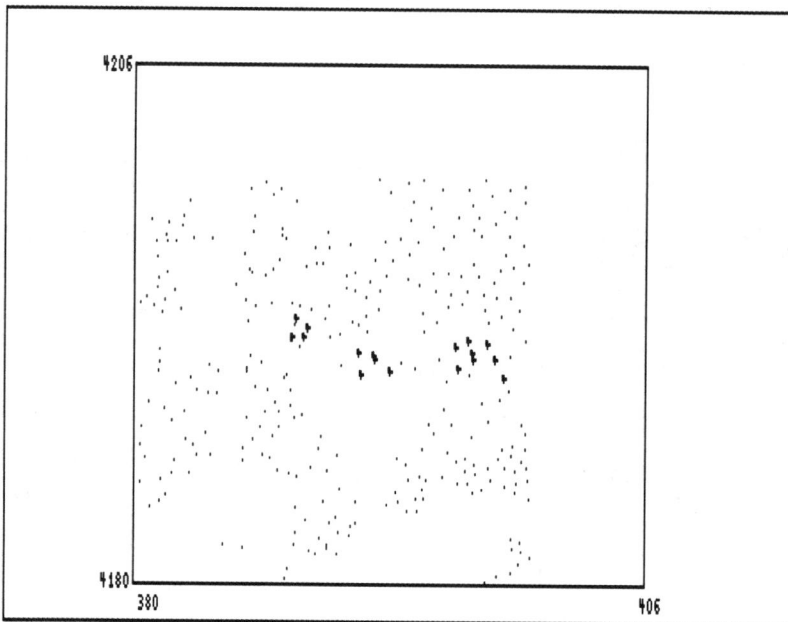

Figure 5. Map showing all wells, with those falling within box in Figure 3 drawn with bold symbols. Geographic coordinates are in meters (UTM projection).

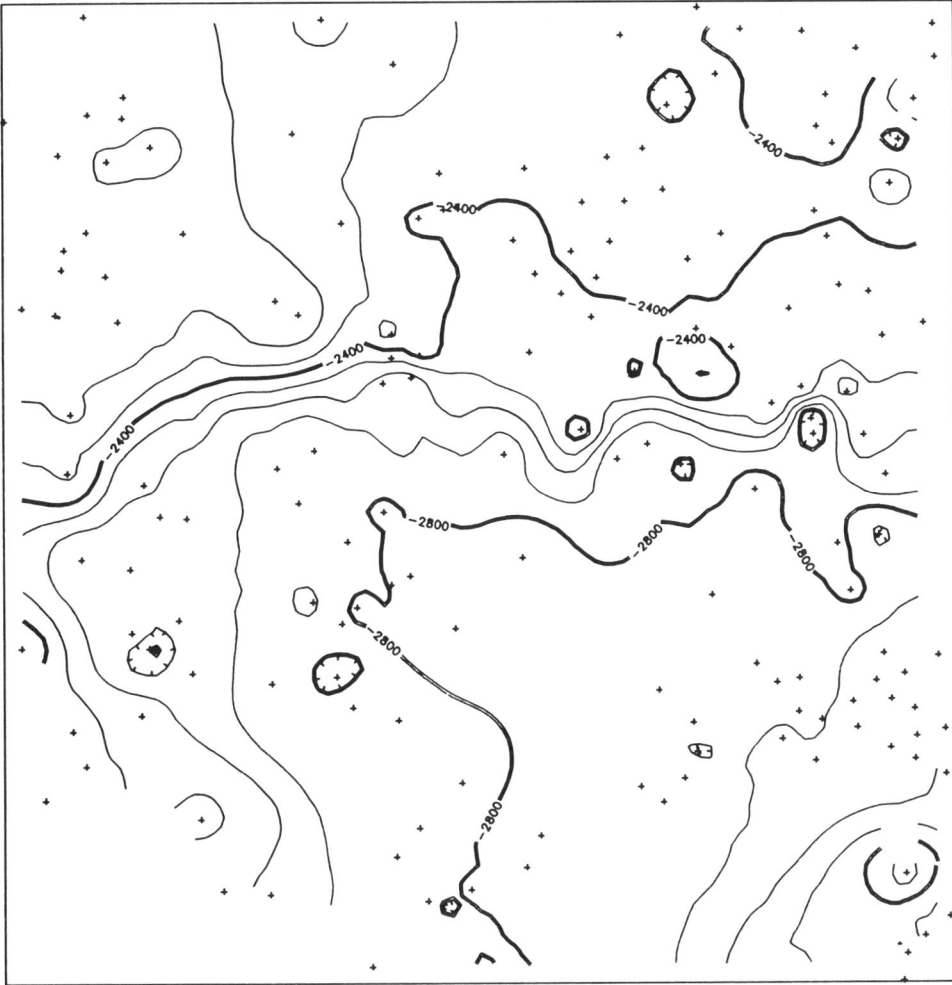

Figure 6. Contour map of subsea elevation of base of Huron Member in southwestern West Virginia.

variability exists. A small set of wells was selected in the area of the Burning Springs Anticline, a major north-south structure that controlled deposition in pre-Devonian time and was the site of some of the earliest oil wells in the world (Filer, 1985).

The h-scattergram for these data (Fig. 7) is in stark contrast to the previous example (Fig. 1), in part because of the poor well control. Few wells are drilled through the entire Devonian shale section in this area. Despite the scatter, some points seem to diverge an inordinate amount from the trend. The box in Figure 8 delineates one such point. Filer (1985) showed that the Devonian shale section in one well of the pair corresponding to this point is overthickened because of thrusting along the anticline. Checks made on other wells in the data set confirmed that thicknesses were accurate, not the result of errors made during data entry, which was suspected at first. If the purpose in mapping total shale thickness is to reconstruct depositional and paleoenvironmental patterns, the geologist would want to trim these wells from the data. Note that some of the divergent

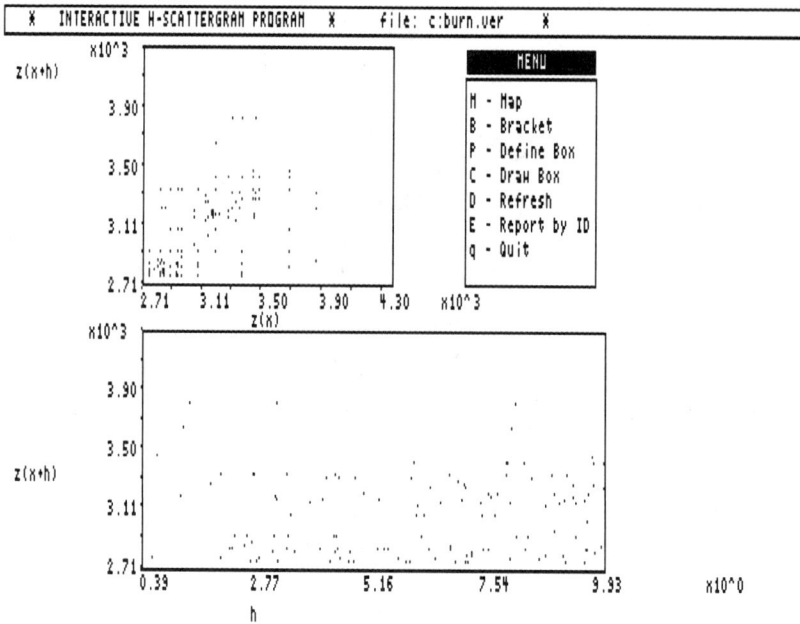

Figure 7. Scattergram of thickness of Huron Member for all values of h less than 10 kilometers.

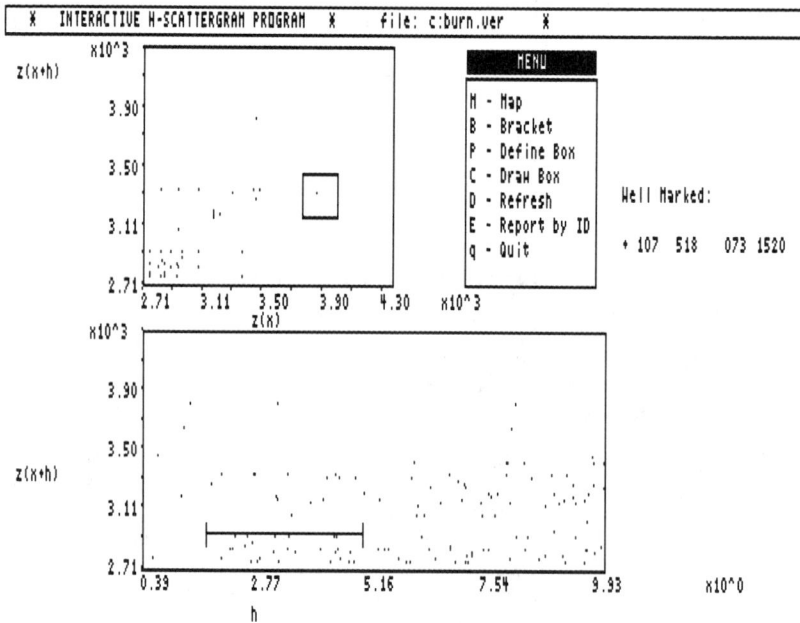

Figure 8. Outlier has been identified, and results from large value for Huron thickness in well 518 in Wood County (107), or small value in well 1520 in Pleasants County (073). Former turned out to be correct here.

points correspond to a thickness that does *not* lie at the extremes of the data distribution (Fig. 9). Because a single well is compared to many to generate an h-scattergram, unusual values manifest themselves as rows or columns of points. Examination of Figure 7 shows that elimination of only three or four wells yields a cleaner scattergram, and a variogram that more accurately represents the spatial continuity of Devonian shale thicknesses.

In small data sets, a few points have a large effect on the variogram. Although sample size is larger for most geostatistical applications, a number of studies have demonstrated the disproportional effect that a few outliers have on the variogram even for large sets of data (Krige and Magri, 1982; and Hohn, 1988).

Comparing Two Variables

In a departure from the previous examples and the h-scattergram described by Journel (1983), the present example uses SCATPKMP to compare two regionalized variables; the first, thickness of the Huron Member of the Ohio Shale in western West Virginia as measured from driller's logs; and the second, values of Huron thickness estimated from gamma-ray logs in nearby wells. Because the Huron is a black, organic-rich shale within a gray shale sequence, gamma-ray logs are reliable for picking the base and top of this unit. The top is defined as the uppermost black shale in the section; the Huron grades upward into gray shales. The top of the Huron is difficult to pick from driller's lithologic logs, and use of these logs for mapping is sensitive to the different degree of resolution reported from well-to-well, and the experience of the geologist. By comparing a kriged estimate calculated from gamma-ray logs with the driller's call for each well, one can spot bias in the driller's tops, and unusually thin or thick determinations which might be in error. Rather than a distance h separating wells, the abscissa on the bottom window is estimation variance (Fig. 10), which is related to the average distance and orientation among control wells and the location of the estimate.

The h-scattergram shows the presence of two populations of wells (Fig. 10). Unlike the first example, a discontinuity in the regionalized variable does not account for the two groups of wells, but rather, wells are clumped within two north-south oriented fields, one eastward of the other (Fig. 11).

More importantly, most points lie close to the line of unit slope. Outliers generally fall above the line, indicating that estimated thicknesses exceed the driller's call for these wells. Apparently, the geologist at the well picks the top of the Huron with the first appearance of a distinct black shale, whereas the log interpreter can select subtle kicks on the gamma ray log. In some situations, the difference is more than 200 feet. One obvious outlier in which the estimated thickness was less than the driller's thickness (Fig. 12) corresponds to a well that was flagged in the database as having a questionable thickness. Whereas the estimated thickness was 390 feet, the driller's log reported a thickness of 490 feet, suggesting an arithmetic error.

These outliers could be determined from a histogram of differences between observed driller's thicknesses and estimates. However, the scattergram also permits one to see that the two groups of wells differ in the amount of scatter about the line of unit slope. The values from the eastern field are not as tightly bunched about the line, particularly when one considers small values of estimation variance (Fig. 12). This observation follows from the fact that as the black shales of the Huron Member thicken to the east, this interval includes more gray shales, the black shales become less organic-rich, and the uppermost Huron becomes difficult to distinguish from overlying gray shales.

Initial Potential of Gas

Final open flow can be used as an early predictor of well performance, and in states such as West Virginia, is available more readily to the public than actual production figures.

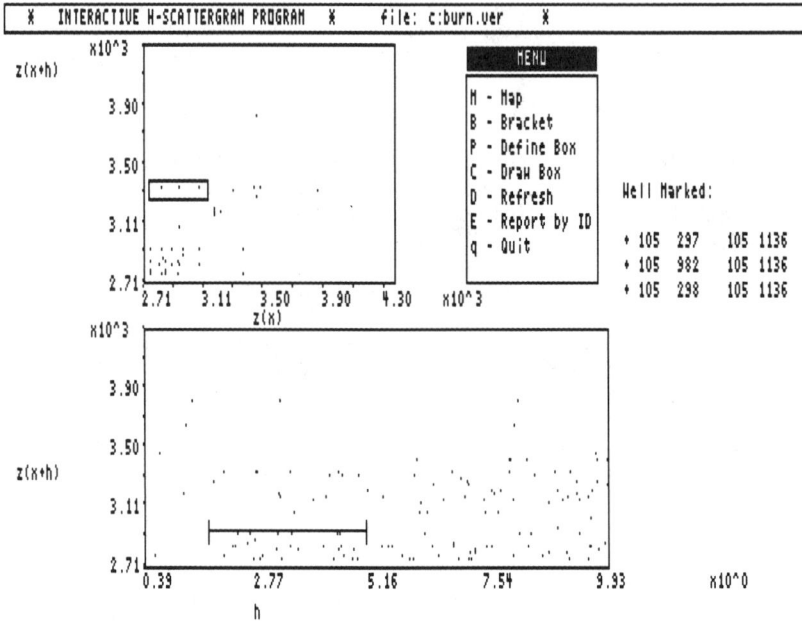

Figure 9. Well 1136 in Wirt County (105) seems to have unusually thick value for Huron thickness when compared with surrounding wells.

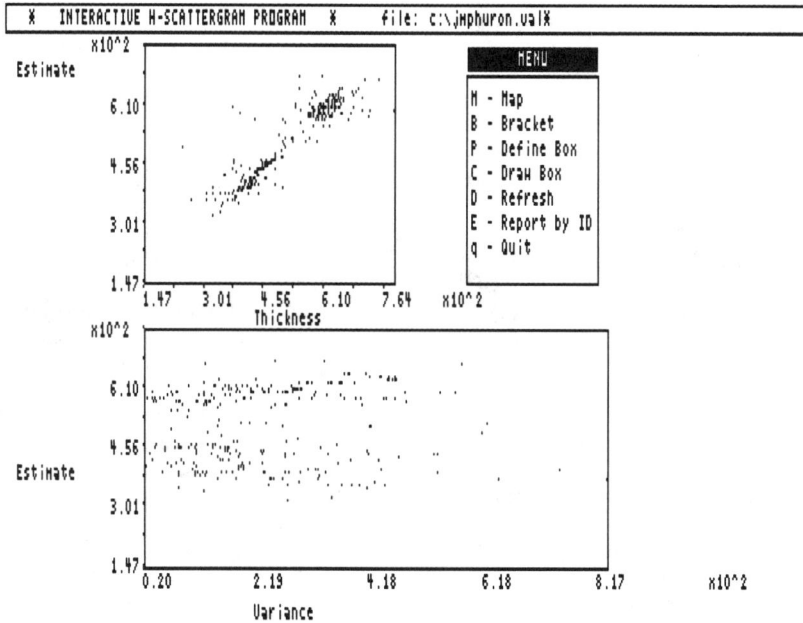

Figure 10. Scattergram comparing thickness of Huron determined from driller's lithologic logs with thickness estimated from gamma-ray logs representing nearby wells.

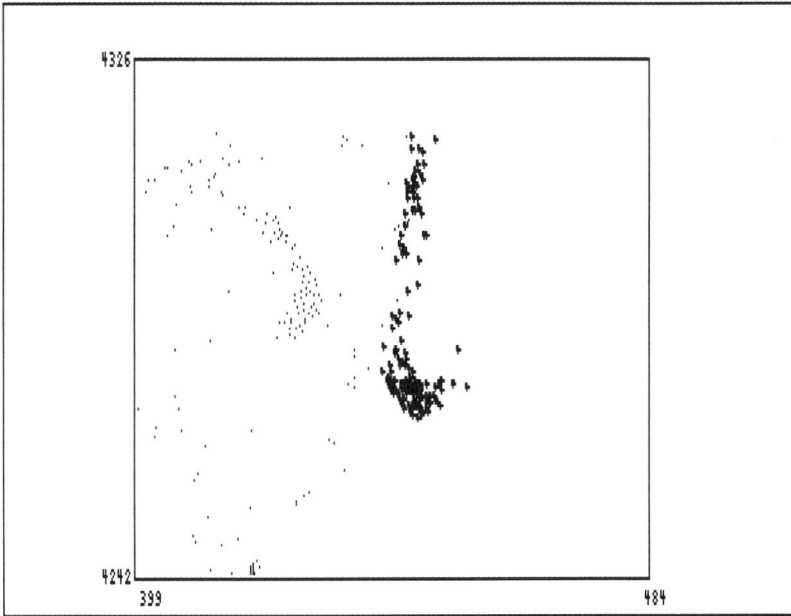

Figure 11. Map showing that wells with measured Huron thickness fall into two main fields. Wells with thicker Huron are highlighted.

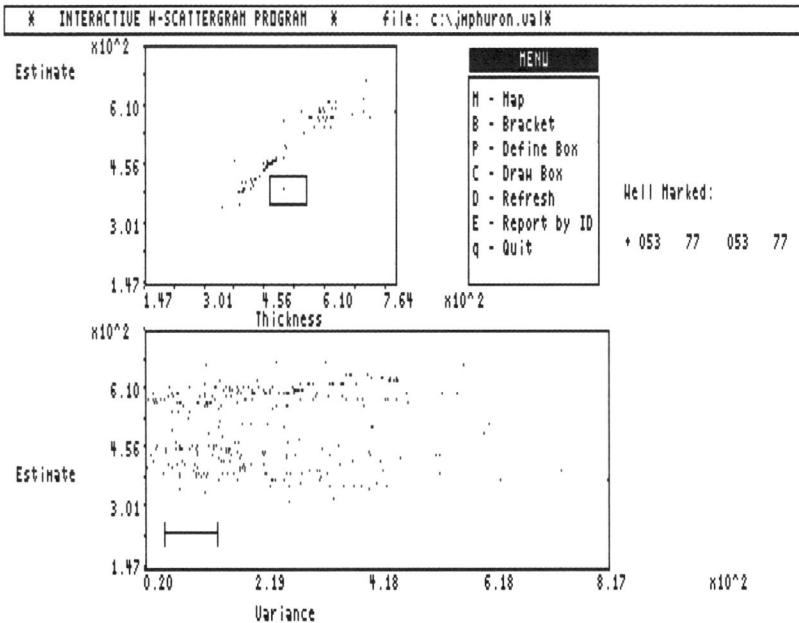

Figure 12. User has spotted problem well.

Unfortunately, standard practices in obtaining initial potential differ among companies, conditions under which open flows were measured go unreported, and methods for standardizing measurements are unavailable. Furthermore, little inducement exists for operators to report these figures accurately.

Calculating a variogram is an important step in mapping initial potential because the nugget effect usually is large relative to the sill. Therefore, the estimates must be smoothed through kriging. One set of data analyzed by Hohn (1988, chap. 1) comprises wells from a gas field in northeastern West Virginia; production comes from several siltstones in an Upper Devonian clastic sequence. Initial potentials were converted to logarithms before analysis.

The h-scattergram for pairs of wells separated by up to 2.35 kilometers (Fig. 13) fits a line of unit slope poorly, reflecting the low continuity in initial potential among wells. The scattergrams demonstrate that the large nugget effect does not result from a few outliers. For restricted values of h, the fit to the line is better (Fig. 14). A box has been used to pick out pairs of wells with the highest initial potentials which, when mapped, define a trend across the field (Fig. 15). In this way, the geologist can set h to focus in on areas with good well control, and adjust the box to highlight wells with an initial potential above a certain threshold. In Figure 15, not all wells occur above the threshold, only those within 0.5 kilometer of another well with an initial potential above the threshold. Thus, the map emphasizes low-risk, high-potential areas.

SUMMARY

An interactive graphics program to display Journel's h-scattergram is useful in (1) teaching what a variogram shows about a regionalized variable; (2) locating and identifying statistical outliers; (3) detecting spatial discontinuities; (4) observing subpopulations in the regionalized variable; (5) comparing different ways of observing the same variable; and (6) visualizing the effects of outliers, spatial discontinuities, and subpopulations on the variogram.

As currently implemented, SCATPKMP requires a data preprocessor to create the file with pairs of points, values of separation distance, and values of the regionalized variable. A program to compute variograms was modified to do this preprocessing on a DEC VAX 11/750.

At present, the lower window is under-utilized, only providing a method to draw the bracket that defines the value of h to use in selecting pairs for the upper graph. The user also can use this display to see the number of pairs separated by a given range of distances.

If SCATPKMP were to be used in modeling variograms, the lower window could be modified to show the variogram and the user could use the upper window for selecting points for deletion. As samples and their corresponding pairs are removed, the user could observe the effect on the variogram.

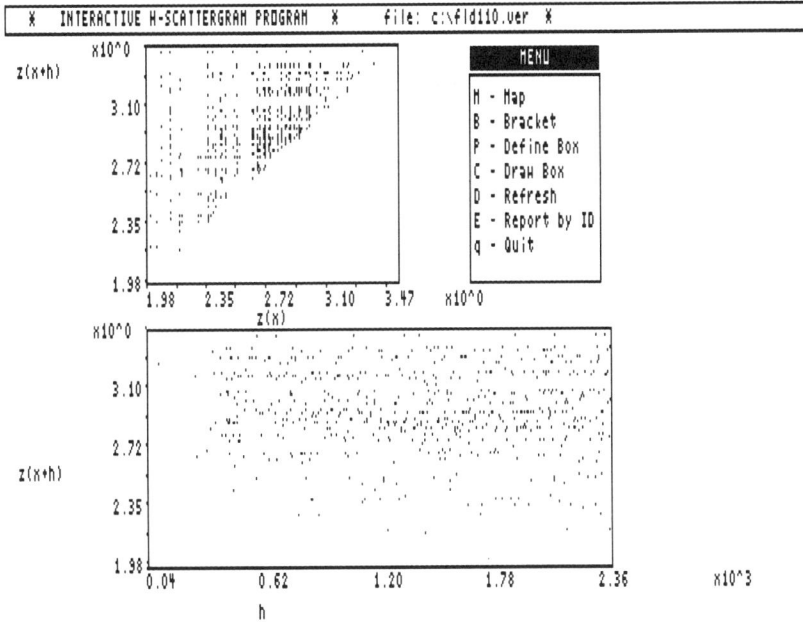

Figure 13. Scattergram of initial potentials from Devonian gas field in northeastern West
            Virginia. Data were sorted in ascending volumes before pairs of values were
            calculated; hence, if $z(x + h) = z_j$ and $z(x) = x_i$, then $z(x + h) > z(x)$ for $j > i$.

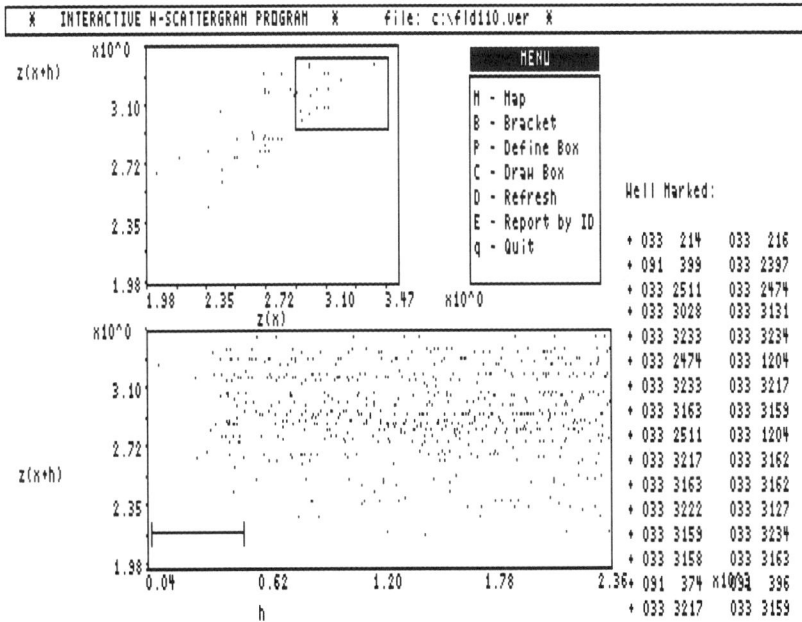

Figure 14. H-scattergram of initial potential for small values of h. User has selected pairs
            of wells where both members of each pairs have relatively large value of
            initial potential.

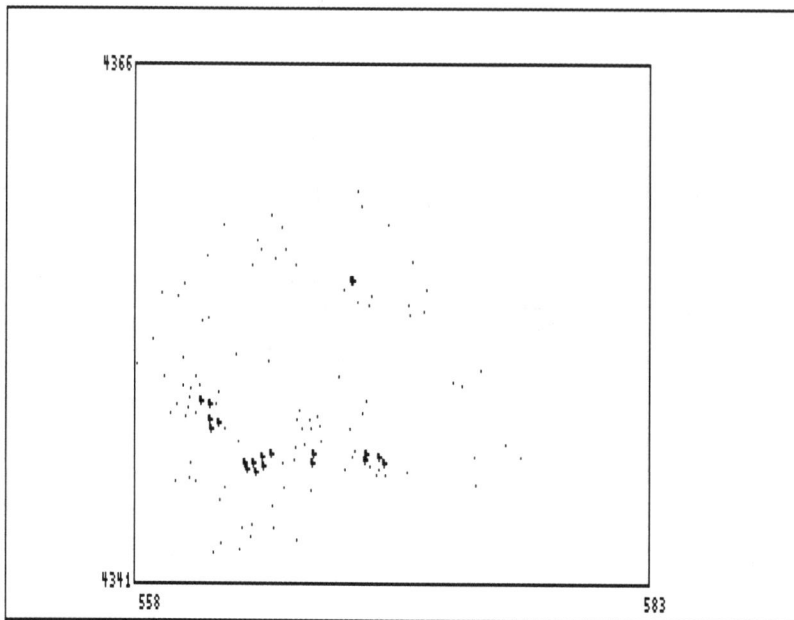

Figure 15. Map showing wells selected by user in Figure 14. Geographic coordinates are in meters (UTM projection).

REFERENCES

Filer, J. K., 1985, Oil and gas report and map of Pleasants, Wood, and Ritchie Counties, West Virginia: West Virginia Geological and Economic Survey Bull. B-11A, 87 p.

Hohn, M. E., 1988, Geostatistics and petroleum geology: Van Nostrand Reinhold, New York, 264 p.

Journel, A. G., 1983, Nonparametric estimation of spatial distributions: Jour. Math. Geology, v. 15, no. 3, p. 445-468.

Krige, D. G., and Magri, E. J., 1982, Studies of the effects of outliers and data transformations on variogram estimates for a base metal and gold ore body: Jour. Math. Geology, v. 14, no. 6, p. 557-564.

Neal, D., and Price, B., 1986, Oil and gas report and maps of Lincoln, Logan, and Mingo Counties, West Virginia: West Virginia Geological and Economic Survey Bull. B-41, 68 p.

Cross Sections and Volume Measurement of Stratigraphic Units

Michael M. Kimberley
North Carolina State University

ABSTRACT

A graphics-oriented microcomputer such as the Macintosh is a potentially powerful tool for sketching cross sections and for calculating either the total or fluid volume between cross sections. General purpose (nongeologic) volume-calculating programs are of limited use because of the peculiarities of geologic data, for example, strike-and-dip data. A program described in this paper is applicable to a wide variety of geologic applications.

Stratigraphic boundaries are interpolated between data locations, for example, boreholes, as smooth curves instead of unrealistic straight lines. The cross-sectional area is calculated for any stratigraphic unit and the land area is calculated between cross sections.

INTRODUCTION

Many geologists plot cross sections. A cross section can depict an orebody, rockbody, waterbody, soil profile, glacier, or landfill. Some geologists continue to rely on manual drafting of cross sections but this is difficult to justify, given the versatility and high resolution of inexpensive microcomputers. Short-comings of manual drafting include inflexible scaling, unnatural linear spline (straightline) interpolation, and the inconvenience of redrafting if additional data points become available. A Macintosh Pascal program is described here for the plotting of cross sections and the calculation of strata volume between cross sections. Topography and strata boundaries are interpolated between scattered data points with a cubic spline (Fig. 1).

Geologists who rely on linear splines realize that they rarely represent the boundaries of a stratigraphic unit accurately. However, no two geologists would select manually identical nonlinear interpolations between data points and so unrepresentative linear splines continue to be used widely. A linear spline is the only manual method for consistent outlining of a stratigraphic unit. An outline must be determined before the cross-sectional area can be calculated. Cross-sectional areas are a prerequisite to the calculation of volume. The program offers both the smooth outline which typically characterizes an actual stratigraphic unit and the consistency of calculation which is required for objective measurement.

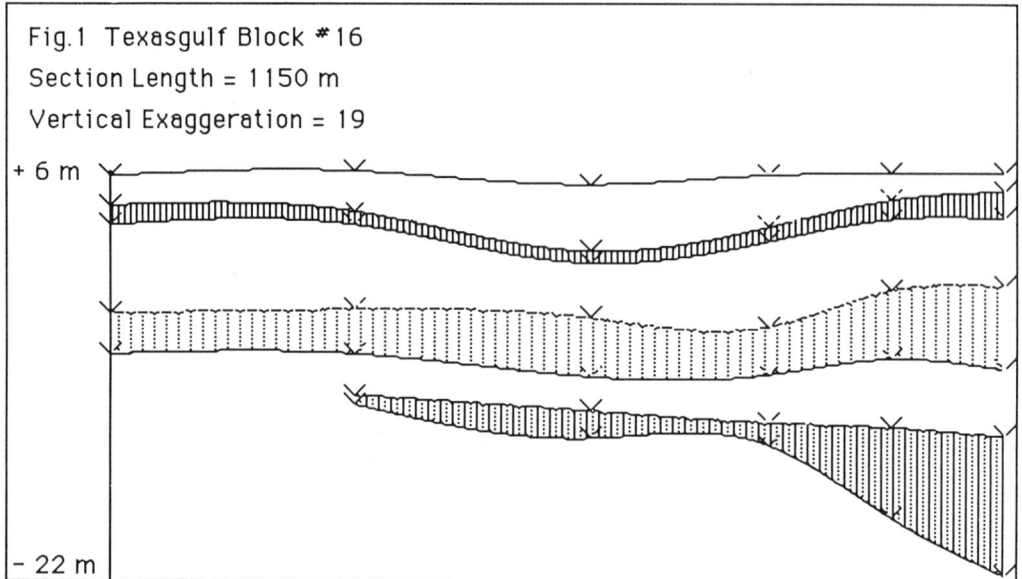

Figure 1. Cross section through part of block #16 at Texasgulf phosphate mine near Aurora, North Carolina. Elevations are relative to sealevel. Lowermost shaded bed is edge of buried river channel.

General-purpose programs are available commercially for the construction of a cross section bounded by known points and for the calculation of a volume bounded by such cross sections. These programs are useful to engineers but do not handle the strike-and-dip measurements which characterize geologic data sets. Geologists require the option of interpolation along a contact which is partly or entirely defined by points where the planar (dip-and-strike) orientation is known locally. For example, consider a geologic contact which extends from an area of outcrop, where strike and dip are observable, to an area of exclusively borehole data where only the depth beneath the land surface is known. The computer program allows for any mixture of these two types of data. Moreover, the program allows data to be entered in any format generally used by geologists.

The three-dimensional Macintosh program is an outgrowth of a two-dimensional Apple II program which was presented in the previous edition of this book (Kimberley, 1986). Both programs rely on cubic splines instead of Lagrangian or other interpolation methods, as justified by Kimberley (1986). The greater length of the updated program (1100 lines) prohibits listing or detailed description here. Both programs have been written in Pascal because Pascal is considered to be the most readable available language. The flow of control through a Pascal program is easy to follow and the number of meaningful characters in any variable name is sufficient that each name clearly indicates the role of that variable.

OVERVIEW OF THE PROGRAM

The following description of the program is limited to suggestions for utilization and an overview of the algorithm. To begin, the user selects the longest cross section and starts entering elevations and corresponding locations along its topographic profile. Elevations and any strike-and-dip data then are entered along the uppermost geologic contact, followed by successively lower contacts. All input data are printed and the user can correct any mistakes after examining the printed record. The program calculates the vertical exaggeration which would result from filling most of the screen with the longest cross section. The user is offered the option of decreasing this vertical exaggeration. There is no limit to decreasing the vertical exaggeration but the maximum permissible increase is roughly 10%. Once determined, the vertical exaggeration remains constant for all subsequent cross sections.

The topographic profile is interpolated as a cubic spline through all input elevations, using an algorithm modified from Cheney and Kincaid (1980). Geologic contacts also are interpolated as cubic splines but any geologic contact which extends above the topographic profile is erased automatically ("eroded") until it returns below the land surface. Strike-and-dip data are used to guide geologic contacts for a portion of the distance to the next data location along the cross section. The portion is decreased automatically for steeply dipping contacts to avoid excessive gyrations along the interpolated spline.

Cubic-spline interpolation between adjacent locations may produce a curved contact which extends well below the lowest data point. It is difficult to predict the degree of curvature which may result from cubic-spline interpolation. The user therefore is asked to specify the lower end of the graphed cross section explicitly rather than rely on an algorithm. If the user sets this limit too high, the lowermost contact may become truncated and the program may have to be run again with a lower bottom limit. The upper limit of the vertical (y) axis automatically becomes the highest elevation.

If the user has no interest in volume measurement, there is no need to know the geographic (Cartesian) coordinates for points along the topographic profile. It is only necessary to know distances along the cross section and any change in the compass orientation (azimuth) of successive straightline segments. The angle between the local line-of-section and each strike line is needed to calculate apparent dip from each true dip. The program plots the apparent dip with the appropriate modification for vertical exaggeration.

Volume measurement requires calculation of the land area between sections and this measurement is performed most readily if Cartesian coordinates are known for the ends of all straightline segments along all sections. The user is given the option of entering these Cartesian coordinates directly or having them calculated by vector addition along the section. In either situation, the program automatically prints the Cartesian coordinates of each data point. Moreover, it prints the total distance along the cross section, whether the data are entered as distance-azimuth locations or as Cartesian points.

Each cross section typically illustrates several beds. The cross-sectional area of each bed is calculated by summing the lengths of a succession of vertical lines drawn between the top and bottom of the bed. The top of each bed is the cubic-spline interpolation along the previously entered geologic contact, except in the situation of the uppermost bed where the overlying cubic spline lies along the topographic profile. All pixel locations of the previously interpolated line are stored in memory to become the upper limit of the vertical lines. Only one previously interpolated line can be stored in memory because of the limitation of variable dimensioning which is inherent to the Macintosh Pascal version (2.0) which was available. If all interpolated lines could be stored, the program could be modified to provide volume calculation for various groups of beds instead of just individual beds.

On a monochromatic Macintosh, successive beds within each section are differentiated by a sequence of four types of vertical lines (solid, dotted, etc.). This differentiation could be modified readily to a series of colors on a Macintosh II or other computer which supports color. The user is offered the option of leaving the area above any contact blank and thereby foregoing the calculation of cross-sectional area. This option is appropriate for any contact which does not extend the full length of the cross section and which underlies a contact that does extend the full length. The area above such a short contact would constitute only part of a bed. A short contact overlies the lowermost shaded bed in Figure 1.

To enhance visual differentiation among the four types of monochromatic lines, vertical lines are drawn at every third horizontal pixel location rather than every horizontal pixel. The precision of the corresponding area measurement would be enhanced if colored vertical lines were drawn at every horizontal pixel. Precision also would be enhanced if bed thickness could be expanded vertically across more pixels. This expansion can be achieved by including only one or two beds per section and then stacking these sections vertically to produce a composite cross section. If sections are stacked, it is important to maintain a constant length among the vertical axes on all sections.

Upon completion of the longest cross section, the topographic profile is entered for the adjacent section. The land area between the two sections then is calculated before any geologic contacts are entered for the second section. Calculation of bed volume requires knowledge of the area between adjacent cross sections as well as the cross-sectional area of the bed in each section. Unfortunately, the calculation of land area between jagged cross sections is not straightforward if the azimuth changes radically between adjacent linear segments along either section.

The accompanying program calculates land area between cross sections by division of the intersection area into a series of triangles and summation of the triangular areas. Each triangular area is calculated with the algorithm of Petersen (1955) which contains the Cartesian coordinates as matrix elements. Intersectional areas, cross-sectional areas, and corresponding volumes are printed as soon as they are calculated.

Manual selection of an appropriate set of nonoverlapping triangles is easy but it is difficult to write an algorithm which consistently avoids both overlaps and gaps between triangles. The reader is encouraged to examine this potential problem by sketching triangles between a pair of highly jagged lines. To circumvent this problem, the accompanying program offers the user a choice between manual and automatic selection of triangles. In manual selection, the user specifies the ordinal positions of the points which become corners of the triangles.

The algorithm for automatic selection makes no attempt to check for overlapping triangles but works well in most situations. In automatic selection, each linear segment of each cross section becomes the base of a triangle. The third corner of each triangle advances along the opposite cross section. If the opposite section contains fewer points than the section with triangle bases, the ultimate point of the opposite section becomes the third corner of all remaining triangle bases. A crude sketch of adjacent lines-of-section generally suffices to indicate if this automatic selection of triangles is a safe selection. The program prints the area calculated for each triangle, whether selected manually or automatically.

As geologic contacts are added to subsequent cross sections, a bed which was present in a previous section may prove to be missing. The program allows the user to skip missing beds so that the correct beds become correlated among the cross sections. Bed volume is calculated for correlative stratigraphic units.

Successive cross sections are entered and plotted on top of the previous sections, producing a superimposed image. Given the inability of the program to refresh the screen, images of individual cross sections cannot be recorded permanently during a program execution which is calculating volume. During volume calculation, all sections are plotted as if the x-coordinates of the leftmost points were identical. The true Cartesian coordinates are retained for these points and are used in the volume calculation but this artificial shifting of sections is required to facilitate plotting. The composite image can be printed or saved on a disk.

If an image is saved on disk, it can be reloaded and modified with a standard graphics program, for example, MacPaint or Super Paint. Image modification is necessary because the Macintosh automatically saves the border of the screen, complete with menu names, along with each cross section. This extraneous material can be erased with any graphics program and the image thus prepared for laser printing, as shown by Figure 1.

ADJUSTING DATA SETS TO PROGRAM LIMITATIONS

The maximum numbers of data locations, contacts, and cross sections are assigned as constants at the beginning of the program to facilitate comparison with the user's data set. This comparison should precede any attempt to run the program. If the user's data set exceeds any of the assigned limits, the corresponding constants should be increased accordingly and any other constants which do not need to be so large should be reduced. If an error message subsequently records insufficient memory, the data set exceeds the capacity of the user's Macintosh Pascal software and the user may consider implementing the program on a mainframe.

Conversion between microcomputer and mainframe dialects of Pascal usually requires translation of several graphics commands and so the conversion is not trivial even if the Pascal text can be moved from one computer to another as an ASCII file. A simpler way to handle a large data set would involve segmenting the data either vertically or horizontally and stacking or splicing the graphic output to produce a composite cross section with a Macintosh. Vertical segmentation would involve subdivision of the stratigraphic units whereas horizontal segmentation would involve subdivision of the geographic area.

UNITS OF MEASUREMENT AND ANGLE FORMATS

Once it is determined that the program can handle the intended data, little other preparation is needed. The program is designed to minimize the organization of data prior to execution. Elevations can be entered in either feet or meters. Once the user selects between these two, the program automatically adds the appropriate abbreviation to all printed calculations (abbreviating feet to "f"). Distances can be entered in any units, for example, miles or kilometers. If the units of elevation differ from those of distance, the program asks for the conversion factor between them. All calculations are performed in the units of elevation. Azimuths can be entered either in full-compass degrees (0 to 360) or as a deviation from north, for example N30W. Dip is entered in degrees. Given that the dip direction is known to be perpendicular to strike, the user simply notes whether the dip lies to one side or the other of the strike line.

SELECTING BETWEEN INTERACTIVE DATA ENTRY
AND DATA STORAGE WITHIN PROGRAM

The user should select carefully between interactive data entry and permanent storage of data within a procedure (subroutine). Data stored within the accompanying program were used to produce Figure 1. Data should be entered interactively if the user is sure that subsequent modification never will be required. A single interactive entry of a data set is quicker than writing the data into a procedure but the latter option is quicker than two interactive entries. Any desired modification of an image would require interactive re-entry of all data.

A desire for data modification generally results from either of two conditions, that is new geologic observations in the area of interest or the realization (after initial running of the program) that the user already knows more about the area than was recorded in the input data. An example of the latter condition would be insufficient data entry along a topographic profile which includes sharp discontinuities at the edges of a cliff. The cubic-spline algorithm inherently tries to avoid abrupt turns and so the image of a cliff will be too smooth unless the user enters closely spaced topographic data near cliff edges.

Data sets which include faulted contacts should be stored within a procedure because acceptable representation of a fault generally requires multiple program executions. Between each execution, the user adds information which was known previously (at least vaguely) but not specified. Initially, a typical user fails to provide any information about the extension of a contact along a fault plane, given the notion that contacts do not extend along fault planes. However, successful representation of a fault with the accompanying program requires acknowledgment that contacts do extend along fault planes.

Contrary to the usual textbook representation, a fault generally smears remnants of beds along its plane. A fault plane may contain several contacts superimposed upon each other. A fault therefore is a place where the orientation of a contact changes dramatically and where several contacts can be superimposed. If either the position or orientation of a fault plane is known poorly, the user generally must hypothesize strike-and-dip data for that plane and experiment with the effect on the calculated cross section. Such experimentation would be time-consuming if the fixed observational data were not stored within the program.

One type of unfaulted section consistently requires two program executions. This involves calculation of the cross-sectional area (hence also volume) of both a bed and a laterally discontinuous member within that bed. Both the upper and lower contacts of a discontinuous member should be omitted during the first execution. During the second execution, the host bed should be left blank (unlined) and only the discontinuous member measured (lined). If all data points are being stored within the program, it is advisable to create two versions of the program, one with and one without the contacts for discontinuous members.

Storage of stratigraphic data in a Pascal program is best done within nested case statements. Nesting of case statements is required because the location (x) and elevation (y) of each point are dimensioned doubly to record the ordinal point and line. The ordinal number of the cross-section line is the selector of the inner case statement whereas the ordinal number of the point is the selector of the outer case statement. A separate nested case statement is required for each geologic contact in the cross section.

Case statements are executed more slowly and require more lines of code than would a series of simple statements which each assign a value to an element of an array. However, simple assignment statements are inconvenient for the subsequent insertion of points within an array. Convenience of data insertion is the motivation for storing data instead of entering it interactively and so this convenience should be optimized. The difference in execution time is trivial.

In a simple assignment statement, the ordinal position of each array element is a fixed ordinal number. All ordinal positions which follow any subsequent insertion within an array have to be increased manually by one. Within a case statement, the ordinal positions themselves are variables which become assigned during program execution. Insertion within such an array requires manual renumbering of subsequent case-statement labels but this task is relatively easy. Moreover, it is quicker to produce a new procedure with case statements than one with explicit assignment of array elements despite the fact that the latter type of procedure has fewer lines of code. New procedures of either type are produced most conveniently by replication (cut-and-paste) of an existing procedure, followed by replacement of data values. Fewer replacements are required if case statements are used.

If the user selects to store data within the program, it is advisable to study the accompanying example of such storage. This example does not include strike-and-dip data but the addition of such data would be straightforward after studying the way strike-and-dip data are entered interactively. Each cross section should be stored as a separate procedure. The program illustrates how the section-storing procedures are called in correct order from the main block at the end of the program.

DATA INPUT AND NORMALIZATION

The program interpolates curved lines between locations. Each location can be entered either as a pair of Cartesian coordinates or as a distance along a given azimuth. If the user selects the distance-azimuth mode, the program asks for the Cartesian coordinates of the leftmost point on the initial cross section and then calculates all other Cartesian coordinates by vector addition. The user specifies the location of the leftmost point of subsequent sections by a distance and azimuth from the left end of the initial section. Points along all sections are thereby normalized to a single Cartesian system.

The user has considerable freedom in data entry and the maintenance of this freedom contributes to the complexity of the program, for example nested if-then-else statements. Classification of the data set is updated continually. Questions posed to the user correspond to the current classification and the number of such questions is minimized constantly.

Whether multiple cross sections are stored or entered interactively, the initial section should be the longest. If the longest cross section is surrounded by shorter sections, the program should be run twice to calculate bed volume, repeating the longest section. The program informs the user of this requirement. The requirement stems from the dual purpose of the program, that is production of both a useful image and useful area-volume calculations. This duality is complicated by the lack of screen refreshment throughout the program. Successive cross sections become superimposed upon the initial section and so the normalization of data distances to screen distances cannot be updated. Any attempt to plot a subsequent cross section which is longer than the initial section would result in a run-time error.

USER INTERACTION WITH THE PROGRAM

Before running the program, the user is expected to select the Drawing Window and overlay the Text Window. All interactive text is designed to fit within the default size of the Text Window. For long messages, this is achieved by requiring the user to advance the text by typing any letter. Otherwise, some of the text would immediately scroll out the top of the small Text Window. The user should note that the required keypress is a letter (as stated in the displayed message) and that depressing a nonalphabetic key will not advance the screen.

The Text Window should not be expanded beyond the default size or else it would hide part of the cross section on the underlying Drawing Window. Any part of the Drawing Window which has been covered by the Text Window will remain blank even if the Text Window is removed subsequently. At the end of program execution, the user is expected to remove the Text Window and then print the image or save it onto a disk.

All single-character interactive entries, for example (Y/N) for (Yes/No), are acceptable in either upper or lower case. The program could be simplified by demanding that the user enter all characters in one case or the other but this would be inconvenient in a program where the user also is entering labels which generally consist of a mixture of upper- and lower-case letters. All single-character entries are read immediately (without requiring either the Return or Enter keypress) whereas any keyboard entry which could involve more than a single keypress, such as a number, does require pressing the Return or Enter key. Subsequent correction is allowed for most numerical entries, hence avoiding the necessity of restarting the program because of a typing mistake. Entered data are printed to facilitate checking for typographical errors. Printing recurs after the correction of any error.

The program offers five labels for the image, that is, a title, subtitle, vertical exaggeration, and both upper and lower labels along the vertical axis. Any of these can be omitted by entering nothing in response to the interactive request for a label. The subtitle is intended to include information about the length of the section, for example "Section A-B: Distance of 10.5 km." No additional labels have been added to Figure 1 but the user may want to provide additional information about the horizontal and vertical scales after reloading the image with a standard graphics program such as MacPaint. Regularly spaced tick marks with corresponding distance labels would be a useful enhancement.

One reason that regularly spaced tick marks are not included in the present program is that such increments should be simple numbers such as 10, 50, or 100 length units but it is difficult for a program to select such simple numbers automatically. The program could be modified readily to add tick marks at fixed proportions of the cross-section length, for example every tenth of the total length, but the intervening distances generally would not be simple numbers.

MODIFICATION OF THE PROGRAM FOR AN ALTERNATIVE COMPUTER

Successful implementation of the foregoing algorithm on some computer other than a Macintosh would require reference to the complete program listing. To facilitate such implementation, all values within the program which are unique to the Macintosh are assigned as constants at the start of the program. These values are related to the dimensions of the Macintosh screen and easily could be modified for a computer with different resolution. Two peculiarities of Macintosh Pascal could require more sophisticated modification, that is the conventions that positive y extends downwards and that zero degrees extends toward the right. The program automatically converts all input so that both the zero azimuth and positive y extend upward on the page. In another dialect of Pascal, the corresponding lines of code may require sign changes.

MODIFIED USE OF THE PROGRAM WITHOUT PROGRAM MODIFICATION

Modification of the accompanying program for specialized applications is facilitated by detailed comment statements within it. Even programmers with limited knowledge of Pascal can make simple modifications. However, some users may prefer to transform their data mentally rather than modify interactive program messages. For example, the program is designed to calculate the fluid volume between cross sections and offers no apparent option for the calculation of ore tonnages. However, these two applications are so similar conceptually that the user could enter an ore grade in response to a question about porosity and the resulting calculation of fluid volume then would correspond to ore tonnage.

CONCLUSION

Several types of geologic investigation require estimation of some type of volume or mass within a stratigraphic unit, for example rock volume or the fluid volume within a unit of known average porosity. The accompanying program measures either total volume or some related property, for example, fluid volume or ore tonnage, across multiple cross sections. Each cross section may consist of a jagged sequence of linear segments, that characterizes input data from scattered outcrops or boreholes. Cross sections are interpolated between dimensionless points or between points at which strike-and-dip are known. The program produces both quantitative area-volume measurements and an image which allows the user to decide if those measurements are being made in an acceptable manner.

REFERENCES

Cheney, W., and Kincaid, D., 1980, Numerical mathematics and computing: Brooks/Cole Publ. Co., Monterey, California, 362 p.

Kimberley, M.M., 1986, Sketching a cross section of folded terrain, *in* Hanley, J.T. and Merriam, D.F., eds., Microcomputer applications in geology: Pergamon Press, Oxford, p. 165-187.

Petersen, G.M., 1955, Area of a triangle and the determinant (abst.): Am. Math. Monthly, v. 62, p. 249.

A Simple Pascal Procedure for Outline Tracing in Image Analysis

Ulf Nordlund
University of Uppsala

ABSTRACT

Tracing of an object's outline is one of the fundamental procedures in image analysis. It is required for different techniques of shape analysis, area computations, among others. A Pascal routine for outline tracing on an IBM PC equipped with an inexpensive video digitizer ("frame grabber") is presented. The routine uses differences in intensity to recognize the outline. Output consists of the x/y coordinates of each pixel along the outline.

INTRODUCTION

With the increased use of personal computers equipped with inexpensive "frame grabber" video digitizers, it has become possible to write simple image-analysis programs designed for special tasks. A fundamental function in computerized image analysis is the definition of outlines of objects, such as fossil specimens and sediment particles. This, for instance, is necessary in different techniques of shape analysis.

In this following paper, a simple routine for the detection of outlines is presented. The programming language used is Turbo Pascal. The method works satisfactorily in both incident and transmitted light provided that there is sufficient contrast in the image and evenly distributed lighting.

SHORT DESCRIPTION

The procedure begins by searching for a starting point on the outline. This is done by tracing a straight line from the margin of the screen towards the center, while reading the intensity for each pixel (picture element) encountered. The starting point is set at the first pixel which has a value less (darker) than a preset threshold value. In order to locate the next point in the outline, the eight pixels surrounding the starting pixel are examined in a clockwise direction, starting with the pixel next to the one read previously. This is repeated for all pixels in the outline, until the pixel read is located next to the pixel of the starting point.

171

THE PROCEDURE OUTLINE

A listing of the procedure, included in a demonstration program, is presented in Figure 1. In the following text, words in capitals refer to the procedure.

The THRESHOLD variable and the variables which contain the direction data (DX and DY), are assigned values in the main program prior to running procedure OUTLINE. The first loop in OUTLINE searches for a starting pixel. The search begins at the position defined by the coordinate-variables X and Y, and continues by increasing X (moving towards the center of the screen) while reading the INTENSITY of the pixels encountered. When INTENSITY receives a value less than THRESHOLD, the margin of the object is reached.

After locating the X,Y coordinates of the starting pixel, and after setting the direction counter, J, to zero, the rest of the pixels in the outline are detected in the second loop. This starts by increasing the value of the point counter, I, and setting J1 to the starting value of J (in order to be able to recognize when all eight surrounding pixels have been read). A small loop then reads the eight pixels surrounding the current one. If INTENSITY becomes less than THRESHOLD the loop is abandoned and X and Y are assigned the coordinates of the pixel last read. This is repeated until the starting point on the outline is approached, or something has gone wrong (RETRY is set to "true"), in which situation the procedure is repeated.

The variables, include-files, and commands which are specific for the CORECO's OCULUS-200 frame grabber video digitizer used here, are:

- variables LUT, FIELD, MODE, and BIT7

- include files OC200.H and PASGRAY.P

- variable MAXY

- variable BASEADDRESS

- initiation command INITGRAY

- command GRCLO (erases previous image)

- command GRRPNT (reads intensity of point)

- command GRGRPH (draws a point in graphical plane)

These should be exchanged for the ones applicable to the digitizing hardware used.

The procedure presented here is included in programs for eigenshape analysis (Lohmann, 1983) and area computations at the Paleontological Institute, University of Uppsala, and has been used routinely for a longer period. The equipment is a B/W IKEGAMI video camera with a monitor, connected to an IBM AT personal computer equipped with an OCULUS-200 "frame grabber." The size of the image is 512x512 pixels, each of which can display 128 levels of gray. A level of 70 has proven to be useful in most situations when transmitted light is used, and therefore, has been selected as a default threshold value. If the threshold value needs adjusting, this usually will be necessary only once, when beginning. Furthermore, the procedure executes rapidly, which indicates that several runs with different adjustments can be made for objects with a blurred or otherwise interfering background. In order to achieve a more exact positioning of the starting point, the computer can be equipped with a mouse with which it is possible to mark the starting point with a cursor.

```
(  THIS PROGRAM DEMONSTRATES THE PROCEDURE "OUTLINE" WHICH DETECTS THE OUTLINE OF OBJECTS IN TRANSMITTED LIGHT  )
(  THE X- AND Y-COORDINATES OF THE POINTS IN THE OUTLINE ARE WRITTEN TO THE SCREEN                             )

program draw_outline;
var    i                     : integer;
       threshold,intensity   : integer;
       dx,dy                 : array [0..7] of integer;
       answer                : char;
       lut,field,mode,bit7   : integer;              ( FRAME GRABBER VARIABLES (HARDWARE SPECIFIC) )

($i \pascal\oculus\oc200.h)                          ( FRAME GRABBER FILES (HARDWARE SPECIFIC) )
($i \pascal\oculus\pasgray.p)

procedure outline;
const  maxpoint=2000;
var    j,j1                  : integer;
       x,y                   : array [1..maxpoint] of integer;
       retry                 : boolean;
begin                                                ( START PROCEDURE OUTLINE )
retry:=false;                                        ( SET ERROR CHECK )
writeln;
write('- Gray level <',threshold,'> : ');            ( SET THRESHOLD VALUE )
readln(threshold);
write('  Press any key to freeze image..');
initgray;grclo(1);                                   ( INITIATE FRAME GRABBER (HARDWARE SPECIFIC) )
grgrab(lut,field,mode,bit7);                         ( ACTIVATE FRAME GRABBER (HARDWARE SPECIFIC) )
writeln;
x[1]:=50;y[1]:=255;                                  ( DEFINE POINT AT LEFT HAND MARGIN OF THE SCREEN  )
repeat                                               ( FIND STARTING PIXEL LOOP )
     x[1]:=x[1]+1;
     intensity:=grrpnt(x[1],y[1],ep);                ( READ INTENSITY OF PIXEL )
     grgrph(x[1],y[1],ermwu);                        ( DRAW SEARCH PATH )
until (intensity<threshold) or (x[1]>maxpoint);      ( QUIT LOOP IF INTENSITY BECOMES LESS THAN THRESHOLD OR
                                                                IF NUMBER OF POINTS EXCEEDS MAXPOINT )

x[1]:=x[1]-1;
i:=1;
j:=0;
repeat                                               ( FIND NEXT PIXEL LOOP )
     i:=i+1;                                          ( COUNTER, PIXELS IN OUTLINE )
     j1:=j;
     repeat                                          ( EXAMINE SURROUNDING PIXELS LOOP )
          j:=j+1;if j>7 then j:=0;                   ( COUNTER, DIRECTIONS TO SURROUNDING PIXELS (8) )
          intensity:=grrpnt(x[i-1]+dx[j],y[i-1]+dy[j],ep);   ( READ INTENSITY OF PIXEL )
     until (intensity<threshold) or (j=j1);          ( QUIT LOOP IF INTENSITY EXCEEDS THRESHOLD VALUE OR
                                                                IF ALL SURROUNDING POINTS HAVE BEEN READ )
     x[i]:=x[i-1]+dx[j];                             ( ASSIGN VALUE TO X-COORDINATE )
     y[i]:=y[i-1]+dy[j];                             ( ASSIGN VALUE TO Y-COORDINATE )
     if (x[i]=x[i-2]) and (y[i]=y[i-2]) then retry:=true;   ( ERROR CHECK )
     grgrph(x[i],y[i],ermwu);                        ( DRAW PIXELS IN OUTLINE )
     if j<4 then j:=j+4 else j:=j-4;                 ( SET NEW START FOR EXAMINING SURROUNDING PIXELS )
until ((abs(x[i]-x[1])<2) and (abs(y[i]-y[1])<2) and (i>5)) or retry;   ( QUIT LOOP IF
                                                                OUTLINE IS COMPLETED OR
                                                                IF AN ERROR HAS BEEN DETECTED )

if retry then writeln('  Adjustments necessary...')
         else begin
               write('- Outline OK <Y> ? ');readln(answer);
               if not (answer in ['n','N']) then
                  begin
                    writeln('  Outline contains ',i,' points.');   ( OUTPUT OF COORDINATES )
                    for j:=1 to i do write(x[j]:4,y[j]:4);
                    writeln;
                    end;
                end;
end;                                                 ( END PROCEDURE OUTLINE )

( --------------------------------------------- MAIN PROGRAM ------------------------------------------------------ )
begin
threshold:=70;                                       ( SET DEFAULT THRESHOLD VALUE )
for i:=0 to 7 do case i of                           ( SET DIRECTION VALUES )
                0: begin dx[i]:=-1; dy[i]:= 0; end;
                1: begin dx[i]:=-1; dy[i]:=-1; end;
                2: begin dx[i]:= 0; dy[i]:=-1; end;
                3: begin dx[i]:= 1; dy[i]:=-1; end;
                4: begin dx[i]:= 1; dy[i]:= 0; end;
                5: begin dx[i]:= 1; dy[i]:= 1; end;
                6: begin dx[i]:= 0; dy[i]:= 1; end;
                7: begin dx[i]:=-1; dy[i]:= 1; end;
                end;
maxy:=485;baseaddress:=$F30;                         ( SET FRAME GRABBER VALUES (HARDWARE SPECIFIC) )
lut:=0;field:=0;mode:=0;bit7:=0;
repeat
     outline;
     write('- Continue <Y> ? ');readln(answer);
until answer in ['n','N'];
end.
```

Figure 1: Listing of Procedure

REFERENCE

Lohmann, G. P., 1983, Eigenshape analysis of microfossils: a general morphometric
        procedure for describing changes in shape: Jour. Math. Geology, v. 15, no. 6,
        p. 659-672.

CatTrack: A Pascal Program to Display Ternary Diagrams on a Macintosh Computer

Colin Ong
University of California, Davis

ABSTRACT

A description of CatTrack, a program to plot solute and soil-texture data in a ternary diagram, is presented. The main algorithm converts chemical data to trilinear ratios and then to x-y screen coordinates. The data categories include major cations (Na^+-K^+-Ca^{2+}-Mg^{2+}), major anions (CO_3^{2-}-HCO_3^--Cl^--SO_4^{2-}) and soil texture (clay-silt-sand). Use of this program's graphics is discussed and output is shown for three examples: (i) major anion composition of evaporation pond waters, (ii) changes in major cation composition along a transect from a mountain stream to an estuary, and (iii) results of a soil-texture analysis of evaporation-pond sediments.

INTRODUCTION

Programs for graphing using personal computers and even mainframes offer a wide range of display styles. However, one well-known graphing style that has not been available generally on computers is that of the ternary diagram. Ternary diagrams are a regular feature in many published articles as well as class exercises, and can be used to display a wide variety of information including soil textures and chemical compositions of waters and soil solutions.

CatTrack originates from the need to display data collected from saline evaporation ponds in the Central Valley of California and associated laboratory simulation experiments. The output is based on the figures of Hardie and Eugster (1970) and includes two types that display a chemical composition and path. Essentially, these two types cover all possibilities for displaying the spatially and temporally distributed major cation and anion data collected.

THE PROGRAM

Use of the program actually requires three steps: (1) data preparation, (2) diagram construction, and (3) diagram refining. The program presented herein covers only step 2. Steps 1 and 3 are to be carried out using commercial software. Specifically, step 1 requires a text processor, preferably not a word processor for the reason of avoiding having to select text format over the default format. Step 3 is a manipulation step which requires software depending on the user's need and, as such, is not included in the coding.

Figure 1 displays a schematic layout of the program illustrating the three shells that the program contains. Optional functions are noted in braces.

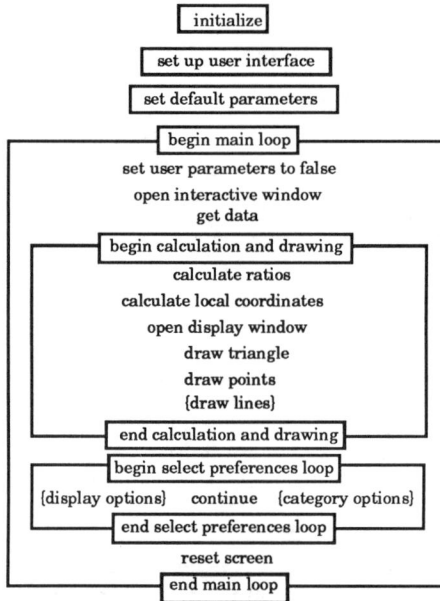

```
                          ┌────────────┐
                          │ initialize │
                          └────────────┘
                     ┌─────────────────────┐
                     │ set up user interface │
                     └─────────────────────┘
                     ┌─────────────────────┐
                     │ set default parameters │
                     └─────────────────────┘
        ┌────────────┤ begin main loop ├──────────────────┐
        │           set user parameters to false          │
        │             open interactive window             │
        │                    get data                     │
        │   ┌────────┤ begin calculation and drawing ├───┐ │
        │   │                calculate ratios            │ │
        │   │           calculate local coordinates       │ │
        │   │              open display window            │ │
        │   │                 draw triangle              │ │
        │   │                  draw points               │ │
        │   │                  {draw lines}              │ │
        │   └────────┤ end calculation and drawing ├─────┘ │
        │   ┌────────┤ begin select preferences loop ├───┐ │
        │   │ {display options}  continue  {category options} │ │
        │   └────────┤ end select preferences loop ├─────┘ │
        │                   reset screen                  │
        └──────────────┤ end main loop ├─────────────────┘
```

Figure 1. Schematic boxed layout of program.

Data files are required to be of the type TEXT and stored in a batch folder with a specific name. Data are stored as one situation per line, each line containing four values. For the version here, there are three categorical options: (1) Na^+-K^+-Ca^{2+}-Mg^{2+}, (2) CO_3^{2-}-HCO_3^--Cl^--SO_4^{2-}, and (3) clay-sand-silt. For options 1 and 2, the line sequence is with respect to the variables as listed in the categories and for option 3, a zero value must reside in the second position whereas clay occurs first, sand third, and silt fourth. The only restraint on the values in the data set is that they are in the same units. There are no facilities to calculate equivalents of concentration or compute other conversions.

To use the application from a floppy disk, the following is needed: (1) Finder file, (2) a System file containing a text-processor desk accessory, (3) a batch data-file folder termed "CatTrackBatchFolder", and (4) the application CatTrack017.

The display includes several options. The symbols used to represent data can be either a black circle or an integer representing the order. Several fonts also are available for selection. The symbols can be joined by straight lines in the order presented in the data. Another option is to overlay a grid with increments of 10 percent.

Note that the variable arrays have been dimensioned to handle up to fifty situations, but can be edited to handle more.

APPLICATIONS AND EXAMPLES

One application has been to display the major anionic composition of waters in an evaporation pond (Fig. 2). In the course of monitoring these chemical species, the effects of precipitation and redissolution sequences can be displayed graphically and trends can be noted. The alternative would be to plot concentration values on a basic x-y graph which makes it difficult to see interactive trends and hence, a pathway is not self-evident. The program also has proven useful in displaying chemical data from different locations along a transect from a mountain stream to an estuary (Fig. 3) as well as soil-textural data from the bottom of evaporation ponds (Fig. 4). The advantage of displaying chemical as well as textural data in a trilinear diagram is that it allows grouping according to several interactive variables. The disadvantage, however, is that the values to be displayed do not necessarily come in threes. Such is the situation, for example, for the major cations in which Na and K concentrations have to be aggregated. The development of a quadrilinear diagram display program would overcome this problem. As evidenced in many publications, the utility of having this program to archive such diagrams on computer media as well as prepare hard copies is an advantage in many areas of the geosciences.

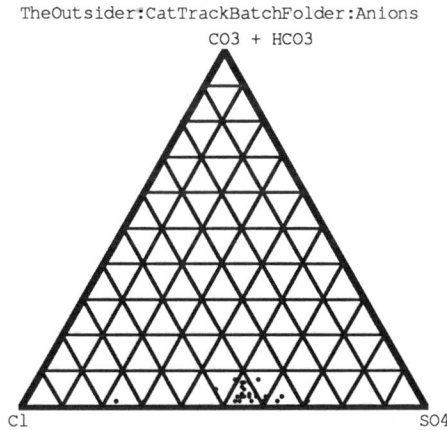

Figure 2. Screen dump output from CatTrack graphic display window. Data are from evaporation pond studies for 2-cell facility during one year. Units are mole%.

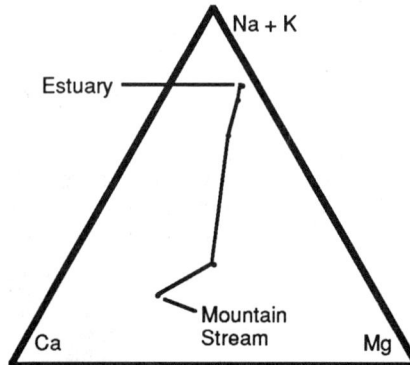

Figure 3. Edited output from CatTrack. Commercial graphics program was used to manipulate elements of diagram. Data are from river transect study and illustrates progression of change of rainwater influence upstream to seawater influence downstream.

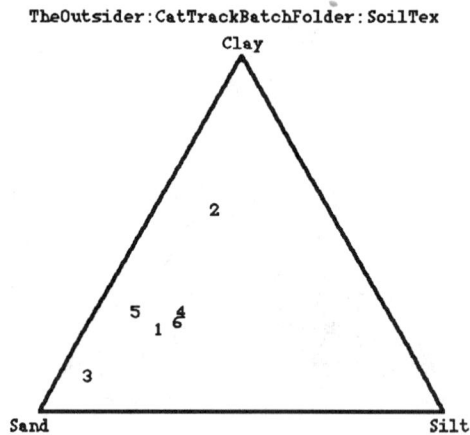

Figure 4. Screen dump output from CatTrack graphic display window. Soil-texture analysis data are from evaporation pond study.

CALCULATION OF LOCAL COORDINATES

The following equations in the procedure CalcLocal are used to convert trilinear coordinates into the coordinates of the window which is bilinear, that is with horizontal and vertical axes:

$$y = 2*[\%(Na+K)] \tag{1}$$

$$x = \%Mg*2.3 + (y/\tan 60°) \tag{2}$$

The equation for the vertical coordinate (1) is derived from a direct overlay of a transition of 0 to 100% on the triangle height of 200 screen bits. For the horizontal axis, the position is determined by adding the x-coordinate of Mg along the base of the triangle and then adding a contribution from the vertical coordinate equal to the base length of the right angle triangle of y. These local coordinates then are used to draw the data points according to the size of the display window.

SUMMARY

This paper presents a program that has utility in archiving and preparing hard copies of ternary diagrams. Extensions to the principle of the idea include quadrilinear diagrams which when combined with trilinear diagrams will afford representations as in the standard Piper diagram form (USGS-GW, Sacramento: 9-260d, Revised March 1952). It takes the versatile graphics capability of the Macintosh computer to be able to design such rudimentary figures and then manipulate them to layout in a desired form.

NOTE

This program (version 017) represents the first working version with a minimum of Macintosh interface code using examples from Chernicoff (1985). It has been tested on Macintosh models from the 512 to the Macintosh II with System 3.2 and Finder 5.3 (and later). It also has been tested as a stand-alone with MultiFinder 1.0. The current release version, 062, includes a variety of standard Macintosh interface functions (from the Print, File, Font, and Scrap Managers) as well as special uses of the standard QuickDraw colors and gray shades. Most notably, this release version supports printing, the clipboard, saving in a PICT format file, use of the compiler's extended number arithmetic type and additional customization routines. Special effort has been put into conforming to the Macintosh Interface Guidelines described in Inside Macintosh (Apple Computer, 1987).

REFERENCES

Apple Computer, 1987, Inside Macintosh: Addison-Wesley Publ. Co., Reading, Massachusetts, v. 1, 550 p.; v. 2, 429 p.

Chernicoff, S., 1985, Macintosh revealed: Hayden Book Co., Hasbrouck Heights, New Jersey, v. 1, 516 p.; v. 2, 626 p.

Hardie, L. A., and Eugster, H. P., 1970, The evolution of closed basin brines: Mineral. Soc. America Spec. Paper 3, p. 273-290.

A MICROCOMPUTER RECONSTRUCTION OF PALEOCLIMATES

Barry L. Roberts, Richard G. Craig, and John F. Stamm
Kent State University

ABSTRACT

We describe here a computer program which solves the climate of the southwestern U.S. on a grid of points with a spacing of 7.5' latitude and longitude. Variables computed include mean monthly values of maximum daily temperature and total precipitation. Annual values for each of these variables also are accumulated. The estimates are made with equations derived with regression procedures. Independent variables include: elevation, slope, maximum and minimum upwind elevations, ocean temperature at the coast, and distances to several key climatic controls. These are solved from boundary conditions that include: sea-surface temperature and dominant wind directions at the seasonal extremes as well as elevation and sealevel. Solutions have been obtained for the modern climate and climate characteristic of the last glacial maximum. The boundary conditions for the latter reconstruction are based upon CLIMAP reconstructions of sea-surface temperatures.

INTRODUCTION

Climatic change in the Great Basin during glacial ages was responsible for the formation of large lakes in basins that now are completely dry (Fig. 1). Some lakes overflowed and fed other systems. Lake Searles was fed by overflow down the Owens Valley (Smith and Street-Perrott, 1983). Other lake systems (e.g. Lakes Lahontan and Bonneville) formed in the Great Basin approximately synchronously. There have been numerous attempts to discover the character of climatic change which was responsible for these pluvial conditions in the southwestern US. Theories include: increased precipitation (Antevs, 1952), decreasing temperature (Mifflin and Wheat, 1979), and decreased evaporation resulting in increased cloud cover (Benson, 1981). A common weakness of these theories is that they fail to represent local climatic change in a context that is consistent with the synchronous regional and global climatic variations. A recent exception is the work of Benson and Thompson (1987) in which a shift in atmospheric circulation is suggested although the exact physical link is incomplete.

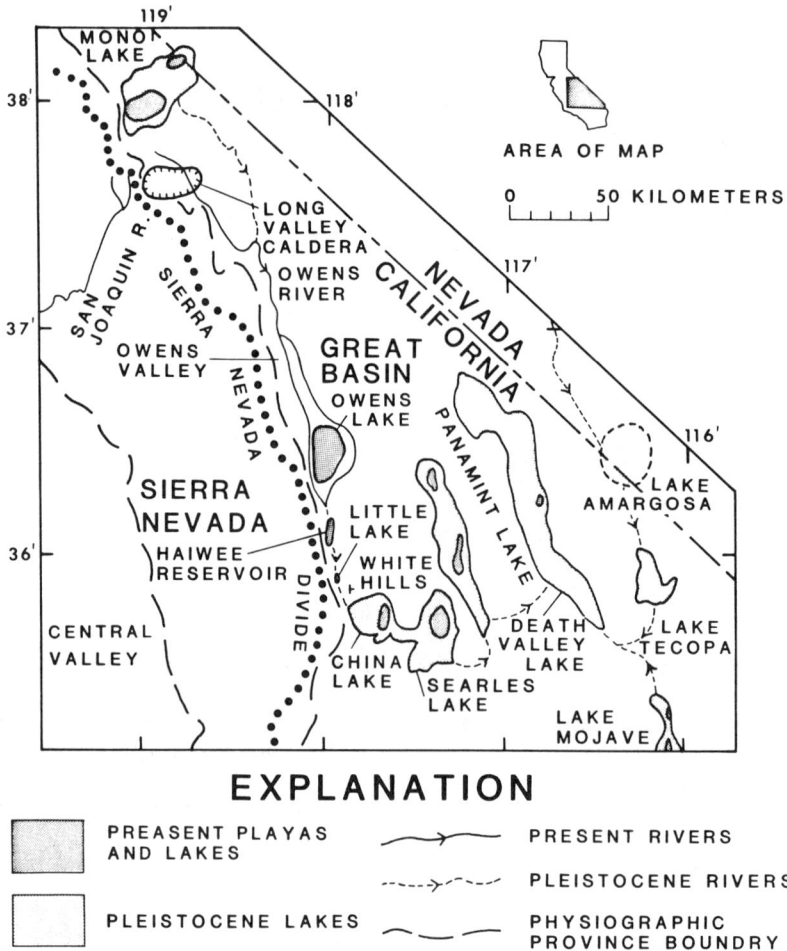

Figure 1.   System of lakes which existed in southwestern U.S. during Quaternary Period
(Smith, 1984).

Existing reconstructions of paleoclimates in the Great Basin are based on field evidence
such as pollen records preserved in lacustrine or alluvial sequences, macrofossils
extracted from preserved paleopackrat middens (Spaulding, Leopold, and van Devender,
1983), or the interpretation of sedimentologic records including shorelines, lacustrine
sediments, or tufa deposits (Smith and Street-Perrott, 1983). Paleoshoreline locations
have been used for estimation of the paleohydrologic balance of individual lake basins
(Snyder and Langbein, 1962; Mifflin and Wheat, 1979). Much of the evidence is difficult to
date precisely and the record suggests periods of rapid change. Thus, it is difficult to
compare one reconstruction to another.

The typical reconstruction is phrased in terms of a constant change from the modern
baseline, either in absolute or relative (percentage) terms (Smith and Street-Perrott, 1983,
Table 10-1; Spaulding, Leopold ,and van Devender, 1983, Table 14-6). Little consideration
has been given to the idea that the changes may have changed both spatially and
seasonally at the time of reconstruction. Equally missing is a clear demonstration of the

impact of even a constant climatic change upon the hydrologic balance. There is no example of applying that constant change to some representation of modern climate and computing the resulting hydrologic balance over an entire series of lake basins. The work we present here is one component of a program to do such computations.

The only digital representation (that we are familiar with) of the spatial pattern of modern climate that is available at a resolution needed for paleoclimate experiments throughout the area was produced by NASA (Labovitz and others, 1983). They digitized each map of the US National Map Atlas at a grid spacing of 20 km. Use of this map allows computation of paleoclimates produced by a constant or proportional change but such reconstructions have not been attempted.

To understand the type of climatic change which created these lakes, we have formulated a model of climate which can solve climatic change as a response to changes in regional boundary conditions. These boundary conditions include the orographic controls on climate, which are so important in this area, as well as sea surface temperature (SST) along the Pacific coast, wind patterns, and sealevel.

The original version of this model (Craig, 1984; Roberts, 1985) was produced for solution on a Hewlett-Packard model 9000 series 500 computer. It was written in HP-BASIC with structured programming extensions similar to Pascal and advanced programming extensions which allowed fast sorts and searches. We have attempted to retain most elements of the original code in this version; however, some parts are not included here, particularly the part to derive the variables for the parameterization of new equations. Without extensive climatic databases, it is not practical to attempt such derivation. We are working currently on several extensions of the equations that will expand the utility of the code and these will be reported separately. The code does allow derivation of climate under a variety of assumed boundary conditions of sea surface temperature, sealevel, topography, and atmospheric circulation patterns. A separate program provides for graphical display of the results.

Hardware requirements for the solution program include 640k RAM, hard disk, IBM PC or compatible (it has been checked on IBM, Zenith, and COMPAQ machines). No math coprocessor is required. The user will need a C compiler if changes are made and recompiling is required. We used the Microsoft C 5.1 compiler for this purpose. As long as no changes are needed, the compiled version can be used.

METHOD

The area modeled covers the entire system of drainage which could be tributary to Death Valley, CA if runoff were sufficient to fill all basins in that area (except Death Valley itself) to overflowing. A rectangular region which includes all of these basins is defined as the level II area in which solutions are obtained. The variables used in the equations are solved themselves from a digital elevation model (DEM) and several additional geographic grids. These grids must cover a larger area (level I), extending west to the Pacific Ocean. The limits of the level I area are selected to include all points upwind (until the coast is reached) of level II points along average wind paths for the months of August and February (Fig 2).

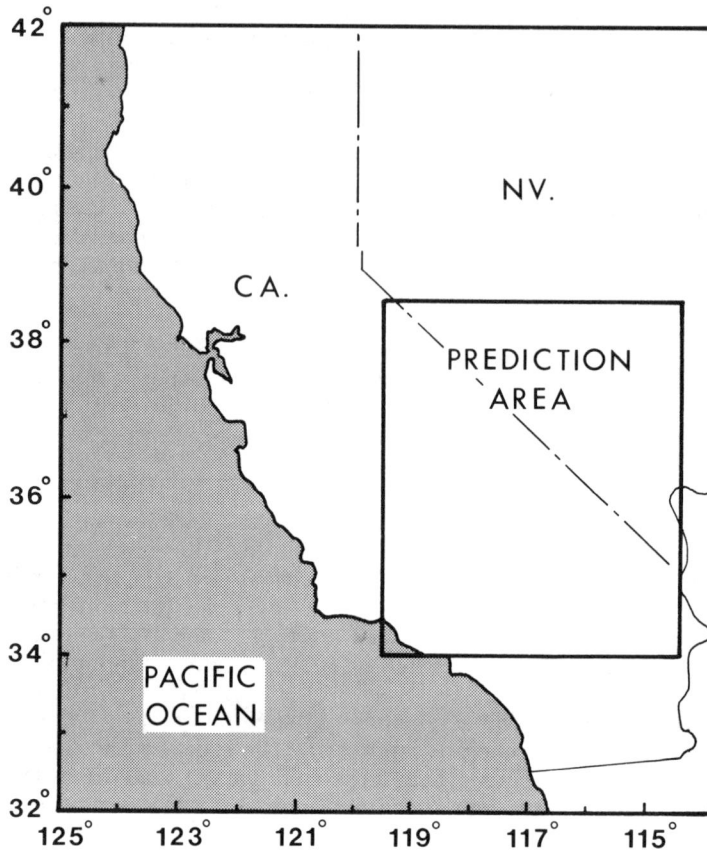

Figure 2. Area within which CLIMATE code can be used to compute mean monthly values of maximum temperature and total precipitation.

The file containing the elevation data is CLIMTOPO.DAT. Coordinates are in units of latitude and longitude and the bounds of the file are:

north -	42°	30'
east -	114°	00'
south -	32°	00'
west -	125°	00'

The grid spacing of elevation file is 7.5 minutes square. Predictions are made within a subset of this region, again at the 7.5' spacing. The bounds of prediction area are:

north -	38 °	30'
east -	114°	30'
south -	34°	00'
west -	119°	30'

These boundaries were selected to include all of the area potentially tributary to Death Valley if each closed basin were to fill and overflow.

The program is based upon a set of regression equations whose coefficients are estimated from modern instrumental records of 180 climate stations. Means for the period 1931-1960 are used (Wernstedt, 1972). Dependent variables are monthly means of maximum daily temperature and means of monthly total precipitation. Thus, there are a total of 24 regression equations. A variety of independent variables are employed. There are eight basic ones and numerous transformations of those. Basic independent variables include: sea-surface temperature (SST), distance to the coast, maximum upwind elevation, minimum upwind elevation (between the site and the maximum upwind point), distances to these maximum and minimum points, upwind slopes, and elevation at the point. Transformations of distances to the natural logarithm of distance and ratios of a variable to its distance also are used.

Wind directions used represent the mean monthly atmospheric circulation pattern for the months of February and August, the same months for which SSTs are available. Although SSTs are available for other months in modern data, such data are more limited in the paleoclimatic record. We have relied upon the CLIMAP reconstructions (McIntyre and others, 1981) for our SSTs and those reconstructions could estimate patterns only for the seasonal extremes. These SST reconstructions are one of the essential boundary conditions that must be specified in a general circulation model of the atmosphere. Such a model can solve for the regional atmospheric circulation pattern, and it is such data that are expected to be used in the climate model described here to represent the needed wind vectors. Thus, the model provides a link between the global models and the field data.

Each of the independent variables is designed to represent some physical control upon climate within this area. They also are selected under the constraint that they must be solvable under boundary conditions representing other climate regimes such as occurred in the geologic past. For the level II area these independent variables are computed from a set of boundary conditions in the larger, level I area. This level I area is large enough to allow tracing wind vectors upwind until they reach the coast.

We used a stepwise regression (Dixon, 1981) using the default values of F-to-enter, F-to-delete, and tolerance level. The order used to predict the dependent variables was determined in a preliminary regression so that climatic variables also were available for use as independent variables once they had been predicted and the order was natural in the sense that months were available when needed.

Once the equation coefficients have been estimated, they are used for solution on a grid of points throughout the level II area. The independent variables have been designed to ensure they could be solved at any point within the area. The spacing used for the grid (7.5') seems fine enough to represent the topographic variability of the area; but is not so fine that microclimatic variability would become significant in defining the climatic variability between adjacent points. Solutions are not limited to this spacing, we have interpolated to a 30" spacing using an elevation-weighted least-squares approach. This limit of detail is given by the point where the inherent uncertainty in the independent variables is important relative to the variability of the dependent variables.

The programs can be used in several ways. We recommend that users familiarize themselves by first generating a standard set of results (e.g. fixed modern). For this, use the default values. A landform data file and the required SST equations are provided. This will solve for modern climate and print values to a disk file in a format compatible with the graphics code. Use the graphics program to view the results. The graphics will show, in color, chloropleths of the variable selected (temperature or precipitation) over the entire level II area. All that will be needed is to specify the file name and give the class limits. Even for these, default values are available. Once the graphics are displayed, you then can move the cursor to individual points. The latitude and longitude of the point will be listed as well as the value of the climatic variable on the grid.

After becoming familiar with the standard modern solutions, experiment with options in the new solutions code. For example, you could select the default last glacial maximum (LGM) values next. Produce the results files and use the graphics program to view those. At this point, one could begin experimenting, such as with LGM sea-surface temperatures but with other sealevels, or modern conditions but some other wind patterns, then LGM with other winds. Such experimentation is a forerunner to serious controlled experimentation using an appropriate experimental pattern.

Some possible modifications of the code include making changes in the computation of SST patterns or changing class limits to demarcate the zero point for temperature. For Monte Carlo simulations, we perturb the DEM by the uncertainty at each point (represented by a normal distribution with mean zero and standard deviation computed for each point by a finer scale DEM). We have made such modifications, and others, in the original code and this is facilitated by the structured organization of the code.

THE PROGRAM FOR CLIMATE SOLUTIONS - USER'S GUIDE

This user's manual is designed to familiarize you with the operation of the computer code which will "solve" the mean monthly values of maximum temperature and total rainfall for a large area of the southwestern U.S. (Fig. 2). The area is bounded on the north by Mono Lake and on the south by Los Angeles. The eastern margin is the Colorado River at Lake Mead and the western edge is again Los Angeles. The computer code is designed to compute solutions at a grid spacing of 7.5' of latitude and longitude. Other grid spacings could be used but the required elevation data must be supplied by the user. Information for that modification is provided later.

The code is designed to solve the regression equations previously calibrated. It works by first computing the independent variables required for the equations. Most of these independent variables are illustrated in Figure 3. These are solved and stored at the set of grid points discussed previously. These independent variables are used in the climate equations to solve the (dependent) climate variables for this same grid. Options are available that allow the user to consider distinct climatic "scenarios." For example, one can compute the climate that would result if the dominant pattern of atmospheric circulation in this area during the winter would shift southward by 10 degrees.

The file containing the prediction code is termed "CLIMATE." To run the program, first make sure that you are in the directory where that executable code is stored. This file requires approximately 250K bytes of storage. During the execution, 26 files are created for each climate scenario selected. These require approximately 14K bytes storage per file. Make sure there is enough storage available for the runs you intend to make. The directory in which the results will be stored is the default directory of the program files.

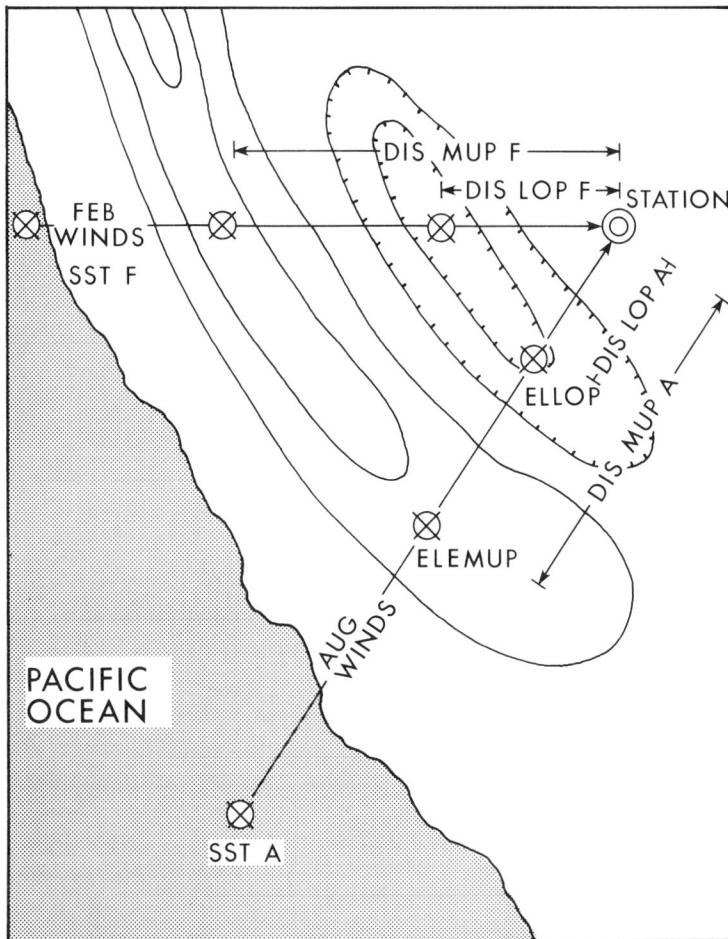

Figure 3. Schematic diagram illustrating independent variables in climate equations of default model.

Program Options

Four options (Table 1) must be set by the user for each run. Default values for these options are built into the code so that the user, especially when learning the code, simply can accept those values, produce output files, and examine them with the accompanying graphics code. Default values are given in braces {modern} in the table. Option four is useful when the code is being modified. In all other situations, the results of the run should be stored. Failure to store the results would lead to errors in predictions because of the bootstrap method that is used in the regression equations.

Table 1. Options available in CLIMATE program.

1. Prediction time period toggle (modern or last glacial maximum) {modern}

2. February wind orientation (direction wind is traveling to; measured in decimal degrees, clockwise from the north) {102.00}

3. August wind orientation (direction wind is traveling to; measured in decimal degrees, clockwise from north) {75.00}

4. Toggle for storing the climate predictions or not {store climate predictions}.

Execution

After starting the code by typing the executable file name "CLIMATE", the program prepares for the calculations. At this time, it will flash the message:

***** INITIALIZING *****

Within a few seconds, it will present the option screen (Fig. 4) from which the user can select items to be modified for the particular climate scenario of interest. These are the variables listed in Table 1. Note that it is possible to make a run without making any changes. This would be useful especially for the first pass through the program.

SELECT MENU OPTION TO CHANGE PARAMETER, OR ZERO TO CONTINUE.

CURRENT VALUES ARE SHOWN BELOW:

0 = **** Values okay, continue program ****
1 = Prediction time period = MOD
2 = February wind direction = 102.00
3 = August   wind direction = 75.00
4 = Store the climate prediction on disk

ENTER DESIRED OPTION:

Figure 4. Options list for CLIMATE program.

The program continues the initialization process. A message is printed on the screen:

READING IN SEA-SURFACE TEMPERATURE EQUATIONS...

---

indicating that the coefficient file for the SST equations has been opened. If this is not successful, a message to that effect is printed and the program will terminate execution and return to the operating system. Table 2 shows the organization of the SST file.

Table 2. Coefficients used to calculate the SST at each point in ocean. Equation is cubic polynomial.

249.978	-0.0585317	-3.780757	0.0	
0.0165434	-0.00501172	0.0000483287	0.0	0.170269
372.268	-4.0357	-3.65132	0.0	
0.0	0.0260343	0.0000905506	0.0000636592	0.558759
136.236	0.0	-1.37717	-0.016392	
0.0	0.0	0.0000777909	0.0000330744	0.439222
346.869	-4.41016	-3.27927	0.0	
0.0	0.026695	0.000150343	0.0000560485	0.539209

These lines contain the 8 coefficients for the sea-surface temperature equations with the ninth element in each row being the standard error for the regression.

ROW 1-2 = FEBRUARY MODERN   ROW 5-6 = FEBRUARY LGM
ROW 3-4 = AUGUST MODERN   ROW 7-8 = AUGUST LGM

ELEMENT 1=INTERCEPT   ELEMENT 6=LAT*LONG
ELEMENT 2=LAT   ELEMENT 7=LAT^3
ELEMENT 3=LONG   ELEMENT 8=LONG^3
ELEMENT 4=LAT^2   ELEMENT 9=STANDARD ERROR
ELEMENT 5=LONG^2

A similar message then will be printed indicating that the elevation data file has been opened successfully.

READING IN TOPOGRAPHY DATA FILE...

Again, if this is not successful, a message to that effect is printed and the program terminates execution and returns to the operating system. The format of the data contained in the elevation file is illustrated next. Each row would extend to include all of the values of the respective row of the data matrix. Because this is the northwestern portion of the elevation grid which is all in the Pacific Ocean, all the points listed are actually bathymetric depths and are reported as negative numbers. The normal continental elevations are reported as positive numbers (except in areas such as Death Valley!).

```
1000  -1000  -1000  -532  -139  -88
1000  -1000  -1000  -499  -128  -38
1000  -1000  -1000  -563  -150  -68
```

The program then begins to extract the independent variable data for the points needed along each of the two wind vector orientations that have been specified. It first does the February wind vector, then the August wind vector. A message is printed to inform the user that this is being done. The message will list the vector orientations that are being used. This can be time consuming, so relax!

Extracting independent variables along a feb wind vector of 102.00 degrees
Extracting independent variables along a aug wind vector of 75.00 degrees

The format that is used for the climate equation coefficient file is listed in Table 3. The data structure is a three-dimensional matrix with 24 rows, 20 columns, and 2 planes. The rows represent the equations for each of the 12 months, for temperature and precipitation. Each row is a sequence of pairs of numbers. There are up to 20 pairs for each equation. The first number is a code indicating the variable, the second number is the coefficient for that variable. The variable code consists of two parts. The integer portion indicates the variable type (Lat, Long, SST,...). The fractional part indicates which month the variable is associated with (0.01=Jan, 0.04=April, 0.12=Dec).

For example:

0.01	35.897	Y-intercept value of 35.897 (month is meaningless)
4.08	3.9999	SST August coefficient is 3.9999
21.5	1.2345	May temperature coefficient is 1.2345

A partial example line of a coefficient file is given in Table 4. The program reads in the coefficients of the climate equations. While it is doing this, a message to that effect appears on the screen.

Reading in climate equation coefficients ...

Table 3. Variable codes used in file used to store coefficients of climate equations.

0 or blank : Y-intercept
   1: Latitude
   2: Longitude
   3: Elevation
   4: SST
   5: Elevation of max upwind point
   6: Distance to max upwind point
   7: Elevation of lowest point since #5
   8: Distance to #7
   9: Upwind slope
  10: Distance to coast
  11: Elevation drop
  12: Elevation gain
  13: Ln(Distance to coast)
  14: Ln(Distance to max upwind point)
  15: Ln(Distance to #7)
  16: SST/Ln(Distance to coast)
  17: Decamax = #5/Ln(#6)
  18: Decamin = #7/Ln(#8)
  19: Decagan = #12/Ln(#8)
  20: Decadrp = #11/Ln(#6)
  21: Previous temperature prediction
  22: Previous precipitation prediction
  -1.0 indicates the end of the equation

Table 4. Example format of coefficient file for climate equations. Only five of 24 rows
         (equations) are shown. For each row only intercept and first coefficient are
         listed.

.010000	16.307631	2.010000	-0.150960
.010000	64.036240	1.010000	-1.099810
.010000	.196840	3.010000	-0.000268
.010000	14.771160	2.010000	-0.108870
.010000	.718010	3.010000	-0.000482

Storing the Results

The next option relates to the output of the results. When the computations are completed
and the code is prepared to write the output, if a file is encountered that has the same name
as the file to be written, the program will ask if that file should be written over.

PREDICTING FEB TEMPERATURE
WRITE OVER THE EXISTING CLIMATE FILES IN THIS DIRECTORY?
0=NO 1=YES

The user response is either 0 (zero) or 1 (one) depending on whether the file should not be erased (0) or should be overwritten (1). If a "zero" is entered, the user is returned to the operating system. If a "one" is entered, the results are printed out, overwriting the previous data. An example illustrating the format of a small part of a prediction file is shown next. This lists the first 5 columns and first 10 lines (rows) of the file for modern March temperature (in degrees Celsius) as forecast with the default options. It should be noted that the precipitation predictions will be stored as the cube root of precipitation except for the annual values which are in centimeters.

0.6260	-3.5303	1.3095	2.3973	2.2044
1.6345	-0.5724	0.2567	-0.5872	0.6351
3.9493	-0.4147	-0.0540	-1.6308	-0.3748
3.7363	-4.2255	-0.8808	2.2539	2.6461
2.7337	-3.5006	-3.2380	2.0652	1.3924
2.2615	-4.2966	-5.4995	-0.4465	-0.8164
1.9031	-3.9991	-4.3651	-2.9468	0.0965
1.3976	-1.2400	0.3124	-1.3719	-5.0434
2.0716	3.3742	2.5474	-1.4411	-2.9872
3.4542	4.5327	0.2089	-0.2426	-1.7384

For each subsequent prediction, files of earlier predictions are opened as needed by the bootstrap procedure. A message indicating which prediction is in progress is printed on the screen.

Predicting Aug temperature

These steps continue for each month of each variable (temperature and precipitation) that is estimated. For each month, the prompt will be given until the predictions are complete. That step completes the program execution. The user is returned to the first option screen for additional runs.

THE PROGRAM FOR GRAPHIC DISPLAY - A USER'S GUIDE

The purpose of this code is to display the results of the program CLIMATE. Two types of graphic displays are available. The first is termed a "normal plot." It displays the climate data for a given variable and given month for the entire area. The second type of plot is termed a "difference plot."

The difference plot displays the difference (change) in climate between two scenarios. The default scenarios selected are the default scenario for modern climate and the default scenario for the climate of the last glacial maximum (LGM). This latter scenario represents that inferred to have existed at about 18ka. Thus, the plot indicates the amount that climate changed between the last glaciation and today. The difference is computed as LGM minus modern. Therefore, if the climate at the LGM was cooler than the modern climate, it will be given on the plot as a negative value. This simplifies the interpretation of changes.

To run this code, you must be sure you are in the proper directory. The program file is termed CLIMPLOT. In addition to that file, the directory should hold all of the climate files that are to be displayed. An example of such a file, and its structure, are listed in Table 6.

To execute the code, type the name CLIMPLOT. After a few seconds, the first menu will appear on the screen. It will look as follows:

```
E ... EXIT PROGRAM
N ... NORMAL PLOT
D ... DIFFERENCE PLOT

INPUT SELECTION NUMBER
```

Table 6. Example format for file containing climate solutions to be displayed by program CLIMPLOT. This is partial listing of file MAR_TMOD. First five (of 30) rows and first six (of 24) columns are shown. Values are given in degrees centigrade.

0.6260	-3.5303	1.3095	2.3973	2.2044	-0.3115
1.6345	-0.5724	0.2567	-0.5872	0.6351	0.4655
3.9493	-0.4147	-0.0540	-1.6308	-0.3748	0.9651
3.7363	-4.2255	-0.8808	2.2539	2.6461	1.4866
2.7337	-3.5006	-3.2380	2.0652	1.3924	0.0268
2.2615	-4.2966	-5.4995	-0.4465	-0.8164	-1.0017
1.9031	-3.9991	-4.3651	-2.9468	0.0965	1.5865
1.3976	-1.2400	0.3124	-1.3719	-5.0434	-2.9465
2.0716	3.3742	2.5474	-1.4411	-2.9872	-4.7306
3.4542	4.5327	0.2089	-0.2426	-1.7384	-5.2713

This option allows you to select whether you will view a "normal plot" or a "difference plot." These were discussed previously. If the EXIT option is selected, the program will terminate execution and return you to the operating system. We will consider the results of selecting each of the other options separately. It is not necessary to press <RETURN>; typing the letter will activate the next option screen.

Normal Plot

After selection of the "normal plot" option, the user is presented with another option screen. This allows selection of the month of the year which is to be displayed. It will appear as follows:

```
                        0 ... ANN
                        1 ... JAN
                        2 ... FEB
                        3 ... MAR
                        4 ... APR
                        5 ... MAY
                        6 ... JUN
                        7 ... JUL
                        8 ... AUG
                        9 ... SEP
                       10 ... OCT
                       11 ... NOV
                       12 ... DEC

              INPUT MONTH NUMBER
```

The first selection is for the ANNual values of the variable. For temperature this would represent mean annual maximum temperature. For precipitation, this represents total annual precipitation. Each of the months is indicated with a three-character code whose meaning should be obvious. To select a month period, type the appropriate number, then press <RETURN>.

Another menu will appear to allow selection of the variable to be plotted, temperature or precipitation. This menu will appear as follows:

```
                        1 ... TEMP.
                        2 ... PREC.

               INPUT TYPE NUMBER
```

After selection of the variable, the next option screen allows selection of the time period to be plotted. There are two options, modern or LGM:

```
                        1 ... MOD
                        2 ... LGM

              INPUT PERIOD NUMBER
```

It is important to realize that, although the time period is termed LGM, this only indicates that the values were solved (in the CLIMATE program) using the sea-surface temperature pattern defined by the CLIMAP project as representing the time of the last glacial maximum. The user could have created a different scenario by modifying the wind directions in that program. This would lead to printing a new LGM file.

This completes the selection of options describing the plot to be produced. At this point, the program will access the disk and read in the data required for the plot. A message will appear in the bottom left-hand corner of the screen to this effect:

READING IN DATA ...

Note: whenever the program attempts to read in a data file, if it cannot locate that file, it will beep and print out a message to that effect. It then will return you to the main menu.

The program displays the data in six colors, each representing a separate class. The class limits for these classes are computed from the range of data values for the selected variable. There are six equal size classes. Before the plot is made, the user is presented with the class limits to be used and asked if these class limits are acceptable. If they are, indicate "Y" for yes and the plot will be made.

If you would prefer to define your own class limits, answer "N" for no. You thenwill be presented with each default class limit and be asked for a new value. You must enter a value, even if it is the same default value. If you wish to define a new class limit, simply type in the numeric value desired and press <RETURN>. The program will prompt for a new value of each class limit but not all need to be changed.

An example where it is useful to redefine class limits is in the temperature data. Especially for the winter months, it is useful to have one class limit be exactly zero (0.0). This makes it easy to see where the model is predicting that above freezing or below freezing values will be encountered.

The option screen for class limits will appear as follows:

CHANGE CLASS LIMITS? [Y/N]

At this point, all options have been specified and the plot will be computed and shown. It will stay on the screen until you type <ESC> to exit. This will return you to the main menu.

A "normal" plot of temperature is illustrated in Figure 5 (see Color Plate 4). This is January temperature under the default modern scenario. Class limits and corresponding colors are listed on the right. Shown on the lower right are values of mean, minimum, and maximum.

Pressing any key (other than <ESC>) allows you to "digitize" individual points to examine the data value at that point. Press a key. To select a point, use the arrow keys to move a rectangular cursor over the screen. The relevant data are printed at the bottom of the screen. These data include:

>   latitude of the center of the 7.5' cell
>   longitude of the center of the 7.5' cell
>   data value

To return to the main menu, hit <ESC>. This returns you to the option screen seen at the beginning.

E ... EXIT PROGRAM
N ... NORMAL PLOT
D ... DIFFERENCE PLOT

INPUT SELECTION NUMBER

Difference Plot

If you select D=DIFFERENCE PLOT the sequence of menus that will be presented virtually is identical to the normal plot option. The major difference is that it does not ask for the time period. This is because it will make use of both files, modern and LGM.

After typing "D" (no return is needed), the menu for selection of the month will be presented. As with the "normal" plots, type the integer corresponding to the month desired and then press <RETURN>. The menu for selection of the variable of interest will be presented. Type either 1 or 2 and press <RETURN>. At this point, the program will begin to read in the desired data. Two identical messages will appear on the screen:

READING IN DATA

Both files (modern and LGM) are needed for this plot and they are read in sequentially. Once both files have been read in, a prompt will appear listing the default values of the upper class limits for each of the six classes of the plot. The user will be asked if these limits should be modified. In this selection it is clearly of interest to have 0.0 be one of the class limits so that both increases and decreases can be distinguished. Of course, this is not needed if only increases or only decreases occur. Thus, it is important that the user review these class limits. Once the class limits have been defined, the plot is produced. Again the option is available to move the cursor across the screen and read off exact individual values at any selected point. Pressing <ESC> once again will return the user to the main options screen for examination of another variable or to exit the program.

Hardware requirements for the plotting program include 640K RAM, a hard disk, IBM PC or compatible (it has been checked on IBM, Zenith, and COMPAQ machines), an EGA graphics card but no math coprocessor is required. The user will need a C compiler (we used Microsoft C 5.1) if changes are made and recompiling is required. Otherwise, the compiled version can be used.

APPLICATIONS

We envision a variety of applications for this code. First, it provides estimates of modern climate at a level of detail not previously available. One can use the climate graphics program to determine the climate at any point of interest. These estimates can form input to hydrologic or hydrogeologic models. They also could be of use to agriculturists, foresters, and botanists. In short, anyone who has use for estimates of modern climate can benefit from this program. These climate grids also could form one component of a geographic information system (GIS).

A larger range of experimental applications is available. The code is designed to allow reconstruction of paleoclimates, particularly those at the last glacial maximum. We have used it to estimate runoff at that time. That estimate of runoff is used as input to another code which can compute the formation of the resulting system of lakes and rivers. We have used the code to estimate the climatic response to other climate scenarios such as the local response to a "greenhouse warming" (Craig and Roberts, 1988).

CONCLUSIONS

The availability of this program makes it possible to compute values of mean monthly maximum daily temperature and total monthly precipitation for each month of the year at any point within the level II area at a resolution of 7.5' of latitude and longitude. This provides a significant improvement upon the usual extrapolation of available records at existing climate stations. The code also allows computation of paleoclimates such as that of the last glacial maximum. Experimentation can be done to estimate the local response of the climatic system to broad scale changes in atmospheric circulation patterns and sea-surface temperature. The quantitative descriptions of climates and paleoclimates will allow more comprehensive examination of the hydrologic impacts of climate change and a better understanding of the spatial and temporal components of climate change in the Southwest.

ACKNOWLEDGMENTS

Preparation of this computer code was funded under contract B-1804 from the Desert Research Institute, University of Nevada, Reno. An earlier version of the program was funded partially by the Department of Energy, Pacific Northwest Labs under Contract B-F7204-A-H. Completion of this project relied upon discussions with M. Singer.

REFERENCES

Antevs, E.V., 1952, Cenozoic climates of the Great Basin: Geologische Rundschau, v. 40, heft 1, p. 94-108.

Benson, L.V., 1981, Paleoclimatic significance of lake level fluctuations in the Lahontan Basin: Quaternary Research, v. 16, no. 3, p. 390-403.

Benson, L.V., and Thompson, R.S. , 1987, Lake-level variation in the Lahontan Basin for the past 50,000 Years: Quaternary Research, v. 28, no.1, p. 69-85.

Craig, R. G., 1984, Glacial climate of the Southwest U.S. from global circulation: a computer model (abst.): Intern. Comm. on Climate, Igls, Austria, p. 60.

Craig, R.G. , and Roberts, B.L., 1988, A sensitivity analysis of Southwestern climate: Fifth Workshop on Climate Variability of the Eastern North Pacific and Western North America, p. 17-19.

Dixon, W.J., 1981, BMDP Statistical Software: Univ. California Press, Berkeley, California, 725 p.

Labovitz, M.L., Masuoka, E.J., Broderick, P.W., Garman, T.R., Ludwig, R.W., Beltran, G.N., Heyman, P.J., and Hooker, L.K., 1983, Experimental philosophy leading to a small scale digital data base of the conterminous United States for designing experiments with remotely sensed data: National Aeronautics and Space Administration, NASA Tech. Mem. 85009, 522 p.

McIntyre, A., Cline, R., Hays, J., Prell, W., Moore, T., Kipp, N., Molfino, B., Denton, G., Kukla, G., Matthews, R., Imbrie, J., and Hutson, W., 1981, Seasonal reconstructions of the Earth's surface at the last glacial maximum: Geol. Soc. America, Map and Chart Series, MC-36, 18 p.

Mifflin, M.D., and Wheat, M.M., 1979, Pluvial lakes and estimated pluvial climates of Nevada: Nevada Bureau of Mines and Geology Bull. 94, 57 p.

Roberts, B.L., 1985, Paleoclimate in the Southwest U.S.: a computer model: unpubl. masters thesis, Dept. of Geology, Kent State Univ., 189 p.

Smith, G.I., 1984, Paleohydrologic regions in the southwestern Great Basin, 0-3.2 my. ago, compared with other long records of "global" climate: Quaternary Research, v. 22, no. 1, p. 1-17.

Smith, G.I., and Street-Perrott, F.A., 1983, Pluvial lakes of the western United States, in Porter, S.C., ed, Late-Quaternary Environments of the United States, vol. 1, The Late Pleistocene: Univ. Minnesota Press, Minneapolis, p. 190-212.

Snyder, C.T., and Langbein, W.B., 1962, The Pleistocene lake in Spring Valley, Nevada, and its climatic implications: Jour. Geophysical Res., v. 67, no. 6, p. 2385-2394.

Spaulding, W.G., Leopold, E.B., and van Devender, T.R., 1983, Late Wisconsin paleoecology of the American Southwest, in Porter, S.C., ed, Late-Quaternary Environments of the United States, v. 1, The Late Pleistocene: Univ. Minnesota Press, Minneapolis, p. 259-293.

Wernstedt, F.L., 1972, World climatic data: Climatic Data Press, Lemont, Pennsylvania.

Microcomputers in Mineral Exploration:
A Database For Modeling Gold Deposits in the
Yilgarn Block of Western Australia

N.M.S.Rock
J.N.Shellabear
M.R.Wheatley
R.Poulinet
D.I.Groves
University of Western Australia and
Western Australian Regional Computing Centre

ABSTRACT

A geologic, geochemical, and isotopic database for gold deposits in the major Archaean gold province of the Yilgarn block, Western Australia MERIGOLD is being set up on an Apple Macintosh II microcomputer using the new relational database-management system (DBMS) 4th Dimension. Detailed consideration was given to the use of mainframe computers and well-tried DBMS such as ORACLE, but the user-friendly Macintosh with its graphic interface proved to have many advantages for a geologic database of this type. Features of existing gold databases, for example, those of the Ontario Geological Survey and the Australian Bureau of Mineral Resources, were adopted in MERIGOLD as far as possible. Although 4th Dimension itself includes simple plotting capabilities (histograms, etc.) data can be exported readily in ASCII format from MERIGOLD into other computer packages or other (e.g. IBM PC, mainframe) computer systems. This facilitates more sophisticated plotting and processing via dedicated graphics and statistical packages. Use of the quasimultitasking Macintosh Multifinder greatly speeds up this process, because database, statistics, and graphics programs all can run concurrently. MERIGOLD at present is fairly small ($\approx$ 1 Mb), with information on about 80 variables for some 170 gold deposits and > 1,000 rock specimens), but rapid growth is not precluded now that 250+ Mb hard disks, 16 Mb RAM, and CD (compact disk) media are available on microcomputers. Whereas existing gold databases contain mainly descriptive information in free text (company names, production figures, rock-types, stratigraphy, mining methods, etc.), MERIGOLD concentrates on information which is either intrinsically numerical (e.g. ore,whole-rock, and mineral chemistry; isotope and fluid compositions), or which can be coded numerically either as nominal or ordinal variables (e.g. mineral abundances as not seen, rare, abundant; metamorphic grade from 1 = lower greenschist to 5 = upper amphibolite). The considerations and problems applicable to MERIGOLD should be relevant to many other users wishing to set up geologic databases of their own: problems such as missing data, sampling validity, interlaboratory comparability and reproducibility of assay data, validation, and duplication. MERIGOLD is being used to assess some of the conflicting hypotheses for the genesis of Archaean gold deposits.

INTRODUCTION: THE AUSTRALIAN GOLD EXPLORATION SCENARIO

A rise in Australia's gold production from 18 tons in 1981 to an excess of 100 tons in 1988 has resulted from a combination of natural-resource endowment, successful exploration, improved mining and metallurgical methods, and—last but not least—the current high price of gold. The majority of this increased production has come from Western Australia's Yilgarn Block (Fig. 1), where approximately 36 new operations began in 1987 and 46 are planned for 1988. The Yilgarn's projected total production of over 85 tons Au now once again exceeds the previous highest production during the Kalgoorlie gold rush at the beginning of the present century.

The resurgence is necessitating more efficient use of the vast amount of geologic information available on gold mineralization. For example, prospecting for gold deposits will become both more technically difficult and more expensive as the current medley of shallow deposits becomes mined out requiring more sophisticated methods to look for deeper, more hidden deposits. Therefore, it is worth spending time and effort trying to identify the best possible exploration areas, using existing geologic knowledge. By analogy with the studies of Hodgson (1983) in Canada, a geologic and geochemical database of Western Australia's significant gold deposits may lead to better understanding of genetic controls on gold mineralization. It can allow comparisons, empirical correlations, and definition of factors common to gold deposits, which then can be used to test existing genetic models or formulate superior ones, and hence develop exploration models more suited to the 1990s. Setting up a database is learning an important lesson from previous "boom and bust" cycles—for example, the nickel boom of the 1970s—where valuable data accumulated during the boom were lost after the bust, because they had never been compiled or archived properly. A permanent gold database ensures that this loss of valuable data will not recur, and thus represents an investment for the future.

This paper outlines the first two stages of an ongoing project which aims to:

(1) compile significant geologic, geochemical, and isotopic information on gold deposits in Western Australia's Yilgarn Block that have greater than 1 ton of contained gold (i.e., reserves + production);

(2) construct a working database of these compiled data;

(3) analyze the data statistically to show any regional or local trends, and hence develop new deposit models.

Stage 3 will be described in later papers. The database is termed MERIGOLD, because it is being sponsored under the aegis of MERIWA (Mineral Industries Research Association of WA) by 22 mining companies.

RECENT HISTORY OF GEOLOGIC DATABASES

A database here refers to a computerized collection of related information stored in such a way that retrievals can be performed by linking various pieces of information together (Date, 1981). It consists of one or more data files, which are collections of related information treated as one unit on a computer. Databases are managed and accessed via software termed database-management systems (DBMS: Hruska, 1976). The combination of hardware, software, and the database itself is referred to as a *database system* (Teorey and Fry, 1982).

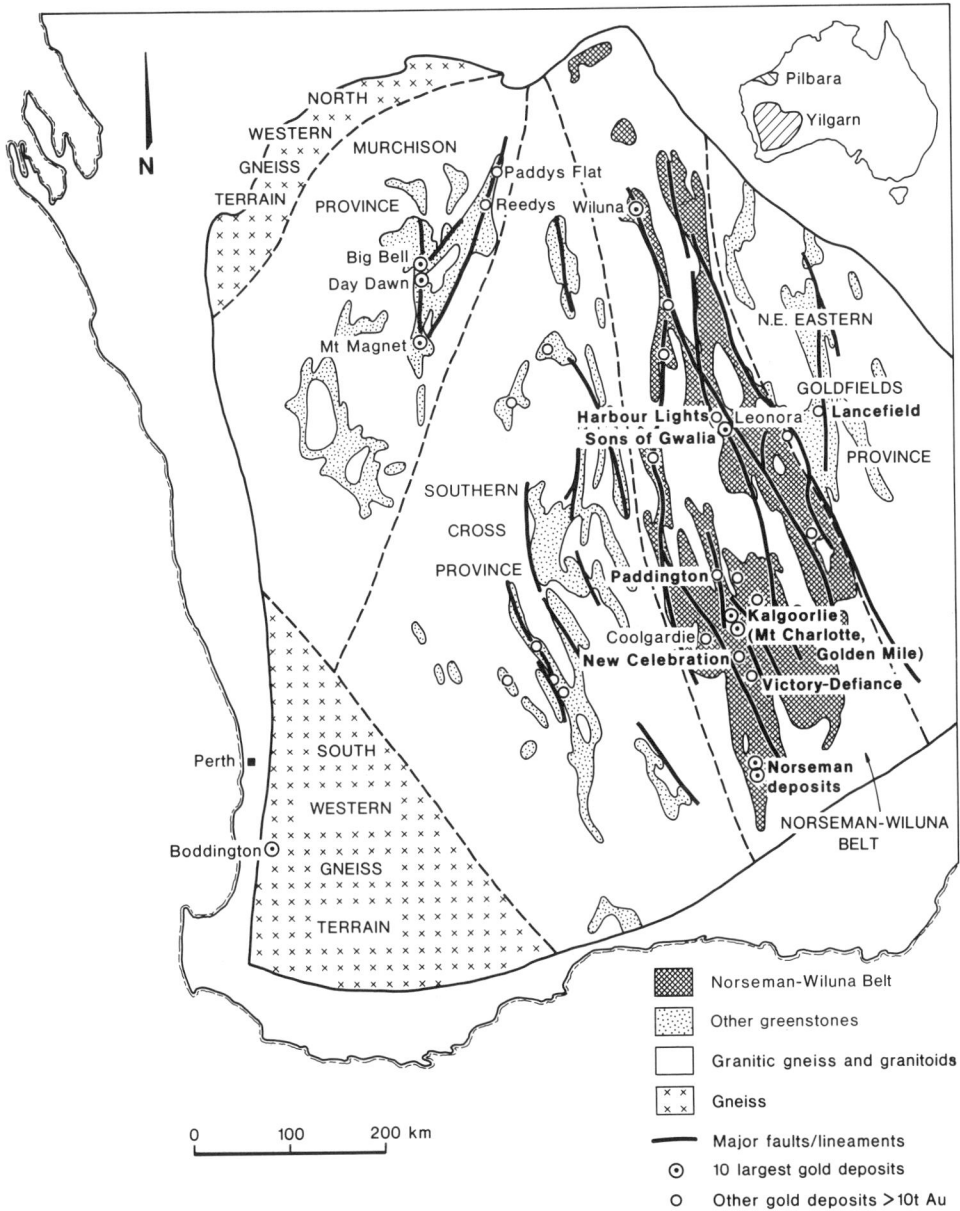

Figure 1. Distribution of gold deposits in Western Australia. Archaean Yilgarn *block* is
divided into several lithotectonic *provinces*, within which *goldfields* comprise
clusters of individual *deposits*. Norseman-Wiluna *belt* is actually only a part of
Eastern Goldfields *province*, but remaining terrain within this province consists
of acid gneisses devoid of gold deposits. Hence, this one *belt* is equivalent to a
province in terms of its gold deposits and is given same hierarchical status in
MERIGOLD's structure.

Databases are divided conventionally into *reference* and *source* databases (Le Maitre, 1982). *Reference* (bibliographic) *databases*, such as GeoRef (USA), Geoarchive (UK), and AESIS (Australia), contain available references (with key-words and some details of data content) on a particular subject (Burk, 1981; Jones, 1981; Rassam and Gravesteijn, 1986). Because these systems are well established, maintained by large commercial support organizations, and contain vast numbers of both published and unpublished references to gold mineralization, MERIGOLD concentrates on different data, and in no way attempts to be a reference database.

Source (numeric, nonbibliographic) databases, by contrast, provide actual data — words, numbers, perhaps even pictures — on a particular subject. Henderson (1986) further distinguishes *numeric* databases as source databases where numbers dominate words and pictures. During the 1960s and 1970s, the history of development of geoscience source databases proved to be somewhat checkered. Some proved to contain an overabundance of irrelevant information (Ackoff 1967), attempting to compile "all" data rather than that specifically required by the ultimate users (Clark, 1976). Others had the converse problem, proving to lack information essential to the development of a project. A third class was unable to cope with increasingly complex demands as the system grew — data simply got "lost." All these problems tended to reflect inadequate consultation between designers and users during the construction phase (Bliss, 1986).

Burk (1981) suggested that the limiting factors have been not technological but rather conceptual and managerial — for example, the general absence of standard or intuitively obvious formats for geoscience data, and even a lack of agreement on what constitutes "data." Accordingly, there can be no universal or all-purpose geologic *source* databases corresponding to the well-established reference databases; each geoscience activity has its own unique set of data requirements. Individual source databases thus will continue to appeal to smaller, specialist audiences than reference databases, even though they may tend to become more grouped in the future, to cover areas of broader interest.

Despite these teething troubles, the number of substantial extant source databases in geology certainly reaches thousands and possibly tens of thousands: Shelley (1985) catalogues just about 250 in Australian governmental organizations alone, and many more have been constructed since 1985. Rock (1988) lists a further range of databases in various geologic disciplines. The tremendous diversity of computer implementations used has led unfortunately to considerable problems in compatability, data interchange, overlap, and hence duplication of effort. Cooperative national and international efforts nevertheless are getting under way, for example, under the auspices of the International Geological Correlation Program (IUGS/UNESCO), leading to prototype *global* databases such as IGBA (IGCP Project 239), which currently contains analytical data for some 12,000 igneous rocks from 1,000 worldwide sources. IGBA builds on previous igneous databases set up by individuals (e.g., Chayes 1982; Le Maitre, 1982). There are some grounds for confidence, therefore, that predominantly negative experiences are a thing of the past.

Unfortunately, these negative experiences do indicate that the geologic literature on databases yields little *practical*, positive advice as to the best ways to set up a system. Nevertheless, specific advice from papers listed in the references (e.g. Aubrey, 1981; Harris, Winczewski, and Umphrey, 1982) has been adhered to as far as possible. On the computer side, advice is equally difficult to obtain, but for different reasons: technological developments are now so rapid that comprehensive expertise is almost impossible to acquire in the first place, and becomes out-of-date almost immediately. A potential user thus can become hopelessly bogged down by the profusion of both hardware and software selections now available. A major aim of this paper therefore is to provide specific advice on selecting computer systems, capturing data, and storing it efficiently, based on our own experience of some 15 years. Readers should realize nevertheless that this advice must be somewhat partisan, because no one can claim to command knowledge of all known computer systems or database packages, let alone the problems peculiar to all types of geologic data!

SELECTING HARDWARE: MAINFRAME OR MICROCOMPUTER?

As anyone working with computers knows, technology has changed beyond all recognition in the mid-1980s. Personal microcomputers now are available which sit on the geologist's own desktop, under his complete control, yet harness the power of the vastly more expensive mainframe computers of only a few years ago. This leads to the question for which there was previously only one answer: mainframe or microcomputer— which now is preferable?

Mainframes offer greater computing speed (in terms of MIPS: millions of instructions per second) and storage capacity (in terms of Mb: millions of bytes of stored information), whereas they also relieve the geologist of day-to-day hardware, software, and file management responsibilities — system crashes, backups, updates, among others. On the other hand, they do, offer the user far less control over his computing environment, and their faster speeds can be cancelled out by far greater access time when multiuser systems become overloaded. Furthermore, it is easy (and a healthy break from staring at a screen) to perform some other task while a desktop microcomputer is performing a lengthy database operation, so that the slower speed of microcomputers is not always a disadvantage. The storage capacity of many modern microcomputers, moreover, already is far beyond that required for all but the largest geologic databases ($1,000-$2,000 hard disks can store the equivalent of 50,000–100,000 A4 pages, $\approx$ 250 Mb, allowing for maps and diagrams, whereas compact disk (CD), "megafloppy," and tape devices are becoming available which store 4-5 times more information ($\approx$ 1 Gb). In these circumstances, local considerations specific to the particular geologist may become far more important than any general advice which could be given on the relative merits of mainframes and microcomputers: factors such as local cost and availability of software and hardware, cost of accessing mainframe systems versus that of purchasing microcomputers, and — of course — who pays for what.

With the MERIGOLD project, a major consideration was making the final data from the database available to the 22 mining companies sponsoring the project, a consideration we suspect is far from unique to this particular project! Unfortunately, these 22 companies already possessed a potpourri of different computer systems (some minicomputers, some mainframes, some microcomputers — mainly IBM PC, and some without computers at all), so that it was impossible from the outset to select one system for MERIGOLD which would be identical to all of the companies' preexisting systems.

Fortunately, rapidly advancing network capabilities makes the particular brand of computer selected for a project less critical. The most popular microcomputers, such as the Macintosh and IBM PC, can be networked fully not only to each other but also to more powerful machines such as workstations, minicomputers, and mainframes. Actually running a mainframe from a microcomputer over a network is already routine, using terminal-emulation packages: the microcomputer merely behaves like an "intelligent terminal", and the user acts as he would running the mainframe from any other type of less intelligent terminal (e.g. Rock and Wheatley, 1989). New developments such as *shells* allow one step further, in which the mainframe is operated using the microcomputer's own operating system: in other words, the user can run the mainframe computer using *the same commands* as if he were running the microcomputer, and thus can operate two distinct computer environments without learning two different operating systems. It also is possible nowadays to run what have been regarded traditionally as mainframe operating systems (e.g. AT&T UNIX) on microcomputers (e.g. Apple A/UX). Overall, these increasing capabilities show not merely that data can be transferred readily from one machine to another, but also that it becomes less-and-less important on which machine a particular database actually runs.

In 1988, the selection of a *microcomputer* system for MERIGOLD was dictated primarily by the following considerations: (1) the availability of a powerful but exceptionally user-friendly, graphics-based microcomputer — the Apple Macintosh II — which is particularly suitable for a graphics-based science such as geology; (2) the small local cost of both the hardware and DBMS software for the Macintosh ($\approx$ A\$1500), relative to that of some well-tried mainframe DBMS (typically $\approx$ A\$20,000 for educational use only; perhaps A\$80-100,000 for commercial use).

Overall, different computer environments are merging more closely together, and their formerly irritating incompatibilies should become more and more transparent to the user as technology improves. The selection of a particular implementation such as that of MERIGOLD now can be made to make best use of local factors and conditions, without isolating the data from users with other systems.

SELECTING A DATABASE MANAGEMENT SYSTEM (DBMS)

A typical DBMS will offer some or all of the following features:

- its own high-level query language (QL), used both to define data relationships and to retrieve and report selectively on the data; SQL (standard query language) is becoming an international standard in this field;

- an interface, enabling other programs to access the data;

- automatic back-up, checking, and validation for duplicates and for internal logical consistency of the data;

- a data dictionary, a centralized repository of information about the data, such as its meaning, relationship to other data, usage, and format; a dictionary also may handle data conversion (e.g. free text to codes or vice-versa, see later), compaction, validation, access, and definition of logical records and files.

Hierarchical, Network, or Relational?

Selecting or developing a DBMS involves a selection of database *architecture*: namely, the basic way in which data elements are arranged and linked within the database structure. Three main architectures are available (Date, 1981):

(1) *Hierarchical* (tree) databases were the earliest developed. Data are arranged in a tree structure (cf. dendrogram from cluster analysis), in which no element can have more than one parent, but each element can have a number of daughter elements at a lower level. Hierarchical databases have two major disadvantages: (a) in order to reach information on a different branch, the system must move to the root of the tree and back along the other branch, which is cumbersome; (b) if the data structure has to be changed, the entire database has to be rebuilt. However, the fixed structure indicates that retrieval is particularly fast, so that if the database is likely to be stable for a long period of time, this type of structure is worth consideration.

(2) *Network* (plex) databases differ from hierarchical structures in that any item can be linked to any other: a "daughter" can have more than one "parent." Many successful geologic databases have been implemented on this model.

(3) *Relational* structures are conceptually the simplest and nowadays perhaps the most popular, particularly in the geosciences (Henderson, 1986; Siegenthaler, 1986). They hold the data in one or more, two-dimensional tables which then are linked together. Critical in relational database structure are the concepts of *primary keys* (links between information in different relations), and *third normal form* (which determines whether the database is set up in the best structure for efficient updating, changing, etc.). Relational databases do not require data relationships to be defined rigidly from the outset. This flexibility has associated penalties, in that some data redundancy may occur and slower access times may result. However, relational databases probably provide the best method for storage of geologic data where:

- data relationships are not well understood or fixed at the time data are entered into the database;

- use of the data may change through the useful life of the database;

- certain well-defined queries are likely to be made frequently;

- access generally will be by geologists rather than by professional computer scientists or programmers;

- speed of access to large volumes of information contained in the database is not critical;

- the database is expected to have a longer lifespan than the computer system on which it is implemented, so that it may have to be moved onto newer systems in the future.

For MERIGOLD, a relational structure clearly was most advisable because most of the described conditions apply.

Buy or Write?

The user in the late 1980s basically has two paths in selecting a DBMS: (1) purchase one of the increasing number of proprietary, commercial packages, for example, ORACLE, or (2) write a "home-grown" DBMS for the particular database in hand. The theoretical advantages of writing "home-grown" database software are: (i) the software remains fully adaptable to the specific purpose; and (ii) experts in the software usually are at hand to rectify any problems. The advantages of using a proprietry DBMS are: (1) the cost to a final product usually is far less, both in time and money; (2) there is generally a far wider user community from which to learn. Considerations such as "user friendliness" and "good user support", do not necessarily weigh in favor of either path. For example, support in theory should be best for a "home-grown" package designed by local staff, but should those individuals leave the organization (an extremely usual occurrence!) support can disappear almost overnight. Support for commercial packages also differs considerably, not only from package-to-package but also from country-to-country for the same package: some DBMS are supported by large commercial organizations with national offices (e.g. ORACLE), whereas obtaining support in Australia for some other packages developed in and distributed from the USA, can be virtually impossible. If "bugs" (all too abundant even in expensive commercial packages) are not eliminated promptly, or updates and newsletters distributed efficiently, a commercial package may prove virtually useless. The size of the user community for a particular package also can contribute to the quality of support: the manufacturer's and distributor's whereabouts and efficiency become less important if there is a large and widely distributed network of intercommunicating users who can help each other. Such a network, of course, is lacking for a "home-grown" product.

On the whole, geologic experience with "home grown" DBMS has been rather salutory: the amounts of time and effort required to generate a working system have been underestimated vastly, and packages have been developed by computer professionals with little input from geologist end-users. As a result, final products all too frequently have turned out either: (a) not to do what geologists want; or (b) to be such nightmares to operate that they give up trying. Some notable examples were abandoned for these and similar reasons after a relatively short useful lifespan (e.g. G-EXEC in the UK: Jeffrey and Gill, 1976). Unfortunately, this same lesson is being relearned in Australia at the present time, where many large organizations are developing their own DBMS (Young, 1985).

In practice, it is better to work around the limitations of a commercial DBMS, or take advantage of a particular feature in order to enhance overall performance. On balance, a commercial DBMS probably is the only effective option for 99% of geologic projects nowadays, and this path was selected quickly for MERIGOLD.

Which Package?

Although this is not the place for a thorough review of commercial DBMS software, a few comments from our detailed examination of several packages are warranted, because reviews in some computer magazines seem to be neither impartial nor thorough (Rock, Brown, and Hattie, 1988). Several well tried and widely distributed commercial DBMS packages are available now, such as ORACLE and MIMER. ORACLE runs on a wide variety of mainframe computers (VAX, IBM, etc.), and although it is a powerful and excellent DBMS, it was ruled out for MERIGOLD by: (a) purchase costs ($\approx$ A\$80,000); (b) hardware considerations (e.g. ongoing computing charges); (c) the complexity of the package itself; and (d) other local considerations (e.g. the concentration of personnel expertise in microcomputers). The microcomputer (IBM-PC) version of ORACLE also was inspected, but proved for present purposes to sacrifice too many of the speed and power advantages of mainframe ORACLE without any concomitant advantages in user-friendliness.

Several microcomputer DBMS were considered as alternatives to ORACLE: dBase III+ (for the IBM PC); 4th Dimension, dBase Mac, and Omnis 3+ (for the Macintosh). The ideal DBMS was required: (1) to implement a standard relational architecture; (2) to interact with external (statistical and graphics) programs, so as to allow manipulation, statistical interpretation and plotting of the gold data. Other major considerations were mentioned in the previous section. Purchase cost differences among these various packages were sufficiently small to be of little concern. Speed also was of little importance: in fact, each of these DBMS is rather slow, Omnis 3+ being fastest. Both 4th Dimension and dBase Mac have promised speedier upgrades, although this may remain an area of inferior performance (relative to, say, mainframe ORACLE) for some time.

As regards a relational requirement, Codd (1982), the original designer of this architecture, has proposed 12 rules for determining how "relational" a DBMS product really is. Although each of the microcomputer DBMS considered in fact fails all of these rules (despite laying claim to being relational), this actually is not unusual, because performance optimization has resulted in many individual microcomputer products targeted at particular application areas. One of the more important failures is that these DBMS all use what are termed *pointer fields* to maintain links between tables (Fig. 2). Efficient, simultaneous data extraction from several tables requires that these pointer fields have been filled out in advance (generally at the data-entry stage). This results in data structures that are less flexible than the "ideal" relational database, because joins cannot be made and broken without large penalties in recomputing links — especially in large databases. The use of pointer fields also results in asymmetrical joins, because the fields are stored in just one of the two linked tables, so that joining table A $\rightarrow$ B is different from B $\rightarrow$ A. Retrievals therefore have to take into account the direction of the join, possibly using different search strategies depending on whether a particular variable is in the linked or linking table.

In *dBase Mac*, the pointer fields are declared explicitly as part of the table definition. They are multivalued fields (taking zero or more values), so 1:*N*, *N*:1, and *N*:*N* links can all be represented. Two-directional links also are possible by joining table A to table B *and* table B to table A. This results in sets of pointers being stored in both tables. *Omnis 3+* employs yet another linking strategy in which tables are built hierarchically with one table designated as the parent and the other as a child. The data structure is rigid, with links established as data are entered. "Ad hoc" joins of data tables are not permitted. This system has little resemblance to the relational model, being closer to a hierachical DBMS in design. In *4th Dimension*, the pointer fields are invisible, with linkages being based on a common variable present in both tables. As the pointer fields are single-valued, only *N*:1 relationships can be represented, not 1:*N* or *N*:*N*. Maintenance of the pointer fields largely is automatic, and the mechanism of table linkages is well hidden at the user level. Of all the Macintosh DBMS considered this one seems the most "relational" in use, although knowledge of the underlying mechanism of linkages becomes important when using the procedural programming language.

In the end, 4th Dimension was selected as the most suitable DBMS for MERIGOLD, not only because of its *relatively* relational form, but also because: (a) it *consistently* has beenwell reviewed in the computer literature (e.g. Stewart, 1988); (b) it is supported fully by Apple Computer, who have so far provided good quality local support; (c) its user community is already large, international, and growing rapidly; (d) it is powerful yet exceptionally user-friendly (students, and geologists even of limited computer literacy, can master simple retrievals from scratch in under 30 minutes!); (e) it makes maximum use of the excellent graphics capabilities of the Macintosh, so that retrievals based on maps and diagrams as well as purely numerical or textual information are possible: a major consideration in geologic databases.

EXISTING COMPUTERIZED DATABASES ON GOLD DEPOSITS

There are several existing gold databases, including two in North America and two in Australia. Given the lack of *general* advice in the geologic literature on setting up databases, MERIGOLD tried to learn as many lessons as possible from these specific databases concerned with gold. The two existing Australian systems were implemented in completely different ways on incompatible media, so MERIGOLD could never hope to interface with both of them, but it was hoped to remain reasonable compatible with one or other. Brief details of their rationale therefore follow.

The Ontario Geological Survey Database

This covers the economic and geologic characteristics, discovery date, discovery method, and extent of exploration of 725 known gold deposits in the Timmins–Kirkland Lake area of Ontario, Canada (Hodgson, 1983). The data were obtained from the published literature and, to a lesser extent, records of the Ontario Ministry of Natural Resources. The objective of the compilation was to answer such questions as: (1) what are the geologic differences, if any, between economically richer and economically poorer deposits? (2) is it possible to define anomalously mineralized areas independently of "exploration intensity" or "percent exposure", using parameters such as the ratio of economically richer to economically poorer deposits, the incidence of certain assemblages of rocks or ore-associated minerals, or the incidence of alteration types in the gold prospects? (3) how important in geochemical exploration is the association of gold with such minerals as tourmaline, scheelite, and arsenopyrite? (4) are there any as yet unrecognized spatial relationships between gold mineralization and particular rock types?

Statistical techniques were used to examine relationships between lithological and mineralogical associations of gold and to relate these to the economic grade (Hodgson, Troops, and Stewart, 1986). This allows for the possibility of making exploration decisions, for example, assigning budgets, on a probabilistic basis, and using quantitative methods for resolving such questions as: how much money should be spent to acquire prospect A, relative to prospect B, given the known differences between them?

The Eastern Washington University Database: GOLDY

This is a dBASE III+ compilation of summary information on 112 North American mining camps with production and reserves>1,000,000 troy ounces gold (Mihalasky, and others, 1987). It includes data on location, geologic age, deposit type, production and reserve figures, ore grades, and tonnages, plus references.

The Australian Bureau of Mineral Resources (BMR) Database: MINDEP

BMR is Australia's federal geological survey organization. A joint project between the BMR divisions of Petrology & Geochemistry and Resource Assessment commenced in 1986 with the object of compiling published information into a MINDEP (MINeral DEPosits) database on all Australian gold deposits. The BMR view MINDEP as the first stage in producing a more comprehensive system for Australian gold deposits to aid mineral exploration and geoscientific research. MINDEP itself is composed of nearly 100 relations, implemented with ORACLE on IBM PC, Data General, and VAX computers. Its contents have been made available to the public as a published report (Mock and others, 1987), together with the actual data as: (1) hard copy; (2) ORACLE format 5.25" floppy disks; and (3) ASCII files on 5.25" floppy disks. Format (3) does not contain ASCII dumps of MINDEP's contents, but ASCII files of *ORACLE reports* generated from MINDEP. This somewhat limits their usefulness for those not possessing ORACLE itself.

MINDEP includes the following descriptive data for 80 Western Australian gold deposits: name of deposit and (numerous) synonyms, names of orebodies within the deposit, operational status in 1987, descriptions of the regional setting, locality, geology and characteristics of the deposits, associated igneous rocks and their host rocks, development history (including year of discovery and mining methods), resource and production data, a brief summary of proposed genetic models, and a selected bibliography. With few exceptions, most of these fields carry free text descriptions, and there is little numerical information in MINDEP which can be subjected to any form of processing (other than simple printouts).

The Australian Mining Industry Council (AMIC) System

Although only in its development stage, this uses a Macintosh with a commercial "graphic file management system", Business Filevision, which has been adopted widely by some large US companies (e.g. Boeing, DuPont, Ford, Sears, Westinghouse). Filevision is not a relational DBMS, and is referred to by its manufacturer as a "hypergraphics" package which "uses graphics as file directories." It however does resemble a simple Geographical Information System (GIS). GIS is a recent techological advance combining a conventional database with a graphics package: that is, spatially distributed data can be analyzed in terms of their coordinates as well as their intrinsic geologic variables (Rock, 1988). Such systems are becoming a powerful force, and are used widely in, for example, the US Geological Survey. Hence, the AMIC system potentially was of considerable interest.

The AMIC system operates as follows: a location map showing five mineral regions of Australia is presented on screen. A scanned image of any region then can be obtained on screen by selecting that region with the Macintosh mouse. In another option, the system presents a map of Australia with dots for the locations of 343 mines. To obtain information about any one mine, the user selects its dot, whereupon a screen of simple information (names of owners, address, deposit type, etc.) is presented. The difference between the AMIC and MINDEP systems is the graphic interface of the former: in MINDEP, the author retrieves information for, say, the Kambalda goldfield by typing the word

"Kambalda" into the system; in the AMIC database, he merely selects the correct dot on the map of Australia.

Existing Databases Versus MERIGOLD

The Ontario database is the closest in both content and intention to MERIGOLD. Although little has been published about its structure or implementation, many of its aspects have been followed in MERIGOLD: for example, the coding of lithologies, and the range of mineralogical and rock-type fields included (Fig. 2).

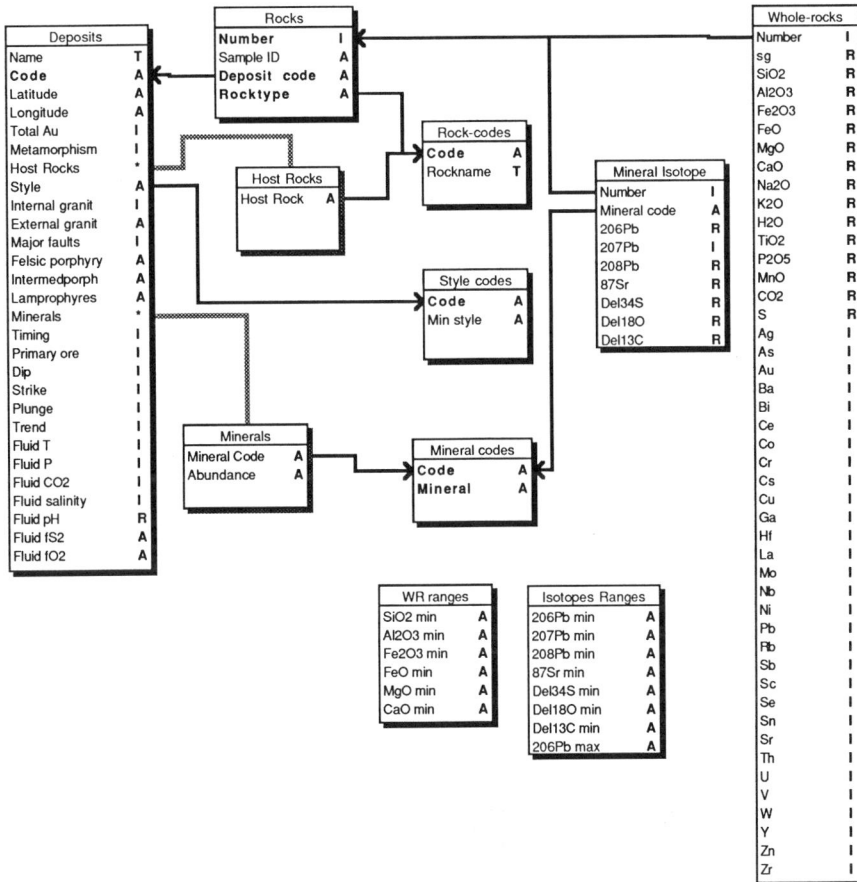

Figure 2. Relational structure of MERIGOLD in DBMS *4th Dimension*. I,R,A,T respectively indicate integer, real, coded, and free text variables, whereas * indicates subfile. Abbreviated names in most situations are dictated by 4th Dimension's naming conventions (e.g. 206Pb/204Pb cannot be used because of "/" character, and so is abbreviated to "206Pb"). Subfile links shown by grey line between two tables take advantage of feature (borrowed by 4th Dimension from hierarchical database structures) for linking any number of subrecords to single record in table. Subfiles could be replaced by relationally linked tables, but result in faster retrievals if number of linked subrecords is small. Contents of most fields and tables are illustrated in Figures 3 and 4 and should be self-explanatory.

Figure 2. (cont.) Following comments therefore elaborate only on less inherently obvious
            fields.

Deposits table

Code. See examples of 4-letter hierarchical codes in Figure 4 and text for further
            explanation.

Total Au . includes both past production and defined reserves of deposit.

Metamorphism. Simple ordinal representation from 1 = lower-greenschist facies
            to 5 = mid-amphibolite facies.

Host-rocks. This subfile records which rock-types have been observed hosting
            mineralization. Use of subfile feature allows number of different hostrocks
            to be specified. Without subfiles, for example, if the maximum number of
            "host-rocks" recorded for all deposits was, say, 6, then 6 "host-rock" fields
            would be required in deposits (or separate) table, but most of these would be
            redundant for other deposits with < 6 recorded host-rocks. Furthermore, if
            another deposit was added subsequently which had 7 recorded host-rocks,
            database would have to be rebuilt.

Style. See Figure 4 for explanation of relevant codes.

            internal/external granite. Records distances between gold deposit and
            internal/external granitoid intrusions (that is, granitoid intrusions within
            or outside greenstone belts).

Major faults. Similarly records distances between deposit and major fault or
            shear systems.

Felsic porphyry, intermed porph, lamprophyres. Records presence of
            corresponding rocktypes as: *abundant, scarce,* or *not seen;* this last is
            different from "no information" (recorded as usual with bullet).

Minerals. This subfile records occurrence of ore and alteration minerals, again as
            abundant, scarce, or *not seen.* See comments on subfiles.

Timing. Records timing of mineralization relative to metamorphism as *prograde,
            retrograde,* or *uncertain.*

Primary ore. Recorded as *free milling* or *refractory.*

Dip to *trend.* These refer to major lode. In due course, more extensive and detailed
            structural information will be incorporated systematically, probably as
            subfile.

Fluid T to *Fluid* fO_2. Record thermodynamic parameters for ore fluid determined
            by fluid-inclusion studies, etc.

Rocks table

sg. Specific gravity of rock example (important in Gresens' type calculations of
            alteration effects).

Ranges tables

These contain acceptable maximum and minimum values for all variables in
            whole-rock and mineral isotope tables, used in data validation procedures.

Interface with MINDEP was our original intention, but use of ORACLE and mainframe
computers was ruled out for reasons discussed earlier, and the AMIC Macintosh-based
system was more suited to our own computing environment. However, Business Filevision
was not adopted because it is not a full-featured DBMS. For example: (1) it cannot handle
relational links; (2) hierarchical attributes such as deposits within goldfields (discussed
later) cannot be defined; (3) the handling of numerical data is cumbersome; (4) display
formats are limited; (5) validation procedures cannot be written; and (5) exporting data to
other (e.g. statistical) packages is difficult.

Overall, the objectives and data contents of the two Australian databases proved to be so
dissimilar to each other and to MERIGOLD that a different system was warranted, but as
much consistency as possible nevertheless was ensured: for example, we have followed
exactly the same classification system for grouping Australian gold deposits into

goldfields, provinces, and blocks as in MINDEP, and reproduced the BMR figures exactly for the few fields which occur in both MERIGOLD and MINDEP.

CAPTURING THE DATA FOR YILGARN GOLD DEPOSITS

The principal steps in building a datafile, according to Gordon and Martin (1974) are: (1) file definition and data capture, (2) data entry, (3) data validation, and (4) updating of the data. This section considers (1), and later sections the remaining steps. Methods of capturing data for igneous rocks have been detailed by Le Maitre (1982) and Rock (1986), and for mineral deposits by Longe and others (1978). The present section therefore details mainly with aspects specific to MERIGOLD, while merely outlining aspects already dealt with in these publications.

Sources

Data for MERIGOLD were obtained from published literature, unpublished theses, company reports, and personal communication with colleagues. A questionnaire was sent to Australian mining companies requesting information, and data on 84 deposits from 54 companies have so far been returned. It must be emphasized that data obtained from mining companies, in many situations, are incomplete. For example, although a particular mineral may be present in an ore environment, it may not have been recognized and therefore will not be recorded in MERIGOLD.

Relevant data were extracted from ASCII report files of the BMR MINDEP database (see previous section). The BMR were not able to supply a complete ASCII dump of MINDEP, although they kindly supplied hard-copies for ASCII dumps of 12 of the 90-odd relations of which MINDEP is composed.

Although several hundred mostly unpublished reports containing gold assay data are available in Australia each year — a prodigious growth over earlier decades — much of the data is either confidential, or only semiquantitative. Retrospective capture therefore can never be complete, but one good guide is the rate of acquisition of earlier data from references in other papers, which slows down dramatically as the limit of available information is approached. This has happened already with MERIGOLD, so that we are fairly confident that the database already contains the vast majority of fully quantitative geochemical and isotopic data generated to date on Yilgarn gold deposits.

Errors

Published geochemical data include an astonishing number of errors, as well as data of doubtful quality. Screening procedures therefore have been erected to handle some of the more important problems, outlined here:

(1) *Discrepancies.* Perhaps the most grotesque problem here is the discrepancy for some 13% of all igneous rock (major element) analyses between quoted totals and sums of the published figures (Le Maitre, 1982, p.216). This problem is acute equally with gold-related data, and a decision usually has to be made between accepting the individual oxide values (rather than the total), or, in situation of extreme discrepancy, rejecting the analysis altogether.

(2) *Poor totals.* An astonishing number of published analyses have totals outside normally accepted limits (99-101%), even where discrepancies of type (1) are absent. This problem is increasing inexorably, as more-and-more published analyses done by machine-based methods (e.g. XRF) are reported either *incomplete* (i.e. without any figures at all for H_2O, CO_2, or LOI), or *volatile free*

(i.e. a separate value for H_2O, CO_2, or LOI is quoted on an analysis which otherwise totals $\approx$ 100%, so that the remaining figures have to be scaled down commensurately to get the actual rock values).

(3) *Inconsistencies between mode and norm.* Blatant examples are not uncommon where, for example, a rock stated to contain substantial amounts of modal feldspathoid in fact shows high normative *hy* $\pm$ *qz* (or vice versa); where such inconsistencies cannot be put down to alteration, they either imply poor analytical data or mismatch between rock description and analysis; in either situation, such data must be eschewed rigorously.

(4) *Miscellaneous errors.* Typographical errors also are abundant, and are most blatant where they yield inherently unlikely values for trace elements (e.g. Nb values of 5,000 ppm).

(5) *Ambiguities.* The ambiguity of "nd" ("not detected" or "not determined"?) is discussed next. Another regrettable ambiguity arises where rock or mineral data are quoted either without any sample identification at all (a usual circumstance), or with identification in some tables but not others. It then becomes difficult or impossible to decide which trace-element or isotope analyses go with which major-element analyses; that is, the links in Figure 2 cannot be established unambiguously. In such situations, it may be possible to deduce the correct links from the context; otherwise "dummy" records must be introduced in the *rocks* table, one for each set of unattributable data.

(6) *Duplications.* The repetition of the same analysis in several publications is general (again, may contain introduced errors!). This has been dealt with by 4th Dimension procedures which sort the data on 2-3 variables, and then compare adjacent records: identical figures for SiO_2, Al_2O_3, *and* MgO almost always indicate a pair of analyses is duplicated. Another problem, not unfortunately open to computer checking, is the publication of new analytical data — usually by a completely different set of authors and occasionally many years later — for samples already analyzed in an earlier paper. If such links are not recognized, superfluous dummy entries in the rock tables are created instead of informative links between the various tables. On the other hand, the number of deposits covered in MERIGOLD is restricted enough that it more often than not becomes obvious when the same samples are being studied by different groups of workers.

ORGANIZATION OF MERIGOLD

Figure 2 shows the structure for MERIGOLD, Figures 3 and 4 display short extracts from the tables of which it is composed, and Table 1 summarizes the major contents of these tables. Some of the more important considerations in setting up the database are dealt with in the following sections.

Structure

MERIGOLD is organized into 11 tables (Fig. 2). The pivotal table, *Rocks* (Figs. 2 and 3), contains a record for every rock specimen for which data are available in other tables. Each rock has a unique sequential identification number (from 1 to, currently, 1035), which serves as the primary key, linking tables together relationally (Fig. 2). The two *subfiles* (*Host Rocks* and *Minerals*) are a useful feature of 4th Dimension which facilitates variable 1:*N* relationships. A separate listing of references is being maintained on the Macintosh but, as mentioned, larger bibliographic listings on gold are available via AESIS (1983, 1986, 1987) so these are not incorporated into MERIGOLD itself.

Rocks table			
No.	Sam. I.D	Deposit	Rock type
1	94104	YNKH	IUFMBA
2	94138	YSFR	IUFMBA
3	94137	YMRA	IUFUKO

Mineral isotopes table					
No.	Min.	206Pb/204Pb	207Pb/204Pb	208Pb/204Pb	87Rb/86Sr
88	SG	14.009	15.085	33.646	•
89	SP	14.003	14.834	33.509	•
90	SG	13.995	15.091	33.702	•

Whole-rocks table (part)							
No.	sg	SiO2	Al2O3	Fe2O3	MgO	CaO	Na2O	K2O
55	2.68	66.510	15.720	2.950	.820	2.860	3.960	3.900
56	2.90	57.520	15.380	8.660	4.780	10.320	1.700	.550
57	2.86	58.880	15.420	7.520	4.450	7.270	2.550	.880

Whole-rocks table (part)										
Ce	Co	Cr	Cs	Cu	Ga	Hf	La	Mo	Nb	Ni
138	3	•	6	•	•	•	54	•	6	13
32	50	•	•	79	•	•	10	2	5	138
37	24	•	1	123	•	•	•	2	4	116

Figure 3. Short extracts from four tables which contain actual geologic and geochemical data in MERIGOLD. Some extracts show only subset of incorporated variables (cf. Table 1) — particularly that shown from *Deposits* table, which also contains other information relevant to gold exploration (Fig. 2).

The *Rocks* table is linked via the primary key (Fig. 2) to two other tables containing geochemical data (one for whole-rocks, one for minerals), and to the *Deposits* table which encodes the geology of each deposit. One entry in the Rocks table may have none, one or, occasionally (where there are duplicate analyses of one specimen), 2-3 entries in the Whole-Rock table, but anything from none to dozens of entries in the Minerals table (i.e. where many grains have been analyzed from that one specimen). The amount of data available for single rock specimens thus ranges greatly, from a minimum of, say, one ∂^{34}S isotopic analysis on containing pyrite, to a panoply of whole-rock plus isotope data.

The number of individual data values for the 44 whole-rock variables also at present differs greatly (Table 1). However, grouping all these data together into a single table greatly simplifies retrievals in 4th Dimension, and does not have a substantial penalty in wasted storage space or efficiency, given the small overall size of the table: for example, there are only some 47 whole-rock records which have major element but no trace-element data, so there is little point in splitting into separate "majors" and "traces" tables. The same applies to the isotopes table, even though most mineral grains have been analyzed *either* for Pb, Sr, *or* stable isotopes, but not usually for all of these.

Province codes	
Code	Province
YM	Murchison Province
YS	Southern Cross Province
YN	Norseman-Wiluna Belt

Deposit codes		Style codes (mineralization)	
Code	Deposit	Code	Mineralization style
YMRR	Rand	S	Shear zone
YNKH	Hunt	Q	Quartz stockworks
YNKV	Victory	L	Large quartz vein

Mineral codes		Host rock codes	
Code	Mineral	Code	Host rock
SP	Pyrite	IVFMBA	Basalt
SG	Galena	IVCFGR	Granite
CC	Calcite	IVPFPO	Felsic Porphyry
CD	Dolomite	S*F*BF	Banded Iron-Formation

Figure 4. Examples of codes used for categorical variables. *Style, rock,* and *mineral* codes form their own tables, but *deposit* codes are incorporated as field in *Deposits* table (Fig. 2). Eventually, these codes will be set up so that user can retrieve information without having to know code first (cf. Fig. 7).

Table 1. Summary contents of MERIGOLD as of October 1988.

	Total records	No. of variables	No. of values per variable within table ranges from
Total size of database (Mb)	≈1	—	—
No. of tables	11	—	—
Deposits table	170	26	Differs because of subfile structure
Rocks table	1035	4	All variables mandatory for all records
Whole-rocks table	248	46	4 (Bi) to 248 (SiO_2)
Mineral isotopes table	787	9	13 (Sr) to 549 (O)

Hierarchical Codes Versus Free Text

Geologic variables in MERIGOLD, as with many other geologic data (Cheeney, 1983), fall into 3 groups: *ratio* variables (continuous, real numbers, such as major element percentages and isotope ratios), *ordinal* variables (ranked orderings), and *nominal* variables (limited numbers of categories). For example, "rock type" is a nominal variable with > 25 categories (e.g. those shown in Fig. 3), which can be represented either as free text (i.e. the actual words "komatiite", etc.), or as some form of code (e.g. K = komatiite, P = porphyry....) Codes A,B,C... are preferable to 1,2,3...., because they do not imply 2 > 1 and similar meaningless relationships: komatiite is in no sense "greater than" porphyry. Mineral phase is a similar categorical variable. Names of gold deposits constitute a third such variable, this time with hundreds rather than just 20 values. On the other hand, metamorphism can best be represented ordinally by simple integers, representing ranks, which indicate that rank "5" (upper amphibolite) exceeds rank "1" (lower greenschist), but not by a ratio of exactly 5 (cf. Moh's scale of hardness, where diamond is *not* simply 10 times harder than talc).

In our view and that of Le Maitre (1982, p.215, etc.), *codes* are preferable nearly always to free text for representing categories in geologic databases. Among cogent reasons for preferring codes here are the following:

(1) *Mechanical errors.* Free text tends to generate more input errors, merely because many more characters have to be typed. Simple coded abbreviations can alleviate this problem, and also help to ensure greater consistency of usage.

(2) *Synonyms, etc.* As anyone who has done a search on one of the large bibliographical databases will know, free text may be ambiguous and inefficient, because of *plurals* ("ore body" may not retrieve "ore bodies" in some systems), variant *capitalization* ("Gold" v. "gold"), *hyphenation* ("ore-body" v. "ore body"), *spelling* ("baryte" v. "barite" or "barytes"), *synonyms* ("idocrase" v. "vesuvianite"), *homonyms* ("spessartite" meaning both a lamprophyre and a variety of garnet), and many other factors. The number of such variants in geology is so large that vast thesauri become essential for efficient free-text retrieval, which tell the computer exact relationships between different terms (e.g. Charles, 1979; AMF, 1987). These can make a microcomputer system totally unmanageable, even though it remains only moderately efficient. Even a thesaurus does not cope with homonyms (a free-text search on "spessartite" will retrieve *inevitably* any irrelevant information on both lamprophyres and garnets!). By contrast, use of simple codes (as in Fig. 3) ensures that all synonyms and spellings variants are given one *unique* code, whereas homonyms receive their several different meanings, so most of the ambiguities are overcome immediately.

(3) *Hierarchies.* Perhaps even more importantly, codes provide the only simple methods of structuring retrievals, or taking account of the inherently hierarchical nature of many geologic variables. Mineral groups and series in MERIGOLD are simple hierarchies: "albite" is a subset of "plagioclase", and "calcite" of "carbonates" but, once again, this is not apparent to a computer from the mere words themselves. Free-text searches on "plagioclase" or "carbonate" therefore probably will overlook data for "albite" and "calcite", unless the user either specifies *all* possible constituent minerals of the group or series individually, or employs a thesaurus which tells the computer that "carbonate" = calcite + dolomite + ... , or "plagioclase" = albite + ... By contrast, the simple codes "CC" and "CD" (Fig.4) immediately give the hierarchical relationships, the two letters corresponding to each level of the hierarchy, while also showing that members of the "carbonate" group are related more closely to each other that members of the "sulphide" group. A retrieval on "carbonate" can be performed using "C@", where @ is the system "wild card", and a retrieval on "calcite" using the full two-letter code "CC."

Exactly the same arguments apply to rock-types, where the top level of the hierarchy is "igneous" versus "sedimentary" and "metamorphic", the second level under "igneous" is "intrusive" versus "volcanic", a third level encompasses "acid", "intermediate", "basic", and individual rock-types constitute a fourth level. The hierarchical codes in MERIGOLD have been devised to include the best features of the two somewhat different systems adopted by Le Maitre (1982) in CLAIR and by Hodgson (1983) in his gold database. For example, "IV" in Figure 4 stands for "Igneous Volcanic", "S*" for "Sedimentary" (the asterisk indicating the lack of a second level of the hierarchy in this example). In this way, retrievals can be made once again at any appropriate level of the hierarchy. Equally importantly, codes can be entered in MERIGOLD which truly reflect the accuracy of available information: rocks described merely as "igneous" or as "ultrabasic extrusive" can be coded less precisely than those described as "komatiite".

A third crucial hierarchy in MERIGOLD, with four levels, categorizes the geography of gold occurrences (Fig. 1). Many individual *deposits* of similar age in one general area form natural groupings, generally termed *goldfields* (in Australia) or *gold camps* (in Canada). Several of these *goldfields* in turn cluster within *belts* or *provinces*, and the provinces themselves group into *cratons* or *blocks*. For example, Victory is one *deposit* of the Kambalda *goldfield*, which lies within the Norseman-Wiluna *belt* of the Eastern Goldfields *province*, within the Yilgarn *block* (Fig. 1). Again, using free text, the geologic/spatial relationship between "Victory", "Hunt", and "Rand" cannot be specified, but using 4-letter hierarchical codes "YNKV", "YNKH", and "YMRR" (Fig. 4), the information is encapsulated that Victory and Hunt are deposits closely related within the same goldfield (Kambalda = YNK), whereas Rand belongs to a different province (Murchison = YM). As before, retrievals can be made at any level of the hierarchy by using the appropriate number of letters: block (e.g. Y@@@), belt/province (YN@@), goldfield (YNK@), or individual deposit (YNKH), where @ is once again the system "wild card." At present, the first letter "Y" applies to *all* deposits in MERIGOLD, but is included to allow for future expansion of the database to cover the Pilbara and other cratons.

(4) *Space.* It is obvious from the previous discussion that codes, being shorter, occupy correspondingly less storage space than free-text equivalents. Although this may not nowadays be a major consideration on mainframe systems, it can be critical for databases reaching towards the storage capacity of microcomputers, where many free-text variables for thousands of records can add up to a considerable space and speed penalty.

The Problem of Missing Data

Missing data are a problem in practically *all* geologic source databases. There are two distinct types, between which any useful database must *always* distinguish: (1) *not determined* values, which are truly missing, with no information value whatever; and (2) *not detected* values (i.e. below machine detection limits), which have a real, if limited, information content. Unfortunately, the manner in which values below detection limits maybe indicated in the literature, for example, "< 5", cannot be used in a database, because "<" is a character, not a number. MERIGOLD uses the Macintosh convention of a special symbol "•" (bullet character) for *not determined*, whereas "0" (zero) is used for *not detected*. Furthermore, all too many references abbreviate both forms to "nd", while many others use "0" for not determined data. Only the dash (−) generally is unambiguous in published papers. Where such ambiguities do arise, values have been interpreted as "•" except where it is reasonably clear from the context that "0" is intended.

Many mainframe databases (such as ORACLE) handle missing data internally. In 4th Dimension, however, all numeric fields in a new record, *and* all bullets, are initialized to zero, because the program removes all nonnumeric characters from character strings before converting them to a number. MERIGOLD therefore requires a special import procedure which traps missing values. This procedure reads the data as a character string, and then parses the separate fields out of the string using the standard Macintosh <tab> characters as delimiter. Each field is checked for equivalence to the missing data bullet before converting to a number. Where the bullet is encountered, a numeric code of −99 is substituted, because *no* variables in MERIGOLD can take such a value. 4th Dimension has extensive formatting capabilities for number display, with separate formats specified for positive, negative, and zero numbers for each field in an output display. Hence the database has been set up to display –99 as a bullet. When outputting data to a ASCII file the same conversion is used, with –99 again replaced by the bullet. This can be altered to any desired code if the data file is to be processed by a program that uses different conventions for handling missing data.

Data Validation

Integrity of data is a vital aspect of any database, usually estimated to take up to 80% of the effort in erecting the system. Because there are all too many errors present in published data, as described in the next section, and further errors are introduced inevitably during input into a database, rigorous quality-control procedures have been established to maintain integrity in MERIGOLD. Each analytical record has an additional field for recording whether the data have been validated. As data are added to the database this field is initialized to "no" (= data not validated). At any time thereafter, general and expandable validation procedures can be run which operate on all nonvalidated analyses, and which fall into two parts. First, each data value is checked against allowable minimum and maximum values for the particular variable (e.g. analytical totals are constrained to lie between 98 and 102%), the permissible ranges being stored within the database for ease of modification (Fig. 2). Second, the validation procedure checks that all codes in the record exist in one of the code tables (Fig. 4), and performs any other specialized checks that need to be done. All analyses which fail validation are displayed through special layouts (Fig. 5) which specify the reason for failure. The data then can be checked for errors and corrected, rejected, or accepted by manual overriding of the validation procedure (for peculiar, but valid, data). The availability of a high-level database language in 4th Dimension allows validation procedures complex enough to trap most likely (and many unlikely!) errors.

SIMPLE DATABASE OPERATIONS

Entering Data into MERIGOLD

There are two alternatives here: (1) *direct entry* , in which the user types the appropriate values into windows such as Figure 6; (2) *import* from data files already created in other Macintosh applications.

Retrieving and Exporting Data

Procedures are being written as required to handle specialized retrievals. In each situation, 4th Dimension presents a menu-driven search editor (Fig. 7), from which the user selects appropriate search criteria. Figure 7A retrieves all volcanic rocks from deposits in the Murchison province. Strings of Boolean commands can be strung together quickly and easily, because the user does not have to master a complex query language as in many other DBMS. After such a search, a window such as Figure 7B is displayed, showing the number of retrieved items against the total searched (which may be a subset of the complete database: successive searches easily can be made to whittle down selections). The user can select one of the buttons on the right to call up the corresponding data for these retrieved records. For example, selecting the *Whole-rocks* button would here call up the 17 retrieved whole-rock analyses for the 19 retrieved rocks. Selecting the *Isotope* button, however, would yield nothing, because there are no such data for these particular rock specimens.

Instead of merely listing retrieved data on the Macintosh screen, MERIGOLD can *export* any of the types of data — for particular rocks or indeed for the entire database — to new ASCII files (Fig. 8). Such files can be imported quickly and easily into other Macintosh applications for specialized processing (e.g. statistical analysis, geochemical plotting), or

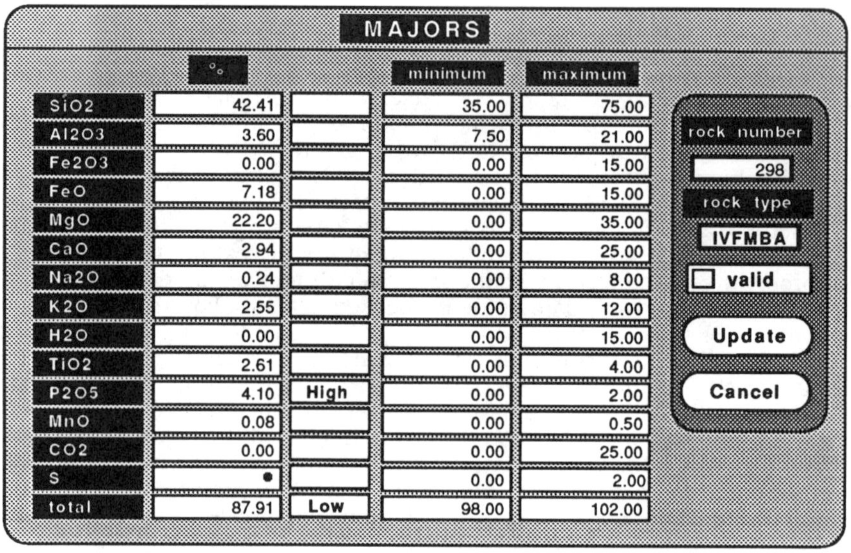

MAJORS

	%		minimum	maximum
SiO2	42.41		35.00	75.00
Al2O3	3.60		7.50	21.00
Fe2O3	0.00		0.00	15.00
FeO	7.18		0.00	15.00
MgO	22.20		0.00	35.00
CaO	2.94		0.00	25.00
Na2O	0.24		0.00	8.00
K2O	2.55		0.00	12.00
H2O	0.00		0.00	15.00
TiO2	2.61		0.00	4.00
P2O5	4.10	High	0.00	2.00
MnO	0.08		0.00	0.50
CO2	0.00		0.00	25.00
S	•		0.00	2.00
total	87.91	Low	98.00	102.00

rock number 298

rock type **IVFMBA**

☐ valid

Update

Cancel

Figure 5. Data validation window. This particular analysis has not passed validation because its P_2O_5 value and total are outside specified ranges (stored in separate tables — Fig. 2). Validation can be overridden manually if, for example, these values in fact are correct. Missing value for S (bullet) is not considered by validation procedure. This window only shows major element part of whole-rock table (Fig. 2), to conserve space.

MAJORS

Rock number	10	Rock type	IVFMPO	Deposit	YNKV
SiO2	54.64	Al2O3	15.42	Fe2O3	5.55
FeO	0.54	MgO	3.76	CaO	6.36
Na2O	4.72	K2O	4.51	LOI	1.17
TiO2	1.26	P2O5	0.98	MnO	0.10
CO2	0.00	S	•		

Done ☒ ok

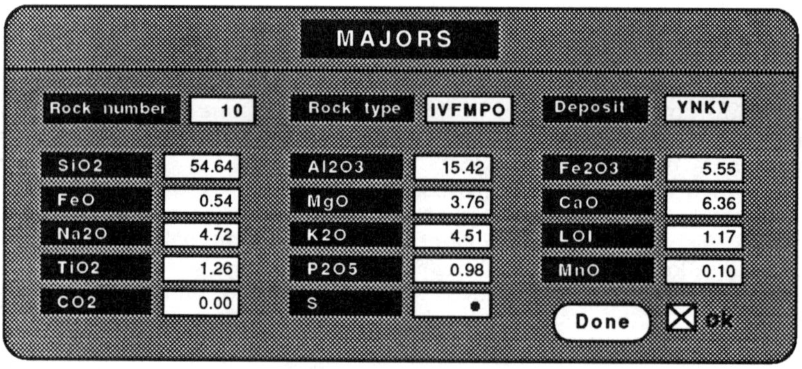

Figure 6. Typical data entry window, for entering data directly into MERIGOLD *majors* table. This window only shows major element part of Whole-Rocks table (Fig. 2), to conserve space.

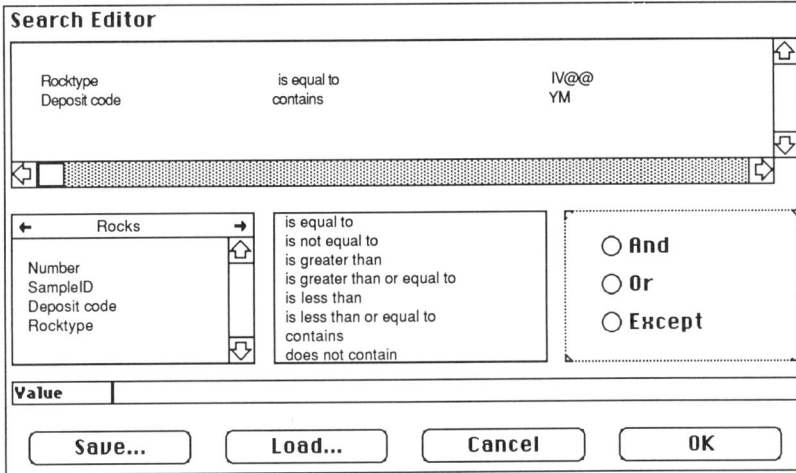

A

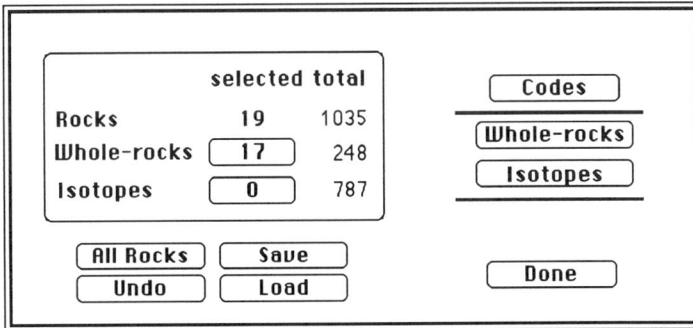

B

Figure 7. Typical retrieval showing search editor (A) and results (B).

indeed ported to other computers (e.g. IBM PC) and read into different DBMS packages running on these machines. The relatively simple structure of MERIGOLD ensures maximum possible portability between these different environments.

The new Multifinder quasimultitasking system on the Macintosh offers tremendous advantages in this context, because it is possible to have 4th Dimension running

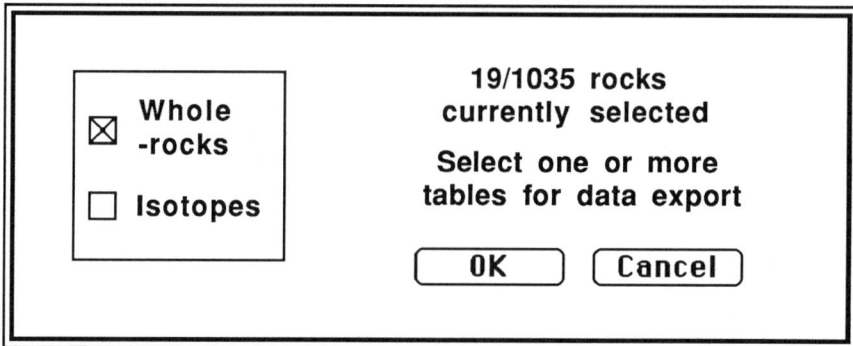

Figure 8. Export window. User selects one or more of tables of data for retrieved set of 19
          rocks; new ASCII files of these data then are created by MERIGOLD. In this
          situation only whole-rocks data are exported, because Figure 7B already shows
          there are no isotope data for particular selection made in Figure 7A.

concurrently with a statistics package, a graphics package, a wordprocessor, and so on,
up to the RAM capacity of the machine. Data thus can be retrieved from 4th Dimension,
processed statistically, the results graphed, and the graph incorporated into a report, all
without returning to the operating system — a quicker and more convenient procedure
than running each package separately.

ACKNOWLEDGMENTS

We gladly acknowledge the support of the following mining companies (names
abbreviated for convenience) through MERIWA (Mineral Exploration Research Institute
of Western Australia): ACM, Asarco, Aztec, Battle Mountain, BHP, Billiton, Carpentaria,
Carr-Boyd, Chevron, Croesus, CSR, Dallhold, Delta, Forrestania, Goldfields, Miralga,
Newmont, Pancontinental, Placer, Reynolds, Stockdale, and West Coast.

REFERENCES

Ackoff, R.L., 1967, Management misinformation systems: Management Science, v. 14,
        p. 147-156.

AESIS (Australian Earth Sciences Information System), 1983, Gold: a supplementary
        list of references from AESIS 1983 (476 references): AESIS Special reference list
        No. 2B/S1. Australian Mineral Foundation, Adelaide.

AESIS, 1986, Gold: a supplementary list of references from AESIS 1985-June 1986 (1558
        references): AESIS Special reference list No. 2B/S3. Australian Mineral
        Foundation, Adelaide.

AESIS, 1987, GOLD: July 1986-July 1987 (788 references): AESIS Special reference list
        No. 2B/54. Australian Mineral Foundation, Adelaide.

AMF (Australian Mineral Foundation), 1987, Australian thesaurus of earth sciences and
        related terms (3rd ed.): Australian Mineral Foundation, Adelaide.

Aubrey, M.C., 1981, The construction of a computerised mineral exploration database, *in* Geoscience numeric and bibliographic data: Australian Mineral Foundation. Seminar Project, no. 154/81, 13 p.

Bliss, J.D., 1986, Management of the life and death of an earth science database: some examples from GEOTHERM: Computers & Geosciences, v. 12, no. 2, p. 199-205.

Burk, C.F., 1981, International review of geoscience source databases, *in* Geoscience numeric and bibliographic data: Australian Mineral Foundation Seminar Project no. 154/81, 20 p.

Charles, R., 1979, Geosaurus: Geosystems' thesaurus of geosciences (3rd ed.): Geosystems, London, 182 p.

Chayes, F., 1982, The rock information system: RKNFSYS: Carn. Inst. Washington Yearbook, v. 82, p. 316-319.

Cheeney, R.F., 1983, Statistical methods in geology for field and laboratory decisions: Allen & Unwin, London, 169 p.

Clark, A.L., 1976, Resource databases — resource assessment: Computers & Geosciences, v. 2, no. 3, p. 309-311.

Codd, E.F., 1982, Relational database; a practical foundation for productivity: Communications of the ACM, v. 25, p. 109-117.

Date, C.J., 1981, An introduction to database systems (3rd ed.): Addison-Wesley, New York, 574 p.

Gordon, T., and Martin, G., 1974, File management systems and geological field data, *in* Computer use in projects of the Geological Survey of Canada: Geol. Survey Canada Paper, v. 74-60, p. 23-28.

Harris, K.L., Winczewski, L.M., and Umphrey, H.R., 1982, Computer management of geologic and petroleum data at the North Dakota Geological Survey: North Dakota Geol. Survey Rept., no. 74, 34 p.

Henderson, W.G., 1986, Implementation of geological knowledge in a relational database: another task for the earth science information centre, *in* Shelley E.P. ed., Proceedings of the 3rd international conference on geoscience information, v. 2: Australian Mineral Foundation, Adelaide, p. 38-51.

Hodgson, C.J., 1983, Preliminary report on a computer file of gold deposits of the Abitibi belt, Ontario: Ontario Geol. Survey Misc. Paper, v. 110, p. 11-37.

Hodgson, C.J., Troop, D.G., and Stewart, J.P., 1986, A new computer-aided methodology for area selection in gold exploration: a case study from the Abitibi greenstone belt, Ontario, Canada (abst): Terra Cognita, v. 6, p. 537.

Hruska, J., 1976, Current data-management systems: problems of application in economic geology. Computers Geosciences, v. 2, no. 3, p. 299-304.

Jeffrey, K.G., and Gill, E.M., 1976, The design philosophy of the G-EXEC system: Computers & Geosciences, v. 2, no. 3, p. 345-346.

Jones, J., 1981. Australian geoscience bibliographic data systems: a review, in Geoscience numeric and bibliographic data: Australian Mineral Foundation,. Seminar Project, no. 154/81, 40 p.

Le Maitre, R.W., 1982, Numerical petrology: Elsevier Developments in Petrology, no. 7, Elsevier, Amsterdam, 281 p.

Longe, R.V., Burk, C.F., Dugas, J., Ewing, K.A., Ferguson, S.A., Gunn, K.L., Jackson, E.V., Kelly, A.N., Oliver, A.D., Sutterlin, P.G., and Williams, G.D., 1978, Computer-based files on mineral deposits: guidelines and recommended standards for data content: Geol. Survey Canada Paper, v. 26, 72 p.

Mihalasky, M.J., Mutschler, F.E., Etienne, J.E., and Gordon, T.L., 1987, GOLDY — a geological and economic database for giant lode gold camps of North American: Eastern Washington University, Cheney, Washington.

Mock, C.M., Elliott, B.G., Ewers, G.R. and Lorenz, R.P., 1987, Gold deposits of Western Australia: BMR datafile (MINDEP): Aust.Bur.Miner.Resour. Rept., no. 3, 34 p.

Rassam, G.N., and Gravesteijn, J., 1986, Factual and bibliographic information in geoscience: the time for integration, in Shelley E.P. ed., Proceedings of the 3rd international conference on geoscience information, v. 2, Australian Mineral Foundation, Adelaide, p. 129-139.

Rock, N.M.S., 1986, A global database of analytical data for alkaline syenitoid, trachytoid and phonolitoid rocks: Modern Geology, v. 11, no. 1, p. 51-68.

Rock, N.M.S., 1988, Numerical geology: a source guide, bibliography and glossary to geological uses of computers and statistics: Springer-Verlag Lecture Notes in Earth Sciences, no. 18, Springer-Verlag, Berlin, 427 p.

Rock, N.M.S., Brown, T.C., and Hattie, J.A., 1988, Macintosh statistical packages an end-users' guide: Wings for the Mind (Newsletter, Australian Apple University Consortium), June 1988, p. 18-37.

Rock, N.M.S., and Wheatley, M.R., 1989, Some experiences in integrating the use of mainframes and microcomputers: Computers & Geosciences, v. 15, no.6, in press.

Shelley, E.P., 1985, Directory of government geoscience databases in Australia: Aust. Bur. Miner. Resour. Rept., no. 269, 180 p.

Siegenthaler, R., 1986, The use of database systems: Terra Cognita, v. 6, no. 1, p. 83-88.

Stewart, J., 1988, 4th Dimension — the Macintosh way of information management: Australian MacWorld, September 1988, p. 60-66.

Teorey, T.J., and Fry, J.P., 1982, Design of database structures: Prentice-Hall, New Jersey, 492 p.

Young, R., 1985, Geochemical databases: the Western Mining experience: Aust. Geosci. Info. Assoc. Occasional Paper, no. 1, 10 p.

MACS: A Macintosh Program For Constructing Marine Magnetic Anomaly Profiles

Eric Rosencrantz
University of Texas Institute for Geophysics

ABSTRACT

MACS, an acronym for Magnetic Anomaly Construction Set, displays and calculates marine magnetic-anomaly profiles on the Macintosh computer. The program is unique in that it allows on-screen editing of magnetic reversals sequences prior to calculating anomaly profiles, so that sequences can be tailored to reproduce suspected gaps, duplications, or other discontinuities within observed magnetic profiles. The application uses a Fast Fourier Transform algorithm for anomaly profile calculations. Application profiles can be printed, saved to disk as a graphic image, or saved to disk for later reworking.

INTRODUCTION

MACS calculates marine magnetic-anomaly profiles from given magnetic field polarity reversal sequences, then compares these synthetic profiles with observed marine magnetic-anomaly profiles, for the purpose of establishing age and spreading rate of the observed profiles. The name "MACS" is an acronym for Magnetic Anomaly Construction Set. The program is designed specifically for the Apple Macintosh desktop computer and makes full use of the menu, window, and mouse features particular to this operating system. The program code is written in Microsoft Basic 3.0 and compiled with the Microsoft Basic compiler. The current version of the program contains about 1800 lines of code, and the compiled application occupies about 100 kilobytes of disk space. The program will run on all current models of the Macintosh.

The process of identifying marine magnetic anomalies as profiled along ship or aircraft tracks is an iterative process of calculation and comparison, wherein plots of calculated (synthetic) profiles with known parameters are matched by eye to plots of observed (real) profiles of unknown parameters. This is at best a tedious process even when some of the parameters of the observed profile are known, such as crustal age or crustal spreading rate. It can be an extremely difficult process when the observed profile displays magnetic anomalies which are not obviously in temporal sequence, such as when the profile transects oceanic crust which includes a spreading center jump, or crosses a fracture zone which juxtaposes crusts of different ages. MACS overcomes these difficulties in two ways. First, it speeds up the overall process of profile comparison by providing the user with on-screen, fast plots of synthetic profiles. Second, it provides the way to alter, on-screen with the mouse, a magnetic-reversal sequence such that it can be tailored to reproduce suspected instances of anomaly gaps, discontinuities, duplications, or local spreading jumps within the observed magnetic profile. The overall process of profile comparison is fast, immediate, and direct.

DISPLAY

MACS' primary display consists of a single plotting window containing all of the application's working elements (Fig 1). These include a scale showing the length of the profiles in km, the plot of the observed profile with an amplitude scale in nT (nannoTeslas), a plot of observed water bottom depth with a depth scale in km, the calculated magnetic-anomaly profile, and a two-part display of the magnetic-reversal sequence. The depth plot is not essential to the anomaly calculations but can be useful both for evaluating possible topographic effects on observed anomalies and for locating probable spreading centers or fracture zones. The magnetic-reversals sequence is displayed in the main window as a pair of "bar codes" separated by a horizontal scroll bar and a pair of scroll buttons (Fig. 1). The horizontal axis describes time. Periods of normal polarity are shown as black, those of reversed polarity as white. The lower of the two sequences displays reversals in the age range 165 Ma to the present, as dated and identified by Harland and others (1982). These are stored in the program as DATA statements. Sequence spreading rate, that is, the length (in time) of the sequence relative to the length (km) of the plot, is set through a dialog box called from the "Parameters" menu. The sequence can be labeled either by time (Ma) or by anomaly identification, can be scrolled across the window, and can be copied to the upper sequence, but cannot be modified otherwise. The upper reversal sequence is a bit-mapped screen image of the lower sequence which can be altered through combinations of menu and mouse commands. Selected portions of the sequence can be copied, erased, stretched or shrunk, moved laterally, or flipped end-for-end, to recreate the variety of magnetic-anomaly sequences which occur in ocean basins.

Program routines not controlled directly from the main window are accessed through the menu bar at the top of the window. These routines include those directly related to the main window display plus those auxiliary to the main plot. The "File" menu controls reading files of observed magnetic profiles, restarts the program, prints plots, saves plots to disk, and shuts down the program. The "Edit" menu controls altering the magnetic reversal display in the main window. The "Parameters" menu calls dialog boxes through which plot and calculation parameters are set, and the "Compute" menu controls the size of computation arrays.

PROGRAM STRUCTURE

Plots within the main window are constructed through two main processes. The first includes reading and plotting observed magnetic anomaly and depth profiles. The second involves plotting and manipulating the magnetic-reversals sequence and calculating the synthetic profile. Although the separate profiles created by these two processes are designed for direct comparison to each other, each can be constructed and manipulated by itself.

Plotting observed magnetic-anomaly profiles: MACS reads magnetic anomaly profiles stored in two standard file formats, the National Geophysical Data Center (NGDC) merged-merged format and the Marine Geophysical Data Exchange (MGD77) format, plus a MACS-specific file format, termed User, comprising three-element records containing distance along profile, magnetic-anomaly value, and depth value. All datafiles are stored in ASCII format and can be edited with standard text editors. NGDC and MGD77 files can

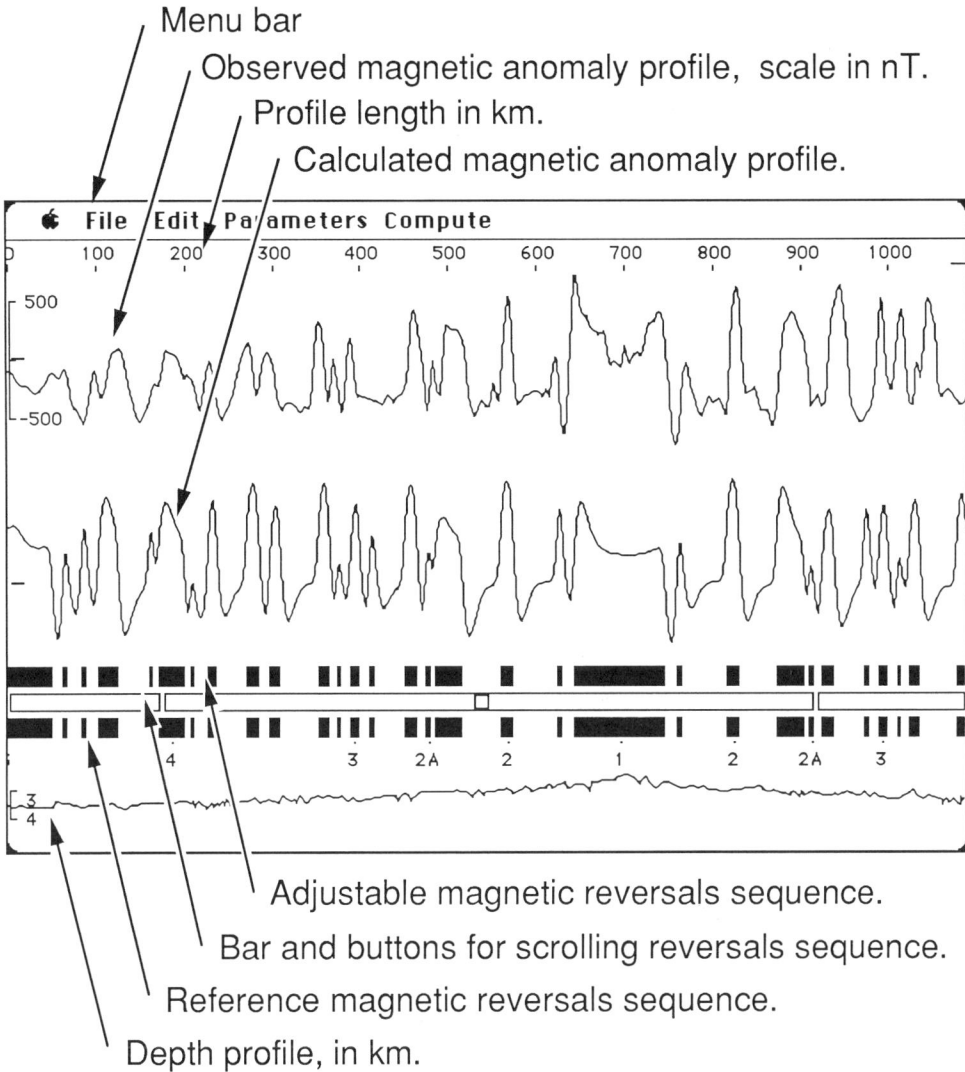

Figure 1: Main window display of MACS, with window elements labeled. Observed magnetic-anomaly profile is drawn from geophysical data collected by R/V Eltanin in southeastern Pacific Ocean, on cruise ELT19. Profile is projected onto vertical plane trending 295°, which is parallel to crustal spreading direction in this area. Profile extends southeast to northwest, left to right. Synthetic profile is plotted with 512 points, is phase shifted by -310° and has been smoothed with application of Gaussian extrusion filter with half-width of 8 km. Thickness of crustal magnetic layer is assumed to be 500 m, and depth to magnetic layer is profile mean water depth. Central magnetic anomaly is centered at 700 km. Calculated spreading (half) rate is 72 mm/yr.

be read in parts delimited by record date and time. The program abstracts latitude and longitude, water bottom depth measurements and magnetic-anomaly measurements from these files, and calculates distance along the track based upon latitude, longitude, and profile azimuth. User files can be read in parts defined by distance. MACS rewrites calculated or read distance, depth, and magnetic values to a small, temporary diskfile, calculates total profile length and ranges of depth and magnetic amplitude, and displays these in a dialog box in which plot-scaling parameters are set.

The second stage of plotting begins with the program setting plot scales as entered through the dialog box, reading the data from the temporary file and drawing the profiles on the screen. All the profiles and scale bars are drawn and saved in memory as objects, rather than as arrays, because this significantly reduces program memory requirements. Profiles can be rescaled and replotted at any time; the program simply redraws the appropriate objects from data saved in the temporary disk file.

Calculating synthetic magnetic anomaly profiles: The synthetic anomaly profile is constructed from the upper magnetic-reversal sequence with a Fast Fourier Transform algorithm similar to that described by Schouten and McCamy (1972). A square wave representing normal and reversed crustal magnetization is constructed from a scan of the edited sequence as displayed in the main window. This function is converted to the transform domain and multiplied by a series of filters, including an earthfilter, phase filter and Gaussian extrusion filter (Schouten and McCamy, 1972). The function then is inverse transformed back into the time domain and plotted. Those parameters controlling the shape of the synthetic curve, such as phase shift, depth to magnetic layer, upward continuation, thickness of the magnetic layer, crustal magnetization, Gaussian extrusion filter half-width, and profile amplitude, are applied through a dialog box called from the "Parameters" menu. The program provides the option of calculating the phase shift and amplitude parameters from appropriate geographic and magnetic-field information. The size of the sample and calculation arrays can be changed from 64 to 512 points. The calculation process is straightforward and fast. A "quick and dirty" profile calculated from 64 points typically takes less than 5 seconds to calculate and plot, and a profile of 512 points takes 20 to 30 seconds.

OUTPUT

The MACS screen plot can be saved to disk both as a data file for later reloading into the program or as a PICT format image file. The screen plot also can be printed. The printout reproduces the profiles and accompanying scale bars and lists the various parameters used in the calculation of the profiles. The image file can be transported subsequently into any Macintosh graphics application which accepts the PICT format.

AVAILABILITY

MACS is available from the COGS public-domain software library. The MACS folder includes the application, instructions, and a sample magnetics profile.

ACKNOWLEDGMENTS

I thank Yosio Nakamura for the use of his BASIC version of the Fast Fourier Transform algorithm. Larry Lawver and Joe Phillips contributed many excellent suggestions for improving the program.

REFERENCES

Harland, W. B., Cox, A. V., Llewllyn, P. G., Pickton, C. A. G., Smith, A. G., and Walters, R., 1982, A geologic time scale: Cambridge Univ. Press, Cambridge, 129 p.

Schouten, H., and McCamy, 1972, Filtering marine magnetic anomalies: Jour. Geophysical Res., v. 77, no. 35, p. 7089-7099.

Theoretical Morphology of Shells Aided by Microcomputers

Enrico Savazzi
Paleontologiska Institutionen, Uppsala

ABSTRACT

Computer modeling of shell morphologies is an important tool in the functional analysis of molluscs and brachiopods, and microcomputers are a choice instrument for this research field. This article describes a set of programs, written in C, that allow interactive modeling of shell morphologies, and of their ontogenetic laws. Different alternatives in the design of the programs are discussed, in order to show some of the problems and approaches characteristic of the microcomputer environment.

INTRODUCTION

Early in the past century, Moseley (1838) observed that coiling of mollusc shells can be approximated closely by a logarithmic, or equiangular spiral. Subsequent analyses of the geometry of mollusc shells concentrated on this principle (Thompson, 1942; Stasek, 1963; Cox and Nuttall, 1969; and references therein). It was not until electronic computers became available, however, that modeling of shell morphologies became feasible as a practical research tool. Raup (1961, 1962, 1963, 1966, 1967) and Raup and Michelson (1965) developed a computer-based procedure for generating graphic images of shells, and applied it to study the range of morphologies existing in regularly coiled gastropods and ectocochleate cephalopods. Modifications of Raup's method were used by McGhee (1978, 1979, 1980a, 1980b) and Savazzi (1985a, 1985b, 1986, 1987) to study the adaptive properties of the shells of bivalve molluscs and brachiopods. A different method was developed by Okamoto (1984, 1988a, 1988b, 1988c) to model the conspicuously allometric morphologies of heteromorph ammonoids. Davaud and Wernli (1974) modified Raup's method to model the geometry of oriented sections of coiled Foraminifera, in order to allow the classification of actual Foraminifera observed in thin sections of sediment samples. Hayami and Okamoto (1986) used computer modeling to study the effects of shell growth on the geometry of sculptures in bivalves. These studies show that theoretical morphology, besides being an interesting field in itself, can be a tool of broad applicability.

The present article describes a set of programs that allow the modeling of shell morphologies on microcomputers. These programs were written with the principal scope of illustrating different algorithms and programming approaches. Although the programs can be compiled and used without modification, their practical usefulness may be enhanced by inserting additional features. In particular, the user interface of the programs has been kept simple deliberately, in order to limit the size of the source code. For the same purpose, some routines and functions have been included in only one program of the set, although they may be useful also in the others.

SYSTEM REQUIREMENTS

Compiling and running the programs requires an IBM PC, XT, AT, PS/2, or compatible computer equipped with a VGA, EGA, CGA, or Hercules monochrome graphic card and appropriate monitor, Microsoft C compiler version 5.0 or later, and the PC-DOS or MS-DOS operating system version 2.0 or later. Any other C compilers that follow the proposed ANSI standard for the C language can be used, provided that (1) it allows single data arrays to exceed 65,536 elements, and (2) the calls to non-ANSI graphic functions are modified as needed. In addition, a text editor is needed to edit the data and source files. Using a math coprocessor will yield a moderate decrease in execution speed. All programs, with the exception of SHELL3D.C, will run in 256 Kb of system RAM. It is recommended that SHELL3D.C be used with a full 640 Kb of system RAM.

PROGRAM FILES

RAUP.C computes a shell model by using Raup's single-helicospiral method (see later). This program produces a simple representation of the shell, and allows it to be oriented and scaled at will. When the desired image is obtained, the shell model can be saved to a file for further processing. SHELLGEN.C generates shell models by using the multispiral method. OKAMOTO.C generates shell models with Okamoto's moving reference frame method. RAUP.FMT, SHELLGEN.FMT, and OKAMOTO.FMT explain the format of the data files containing the initial configuration of a shell model. A different format is required by each program. RAUP.DAT, SHELLGEN.DAT, and OKAMOTO.DAT are actual examples of data files. PLOT.INC contains functions used by RAUP.C and OKAMOTO.C.

SHELLSRT.C and SHELL3D.C read a data file generated by RAUP.C, SHELLGEN.C, or OKAMOTO.C, and use hidden-surface and pseudoshading algorithms to produce a representation of the model that is closer to the aspect of a real shell than a wire-cage drawing. The differences among these programs are described in detail in the following sections.

SHELLGEN.H is a header file containing constants, functions, and definitions for the graphic hardware used by all programs.

HERCULES.C includes low-level graphic routines for the Hercules monochrome graphic card. This file is needed, because the Hercules card is not supported by Microsoft C version 5.0. HERCULES.H includes the function declarations for HERCULES.C.

SETTEXT.C is a short utility program that sets the display in text mode. It is useful when one of the programs terminates abnormally, leaving the display in graphic mode.

INSTRUCTIONS FOR COMPILING THE PROGRAMS

The files HERCULES.H, SHELLGEN.H, and PLOT.INC must be placed in the path specified by the DOS environment variable INCLUDE (see the documentation of the Microsoft C compiler). SHELLGEN.H must be edited in order to adapt the programs to your specific hardware configuration. In particular, one of the symbolic constants HERCULES, CGA, EGA, and VGA must be defined, in order to select the proper graphic interface. This will instruct the compiler to insert in each program only the code required by the select graphic equipment. In some systems, the horizontal versus vertical aspect ratio of pixels (controlled by the symbolic constant ASPECT_RATIO) also may need to be changed.

All files with extensions .C and .DAT are placed conveniently in the subdirectory used for storing Microsoft C source files. If the Hercules monochrome graphic card will be used, HERCULES.C must be compiled first, in order to produce the file HERCULES.OBJ. Subsequently, the object file must be linked with all other programs. All programs are compiled using the small memory model. It is recommended that Microsoft's utility EXEPACK.EXE be used on the compiled programs (in particular, SHELL3D.EXE). This will reduce their size by up to 90%, and considerably shorten their load times.

SHELL MODELING

Raup's method uses a logarithmic helicospiral (Stasek, 1963), that is a spiral that translates along its coiling axis as it develops, as the element controlling shell coiling. If writing a program for a digital computer that uses a raster display, Raup's differential equations are approximated conveniently by finite difference equations in Cartesian coordinates (Bayer, 1977):

$$x' = x + I \left(\frac{\ln W}{2\pi} \, x - y \right)$$

$$y' = y + I \left(\frac{\ln W}{2\pi} \, y + x \right)$$

$$z' = z + I\,Hz$$

in which x, y, and z are the coordinates of the last computed point on the helicospiral, and x', y', and z' are the coordinates of the next point. Thus, the helicospiral is approximated by a set of successive points. The parameter I controls the magnitude of the increment between successive steps. W, the whorl-expansion rate, describes the curvature of the spiral in the coiling plane (the XY plane). H, the helicospiral component, controls the translation rate along the Z axis (i.e., the coiling axis). The origin of the axes coincides with the point reached by the helicospiral after infinite regression. The degree of approximation contained in these equations produces an accumulated error as successive points are computed. If the size and number of growth steps is small, this error remains within tolerable limits. When

$$H = \frac{\ln W}{2\pi}$$

the helicospiral grows isometrically (i.e., is self-similar). In addition, when the generating curve also has a growth rate equal to IH, the shell model as a whole grows isometrically.

The shell aperture is represented by a "generating curve" (typically a circle) centered on the helicospiral, and lying in a plane that includes the coiling axis. The generating curve follows the growth of the helicospiral, progressively increasing in size. The three-dimensional surface swept by the generating curve as it moves along the helicospiral represents the surface of the shell.

RAUP.C implements Raup's method, and produces a simple graphic representation of the shell, consisting of the helicospiral and of the outline of the aperture at each growth step (Fig. 1A). The generated model can be viewed from three orthogonal points parallel to

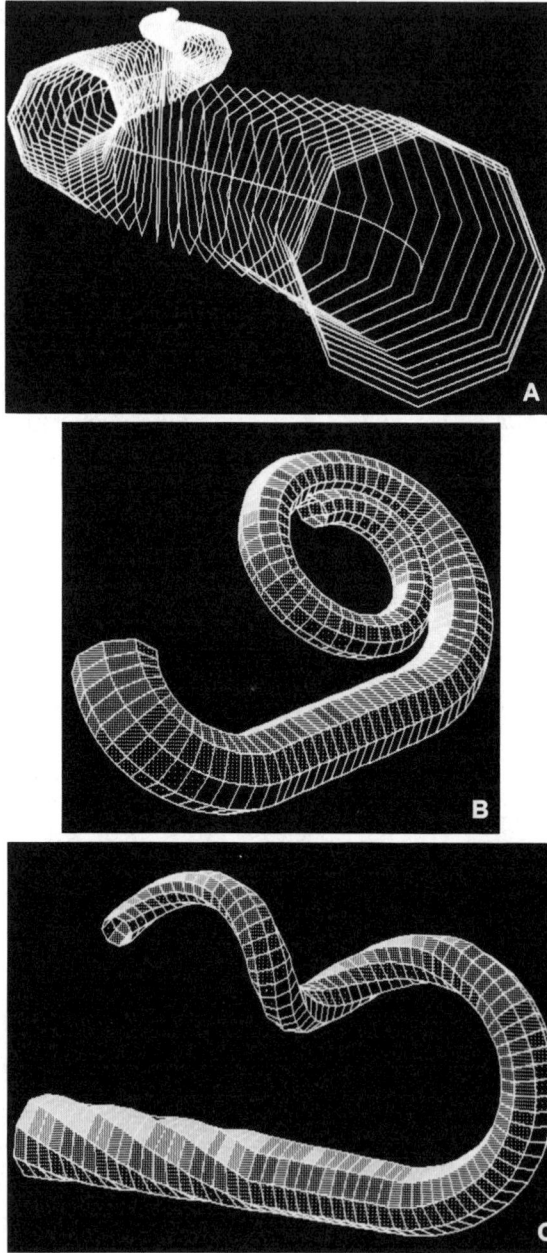

Figure 1. A, Display produced by RAUP.C. Helicospiral and successive positions taken
          by shell aperture during growth are visible. B, Pseudoshaded rendition of
          heteromorph ammonite produced by OKAMOTO.C and post-processed with
          SHELL3D.C. C, Example of shell possessing aperture torsion in its earliest and
          latest growth stages.

each of the reference axes, and can be rotated and scaled at will. Once a satisfactory result has been obtained, the shell model can be saved to a file for further processing by SHELLSRT.C or SHELL3D.C. When saving the shell model, it is recommended that you use the extension .ARY (shorthand for "array") in the file name. This will avoid accidentally overwriting the original configuration file, which has the extension .DAT.

In Raup's model, additional parameters may be used to allow the aperture to be inclined with respect to the generating curve, and to assume a complex shape (this situation is usual in actual shells). These capabilities require simple modifications to the program. A further valuable addition would be the capability of editing the file containing the initial configuration without leaving the program. This would provide the capability of manipulating the parameters, in addition to the display of the shell model, in an interactive way.

McGhee (1978) modified Raup's method by substituting the single helicospiral and the generating curve with a set of helicospirals. In this variant of Raup's method (hereby referred to as the multispiral method), the outline of the shell aperture is formed by joining with segments the points representing the last growth step on adjacent helicospirals. Growth of the model begins from a set of points, each corresponding to the start of a helicospiral. In this way, the initial shape of the shell aperture and its inclination with respect to the axes can be selected freely, without having to introduce additional parameters for this purpose. In practice, it may be useful to have such parameters, in order to be able to change the placement of the shell aperture as a whole (as done by Savazzi 1985a, 1987). The multispiral method also eliminates the need for a parameter controlling the growth rate of the shell aperture. Thus, isometric growth results when all helicospirals have the same values of I, W, and H. In such a situation, the resulting shell morphologies are comparable closely with those produced by Raup's method. If allometry is introduced in the growth programme, however (e.g., by altering one or more parameters during growth), the results of the two methods can be strikingly different. Savazzi (1987) showed that allometric growth is usual in bivalves, and therefore, allometry may be an important factor in successfully modeling the shells of other groups.

SHELLGEN.C implements the multispiral method. At a difference with the preceding program, it produces a wire-cage representation of the shell surface (see illustrations in Savazzi, 1985a, 1985b, 1987). This type of display may make it easier to perceive the shape of the shell. A third alternative would be to represent a shell model by displaying only the helicospirals (and therefore, the outline of the aperture at the end of the growth process). This process can be useful if the model consists of at most a single whorl (such as in bivalves and brachiopods), but results in a confusing image in other situations.

Okamoto (1984, 1988a) developed a method for shell modeling that builds on an assumption different from that of methods based on helicospirals. Instead of describing shell growth from the point of view of a fixed reference frame, the direction of each growth increment is specified relative to the current position and orientation of the shell aperture. As a result, the reference frame is centered on the shell aperture, and follows it during growth. Therefore, shell morphology is controlled by three parameters: (1) relative direction of growth, (2) growth rate along the direction of growth, and (3) growth rate of the aperture. A further difference of Okamoto's method is that, instead of trying to describe shell growth by a set of equations and constant parameters, the values of the parameters are changed freely at each growth step. In practice, the instructions directing each growth step are read from a "script" (i.e., a graph or a data file). This process results in the shell assuming any possible shape (Fig. 1B). At present, Okamoto's method is the only one that succeeds in modeling heteromorph ammonoids and other "irregular" shell morphologies. However, more than the other methods of shell modeling, Okamoto's method must be regarded as an "ad hoc" instrument, rather than as a faithful replication of a morphogenetic programme [Author's note: "programme" is used as a general term, as opposed to "program", which is restricted to computer code]. Okamoto's method is useful as a tool for producing shell morphologies, but it is not an explanation of how the morphogenetic programmes producing these morphologies are carried out in the actual organisms.

There are several possible ways of expressing the fundamental parameters in Okamoto's method (see also the program SNAKY.BAS in Okamoto, 1988a). In OKAMOTO.C, the shell aperture lies onto the XY plane. The direction of the growth vector is given as the angles between the growth vector and the Z axis in the XZ, YZ planes, respectively. The length of the growth vector is specified by a third variable. This way of expressing a vector facilitates writing a data file for use by this program, but any other method (similar to using the components of the growth vector along the X, Y, Z axes) is equivalent. Length and direction of the growth vector and growth rate of the shell aperture in OKAMOTO.C are independent of each other and of other parameters. In some situations, instead, (similar to Okamoto's studies on heteromorph ammonoids), it may be useful to relate growth vector and increment rate of the aperture to the total length of the shell, or to other parameters. This may yield a regular coiling simply by keeping all parameters constant. Introducing such relationships between parameters results in a hybrid between Raup's and Okamoto's methods. It also is possible to add parameters to control further morphological characteristics of the shell: for instance, OKAMOTO.C uses a parameter termed aperture torsion, which expresses a torsion of the aperture around the growth vector. This results in the shell surface twisting around the direction of growth (Fig. 1C). This phenomenon is observed in serpulid worms, vermetid gastropods, and perhaps in some Paleozoic nautiloids.

The most efficient way for generating a shell model (in either a fixed or a mobile reference frame) is to keep the shell immobile, relative to the reference frame of the display. This involves (1) producing the outline of the new shell aperture in a standard orientation, (2) computing the orientation and position that must be taken by the aperture, and (3) moving the aperture to its proper position. In this way, the portion of the shell model that has been built is left unaffected by further growth. OKAMOTO.C, instead, maintains the reference frame at the shell aperture, and moves and rotates the rest of the shell as needed at each growth step. This approach involves a larger amount of computing, but more closely reflects the concept of a moving reference frame. This allows, for instance, the growth process of bivalve shells to be observed (by successively displaying each growth stage), because the two valves remain assembled together at the commissure plane at all growth stages. In a fixed reference frame, the two valves would be seen to grow toward each other, starting from widely distant positions.

The graphic methods used by the described programs are fast and simple, but the images they produce lack realism, and may be difficult to interpret. The first step to produce better images is to introduce routines for removing hidden lines and surfaces (i.e., elements that lie behind others, and therefore should be invisible). Solutions to this difficult problem can be grouped into two main classes: algorithms that operate in geometric space (i.e., the precision of which is limited only by the floating-point arithmetic they use), and raster-oriented (or pixel-oriented) algorithms. The second class of algorithms produces images that are limited in resolution. Raster-oriented algorithms usually are faster in producing images composed of a high number of elements, and use mostly integer arithmetics. Their principal drawbacks are that they are difficult to adapt to vector-oriented displays (such as digital plotters), and that the image must be recomputed every time a change in size or resolution is required. It was decided to use pixel-oriented algorithms for the programs described here, because virtually all graphic interfaces connected to microcomputer monitors are of the raster type.

When using hidden-surface removal, there are two possible ways of displaying shell models. The method selected by Okamoto (see references) uses as surfaces the successive positions of the shell aperture during the growth process (Fig. 2C). In this paper, instead, the true shell surface is displayed (Figs. 1B-C, 2A-B, 3A-C). This type of display more faithfully reflects the actual shell morphology, and allows further improvements on the appearance of displayed models (see later). The data files generated by the programs are organized as lists of trapezoid polygons constituting the shell surface. Therefore, the data can be used immediately by surface-processing graphic algorithms.

SHELLSRT.C is a first approach to this problem. The program uses an algorithm that is faster for images composed of a small number of polygons, and yields correct results in almost all (though not all) situations. The algorithm consists in (1) sorting all polygons from farthest to closest in the line of sight (either of the reference axes can be used as a line of sight), and (2) drawing the polygons in the sorted order. Thus, surfaces that are hidden by elements that lie closer to the observer are "painted out" automatically (Fig. 2A-B). The source of errors lies in the fact that surfaces may be inclined along the line of sight, and therefore, their corners may have a different distance from the observer. This problem is reduced by first selecting, for each polygon, the corner that lies farthest (or closest) to the observer, and then sorting the polygons on the selected corners. This method works best with images composed of small polygons (which is the situation of our shell models). It will not solve cyclically overlapping or intersecting surfaces correctly , but these situations are rare in shell models.

The process used in building the display of the shell models produces the affect of a cross-section view sweeping across the volume of the shell. Thus, one has the opportunity of observing internal details of the shell that normally are hidden, such as the umbilicus and the columella (Fig. 2A). This capability can be a valuable research tool.

SHELLSRT.C uses a simple exchange-sort algorithm. More sophisticated sorting algorithms can be used, but special considerations apply to this example. Comparing two polygons involves a single comparison between two integers, whereas swapping two polygons requires 39 assignment operations. Therefore, all algorithms that move elements by inserting them among other elements (and therefore must move several other elements up or down by one step) are excluded. Algorithms that maximize the number of comparisons and minimize the number of swaps should be efficient in this context.

Because the images generated by RAUP.C, SHELLGEN.C, and OKAMOTO.C are "transparent", they do not allow an observer to distinguish between foreground and background. Therefore, after transferring a shell model generated by these programs to SHELLSRT.C, one may realize that the image is seen from a direction opposite the intended one. To solve this problem, SHELLSRT.C allows the user to reverse the shell model about either axis. In principle, it is possible to add to this program the capabilities of rotating and scaling shell models which occur in the other programs. In SHELLSRT.C, the whole model must be loaded into memory for sorting. Because the shell model is split into separate polygons, the number of corners is about four times larger than the number of points in the original model. Therefore, data from the original model are converted from floating point to integer format prior to recording to a data file. As a consequence, repeated scaling and rotating in SHELLSRT.C would result in a progressive loss of precision. Simply reversing the shell model, on the other hand, only involves changing the sign of variables, and therefore, this operation produces no accumulated error.

SHELL3D.C implements a more general solution to the problem of hidden-surface removal. A large array is maintained in memory. The number of elements in this array is equal to the number of pixels on the screen. As each pixel on the screen records the color of that point of the image, the corresponding element of the array stores the depth of the point along the line of sight. When a new element is to be added to the picture, it is split into pixels, which are processed one-by-one. When a pixel of the new element lies in a position already occupied by a pixel of the picture, the depth of the two points if compared. If the new pixel is farther away from the observer than the old one, it is

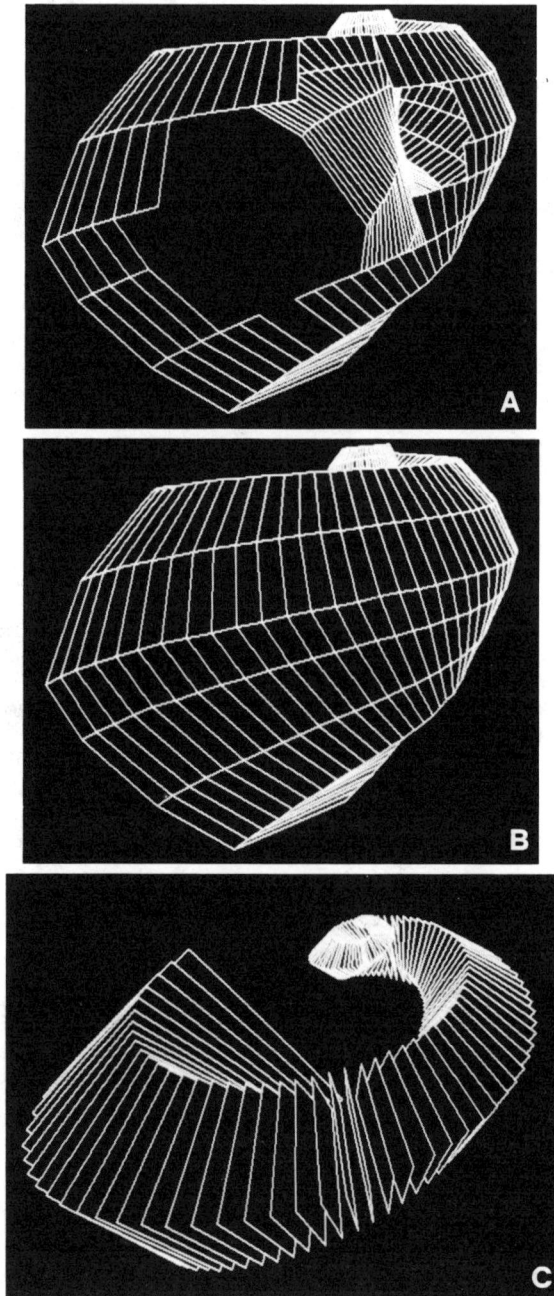

Figure 2. A-B, Intermediate and final stages of process used by SHELLSRT.C for image construction. Note internal details of shell in A. Shell model was generated by SHELLGEN.C. C, Method of shell rendition used by Okamoto.

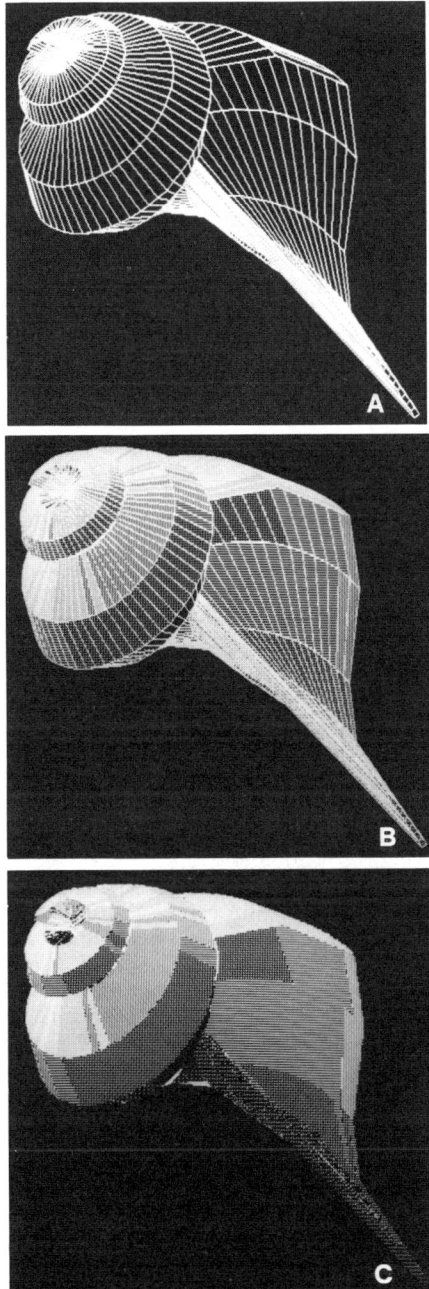

Figure 3. A-C, Post-processing capabilities of SHELL3D.C. In A, only hidden-surface removal is used. In B, both polygon borders and pseudoshading are displayed. In C, only pseudoshading is used. Shell model was produced by SHELLGEN.C.

discarded. If it is closer, it is plotted on the screen, and the corresponding element in the depth array is updated. This approach is known as a z-buffer, or depth buffer.

Several considerations are involved in the design of an algorithm of this type. These considerations will be discussed elsewhere, but one of them is worth mentioning here. According to conventional wisdom, the precision of the depth buffer (i.e., the number of bits of its elements) should exceed the linear resolution of the image. However, even an array of integers (which are 16 bits long in microcomputers using a CPU of the Intel -86 family) would exceed the amount of system memory available under the PC-DOS and MS-DOS operating systems. Placing the depth buffer in a disk file would increase the execution time by several orders of magnitude, and placing it into expanded or extended memory would require equipment not available in many of these microcomputers. Therefore, I was forced to use a character array (8 bits) for the depth buffer (as an alternative, it might be possible to use specially designed variables, 10 or 12 bits long). To partly compensate for the reduced depth of the buffer, the shell model is preprocessed, to resize the Z component of each polygon corner, so that the full range of allowed values (0 to 255 for an unsigned char) is represented in the model. After this preprocessing, problems are only apparent when drawing surfaces that lie at small distances from each other, and that form small angles to each other (similar to those in Fig. 2C). Virtually all models derived from the wire-cage type, instead, will be displayed correctly (Fig. 3A). The algorithm used assumes that the polygons are convex in the line of sight, and that they are flat or nearly flat. These conditions are met always in wire-cage shell models, except in Okamoto's models with a high amount of aperture torsion. In some of these models, the corners of polygons can deviate substantially from coplanarity. In these situations, it would be possible to split each trapezoid appropriately into two triangles, which are of necessity flat and convex.

After removing the hidden surfaces, the next step for enhancing the appearance of a shell model is introducing a form of shading (perspective rendering rarely is needed in shell models, and will not be discussed here). True shading requires ray-tracing algorithms, which probably cannot be executed in reasonable times by microcomputers. A simple approximation to shading, however, is possible. This technique, which here is termed pseudoshading, assigns to each polygon a shade of color, depending on the angle formed by its surface with the direction of incident light. Surfaces that are nearly perpendicular to the incoming light are lighter, whereas surfaces that face away from the light become proportionately darker. The algorithm proceeds as follows: (1) the vector cross-product of two adjacent sides of the polygon is computed, yielding a vector perpendicular to the surface, (2) if this vector points away from the observer, the opposite vector is considered (this allows a polygon to be seen from either side, such as in the situation of the interior of the shell aperture), (3) the divergence of the vector from the direction of illumination is computed, and (4) a shade is assigned to the polygon, on the basis of the amount of divergence. Also this algorithm is discussed in detail elsewhere. SHELL3D.C displays a shell model exclusively in the XY plane. The capabilities of the other programs can be incorporated in SHELL3D.C. In addition, the pseudoshading algorithm can be implemented also in SHELLSRT.C.

Most of the graphic interfaces available for microcomputers of the IBM-PC family provide only a limited number of colors or shades. However, shades of grey may be obtained, although not accurately, even on a monochrome monitor. In this situation, shades are approximated by using a "screening" with dot patterns of variable density. Unfortunately, given the relatively low resolution of microcomputer displays, only about 10 shades can be obtained. This gives reasonably good results in simple shell models (Fig. 1B-C), but fails in models composed of a large number of polygons and possessing evenly curved surfaces (Fig. 3B-C). The EGA and VGA graphic interfaces can yield 16 and 256 shades of gray, respectively (by remapping the color palette). The good range of shades of the VGA is counterbalanced by a poor resolution, making this solution feasible only if the image is to be reproduced at a small size (2-3cm). The palette-remapping technique is not implemented in the program being discussed here.

CONCLUSIONS

Several aspects related to shell modeling have not been explored yet to a sufficient depth. For instance, none of the methods discussed here seems to mimic closely the biological programmes controlling shell growth and morphology in actual organisms. The effects of shell growth and geometry on sculptures and color patterns have received scant attention (Savazzi, 1986; Hayami and Okamoto, 1986). Recent work on regulatory mechanisms in biological systems might suggest new ways of approaching these problems. Because microcomputers possessing both the graphic capabilities and the computing power necessary to carry out this type of work are ubiquitous, exciting developments in this field are to be expected.

REFERENCES

Bayer, U. 1977, Cephalopoden-Septen Teil 2: Regelmechanismen in Gehause- und Septenbau der Ammoniten: Neues Jahrbuch fur Geologie und Palaontologie, Abhandlungen, v. 155, no. 2, p. 162-215.

Cox, L.R., and Nuttall, C.P., 1969, Geometry of shell, *in* Moore, R.C., ed., Treatise on invertebrate paleontology: Geol. Soc. America and Univ. of Kansas, Pt. N., v. 1, p. N84-N91.

Davaud, E., and Wernli, R., 1974, Simulation de sections orientees de foraminiferes planispirales au moyen d'un ordinateur: Eclogae Geologicae Helveticae, v. 67, no. 1, p. 31-38.

Hayami, I., and Okamoto, T., 1986, Geometric regularity of some oblique sculptures in pectinid and other bivalves: recognition by computer simulation: Paleobiology, v. 12, no. 4, p. 433-449.

McGhee, G.R., Jr., 1978, Analysis of the shell torsion phenomenon in the Bivalvia: Lethaia, v. 11, no. 4, p. 315-329.

McGhee, G.R., Jr., 1979, Geometric analysis of biconvex brachiopod shell morphology: ordinal distributions and stability strategies (abst.): Geol. Soc. America, Abstracts with Program, v. 11, p. 44.

McGhee, G.R., Jr., 1980a, Shell form in the biconvex inarticulate Brachiopoda: a geometric analysis: Paleobiology, v. 6, no. 1, p. 57-76.

McGhee, G.R., Jr., 1980b, Shell geometry and stability strategies in the biconvex Brachiopoda: Neues Jahrbuch Geologie und Palaontologie, Monatshefte, v. 1980, no. 3, p. 155-184.

Moseley, H., 1838, On the geometrical forms of turbinated and discoid shells: Phil. Trans. Roy. Soc. 1838, p. 351-370.

Okamoto, Y., 1984, Theoretical morphology of Nipponites (a heteromorph ammonoid): Kaseki, Palaeontological Soc. of Japan, v. 36, p. 37-51; [in Japanese].

Okamoto, T., 1988a, Analysis of heteromorph ammonoids by differential geometry: Palaeontology, v. 31, no. 1, p. 35-52.

Okamoto, T., 1988b, Changes in life orientation during the ontogeny of some
    heteromorph ammonoids: Palaeontology, v. 31, no. 2, p. 281-294.

Okamoto, T., 1988c, Developmental regulation and morphological saltation in the
    heteromorph ammonite Nipponites: Paleobiology, v. 14, no. 3, p. 272-286.

Raup, D.M., 1961, The geometry of coiling in gastropods: National Acad. of Sciences Proc.,
    v. 47, p. 602-609.

Raup, D.M., 1962, Computer as aid in describing form in gastropods shells: Science,
    v. 138, no. 3537, p. 150-152.

Raup, D.M., 1963, Analysis of shell form in gastropods: Geol. Soc. America, Spec.
    Paper 73, 222 p.

Raup, D.M., 1966, Geometric analysis of shell coiling: general problems: Jour.
    Paleontology, v. 40, no. 5, p. 1178-1190.

Raup, D.M., 1967, Geometric analysis of shell coiling: coiling in ammonoids: Jour.
    Paleontology, v. 41, no. 1, p. 43-65.

Raup, D.M., and Michelson, A., 1965, Theoretical morphology of the coiled shell: Science,
    v. 147, no. 3663, p.1294-1295.

Savazzi, E., 1985a, SHELLGEN, a BASIC program for the modeling of molluscan shell
    ontogeny and morphogenesis: Computers & Geosciences, v. 11, no. 5, p. 521-530.

Savazzi, E., 1985b, Adaptive themes in cardiid bivalves: Neues Jahrbuch Geologie und
    Palaontologie, Abhandlungen, v. 170, no. 3, p. 291-321.

Savazzi E., 1986, Burrowing sculptures and life habits in Paleozoic lingulacean
    brachiopods: Paleobiology, v. 12, no. 1, p. 46-63.

Savazzi E., 1987, Geometric and functional constraints on bivalve shell morphology:
    Lethaia, v. 20, no. 4, p. 293-306.

Stasek, C.R., 1963, Geometrical form and gnomonic growth in the bivalved Mollusca:
    Jour. Morphology, v. 112, no. 3, p. 215-231.

Thompson, D'Arcy W., 1942, On growth and form: Cambridge Univ. Press, New York,
    1116 p.

Program to Prepare Standard Figures for Grade-Tonnage Models on a Macintosh

D.A. Singer and J.D. Bliss
U.S. Geological Survey

ABSTRACT

Grade-tonnage models are frequency distributions of deposit tonnage and grades of mineral deposits of a specific type. The program described here allows users to prepare standard figures of grade and tonnage distributions and display the deposit name associated with any of the data points. Titles and scales appropriate for most deposit types are plotted automatically for tonnage, Cu, Ni, Sn, Nb, W, Au, Hg, Mo, Zn, Pb, Ag, Co, Pt, Pd, Sb, Fe, Cr, Mn, and Ba.

INTRODUCTION

Mineral-resource assessments are conducted to meet a wide range of needs by government and industry. Issues that assessments can be used to address include land-use planning, economic planning, exploration, and resource availability. Assessments are useful particularly if they are quantitative and explicitly consider uncertainty. Grade-tonnage models are frequency distributions of deposit tonnage and grades of mineral deposits and are fundamental in translating geologists' resource assessments into a language that economists can use. The availability of grade-tonnage models and estimates of the number of undiscovered deposits allow economic analysis of the value of these sources of potential supply.

An important tool in mineral-resource assessments is the use of deposit models. Models applicable to geologic problem solving are numerous - descriptive, genetic, or exploration models are only but a few. Grade-tonnage models are applicable to many metal and some industrial mineral-deposit types; other models filling the same role as grade-tonnage models are being developed for some industrial mineral-deposit types.

Grade-tonnage models as shown in Figure 1 are frequency distributions. In this figure, the deposit type is hypothetical but realistic. The tonnages and average grades are for well-explored deposits of each type and are used as models for grades and tonnages of undiscovered deposits of the same type in tracts delineated as permissible. The frequency distributions are usually lognormal. The largest published set of grade-tonnage models (60 models) is presented in a bulletin edited by Cox and Singer (1986).

Of the techniques available, the "three step method" or "three part method" (Singer, 1984) meets these requirements as well as permits some parts of the assessment to be modified

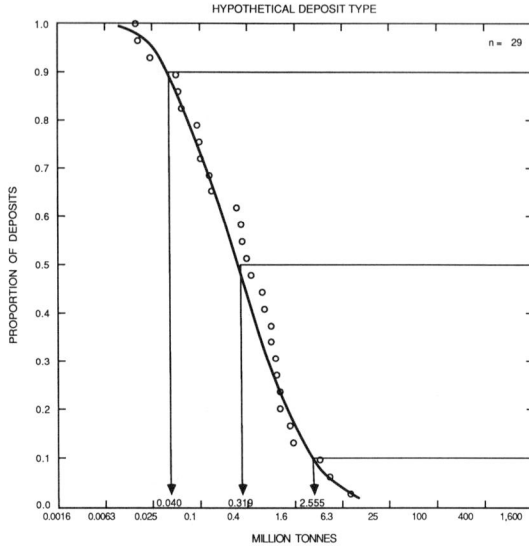

Figure 1.   Grade-tonnage model for hypothetical mineral deposit type. A, Tonnage.

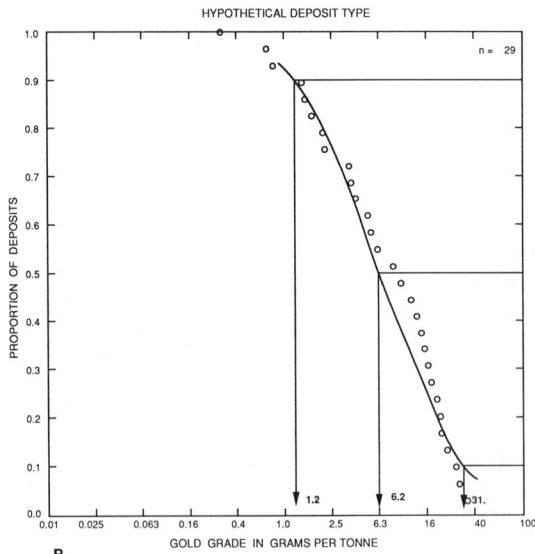

Figure 1(cont.).   B,Gold grades.

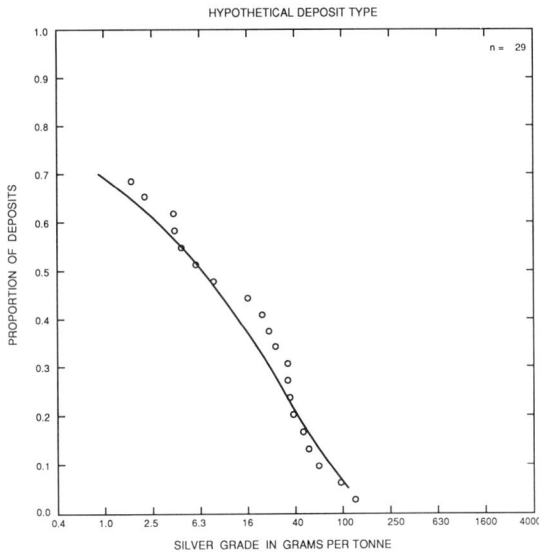

Figure 1(cont.).   C, Silver grades.

without invalidating the others. Both the descriptive and grade-tonnage models are used in the three step method. The steps in the method are as follows:

(1) Areas are delineated according to the types of deposits their geology will permit;

(2) The amount of metal or material and some characteristics of ore are estimated by means of grade-tonnage models; and,

(3) The number of undiscovered deposits of each type within delineated tracts is estimated.

The procedures described here are presented in an executable program. The program makes frequency diagrams identical to those in Figure 1 as well as those in Cox and Singer (1986) and labels the name of one or more of the deposits adjacent to the appropriated data point in the plot.   The program (Grade-Tonnage Plots) does not generate completely finished grade-tonnage models. The program generates MacDraw[*] files which in turn require minor modification.   These figures are not compatible with MacDraw II and its use is not recommended unless the file is opened first and then saved in MacDraw. The computers on which the program works include all Macintoshes tested (MacPlus and others). The program requires a minimum of about 650K memory (including System). Also included on the diskette is the source code for the program.

The program has not been tested exhaustively and no claim is made that it is without error.   Error capture during execution of the program is not available; failure to meet the requirements of the input data will result in system failure during program execution.

---

[*] Any use of tradenames in this publication is for descriptive purposes only and does not constitute endorsement by the U. S. Geological Survey.

Table 1. Text file from EXCEL for use as data input into Grade-Tonnage Plots.

29			
3			
2.2	12.5	0.091	
23	9.5	0.05	
20.7	6.2	0.61	
34	0	0.0974	
15	5.1	1.35	
29	4.6	0.1	
19	3.8	0.911	
16	3.6	1	
9.5	3.4	0.0582	BLACK BEAR
13	3.4	0.44	
27	2.7	3.2	
11.5	2.4	0.29	
8.3	2.1	4.3	
5.4	1.6	0.796	HOOSIER
2.1	0.83	0.13	
6.1	0.58	1	
1.7	0.44	0.14	
3.5	0.39	0.343	
4	0.38	0.65	GOLD QUEEN
3.7	0.22	0.39	
0.71	0.17	0.33	
1.4	0.04	0.88	
5.1	0	0.016	
14	0	0.017	
20	0	0.024	
1.5	0	0.0526	
0.29	0	0.78	LUCKY LADY
17	0	1.5	
0.81	0	7.8	

The program is written in Pascal (Lightspeed v. 1.11) and contains modifications and enhancements to a program published by Kelly and Smith (1988) that determines the roots of a quadratic equation and then generates a MacDraw plot. Because the final figure is in MacDraw format, additions can be made easily .

HOW GRADE AND TONNAGE MODELS ARE CONSTRUCTED AND USED

Construction of grade-tonnage models involved multiple steps, the first of which is the identification of a group of well-explored deposits that are believed to belong to the mineral-deposit type being modeled. The data consist of average grades of each commodity and tonnages which are based on the total of production, reserves, and resources at the lowest possible cutoff grade. These data represent an estimate of the mineral endowment of each deposit so that the final model will represent the mineral endowment of the undiscovered deposits. In practice, the available grades and tonnages seldom are reported at the same cutoff grade and cutoff grades are reported only infrequently. A second consideration at the data-gathering stage is the question of what should be the sampling unit. Grade and tonnage data are available to differing degrees for mines, deposits, districts, and shafts. It is extremely important that all of the data used in the model represent the same sampling unit because mixing data from deposits and districts usually produces a bimodal distribution or at least a nonlognormal distribution and can introduce artificial correlations among the variables.

The next step is to examine the data using plots and statistics to discover if the data contain multiple populations or outliers. For tonnage and most grade variables, a transformation to logarithms is necessary to remove skewness. Histograms, normal probability plots, cumulative-frequency plots, and empirical quantile function plots all are appropriate as is the examination of skewness and kurtosis statistics. Bivariate (scatter) plots of each pair of variables also should be constructed. Deviations from lognormality, outliers, or subgroups are all cause for reexamination of the data. If any of these conditions exist, the data should be checked for correctness of data entry, data reporting, and lastly, correctness of the geology that led to the classification of the individual deposits. If subgroups of data exist, geology of the subgroups probably will be different which suggests that the descriptive model may need reexamination. In most situations, the process of model building is iterative and requires multiple passes. Sources of difficulties include mixed geologic environments, poorly known geology, data-recording errors, incomplete records of production or resources, mixed deposit and district data, and mixed mining methods.

Although it is not possible to guarantee that a grade-tonnage model will never change, several factors ensure its stability. This probably will be the situation if tonnage and grades are not different significantly from lognormal (or normal), at least 20 deposits are used, and there are no significant correlations between tonnage and grade. For some deposit types such as placer Au, a correlation between tonnage and gold grade exists as a result of the effects of different, but inseparable, mining methods having been used. In situations such as placer Au, the model will stand until the effects of mining method can be related to grades and tonnages and the revised model can be linked to geology.

The grade-tonnage models are presented in graphic format to make it easy to compare deposit types and to display the data. All plots show grade or tonnage on the horizontal axis, whereas the vertical axis is always the cumulative proportion of deposits. Plots of the same commodity or tonnages are presented on the same scale; a logarithmic scale is used for tonnage and most grades. Each dot represents an individual deposit (or a district), cumulated in ascending grade or tonnage. Smoothed curves (added in MacDraw) are plotted through arrays of points, and intercepts for the 90th, 50th, and 10th percentiles are constructed. For tonnage and most grades, the smoothed curves represent percentiles of a lognormal distribution that has the same mean and standard deviation as the observed data; exceptions are plots where only a small percentage of deposits had reported grades, and grade plots that are presented on an arithmetic scale, such as iron or manganese, for which the smoothed curve is fitted by eye. The 90th and 10th percentiles are 1.282 standard deviations from the mean (in logarithms).

INPUT DATA

Data to be used by this program must be stored as text files. For example, one option in EXCEL is that the spreadsheet can be saved as a text file. The input data should have the layout as shown in Table 1. To assist in description, columns in the table are identified alphabetically and the rows numerically. A1 (column A, row 1) is the number of observations (29) in the data set which begins in row number 3. In A2 is the number of variables in the data set which should be equal to the number of columns (3). Beginning in row 3 and ending in row 31 are the data for the 29 deposits which are members of this deposit type. In this situation, column A contains gold in grams per metric ton (g/tonne), and B contains silver in g/ton. In the situation of silver, some deposits (B6, B25-B31) did not have silver reported and a zero must be entered. Column C contains ore reported as millions of tons. For example, the tonnage report for the deposit in row 3 is 91,000 tons and therefore is reported as 0.091 (C3). The column order in which the variables occur (Au grade, Ag grade, tonnage, etc.) is not important because the program requests which variable is present in which column. Deposit names always occur along the right edge of the data values (D11, D16, D21, D29). Note that the program does not treat the deposit name as an additional variable (A2). Deposit names could have occurred with all deposits; if names are not desired, a space with carriage return should be placed in each name field. The deposit name must contain no more than 20 characters; leading and trailing blanks are lost during posting of the name on the plot. The maximum number of deposits (records) this program will accept is 250; the maximum number of variables is six (excluding deposit name).

Past work in grade and tonnage modeling suggests that commodities for most ore-deposit types tend to fall repeatedly within the same range of sizes and commodity concentrations. The program automatically scales and labels the X axis accordingly for the following commodities: tonnage (x 10^6), Cu, Ni, Sn, W, Nb, Au, Zn, Pb, Ag, Mo, Co, Pt, Pb, Hg, Sb, Fe, Cr, Mn, and Ba. More details on this are given later.

PROGRAM EXECUTION AND OUTPUT

In addition to the executable program Grade-Tonnage Plots, MacDraw needs to be available. The program units include Misc, MyFileStuff, MyPlotStuff, MyPrintStuff, PlotGlobals, plotter main, Plotter Project, Plotter.R, and Plotter.RSRC, and solve. Grade-Tonnage Plots is executed by a double click. The first screen gives a WaitNewEvent message on which you click OK. This is followed by a screen which allows you to select the name of the text file in which your input data are located. If your input is in the correct form, the top banner will appear with *File*, *Edit*, *Color*, and *Print Options*; because input error detection is not included, errors of input format will lead to a system crash. In the

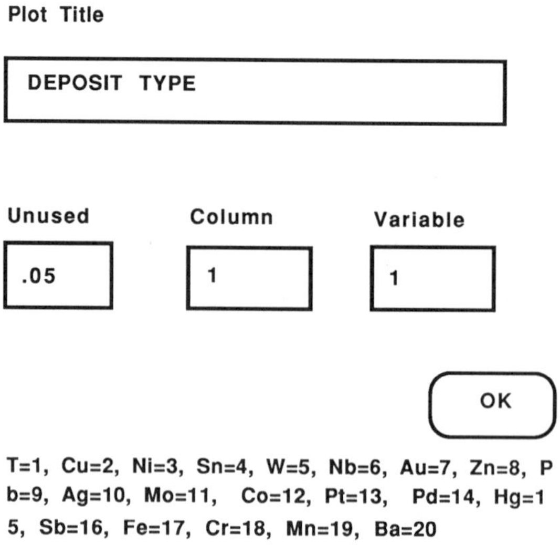

Plot Title

DEPOSIT TYPE

Unused   Column   Variable

.05   1   1

OK

T=1, Cu=2, Ni=3, Sn=4, W=5, Nb=6, Au=7, Zn=8, P
b=9, Ag=10, Mo=11, Co=12, Pt=13, Pd=14, Hg=1
5, Sb=16, Fe=17, Cr=18, Mn=19, Ba=20

Figure 2. Layout of panel exhibited by program on which users make column and variable selections.

pull-down menu for *File* are options *Plot, Save as..., Page setup...*, and *Quit.* Select Plot. The next screen is shown in Figure 2. The plot title is located in the upper box with a default of DEPOSIT TYPE. Change to the appropriate title. The column box allows you to select the column you wish to process in your data set. This requires that you remember which variable (tonnage, Au grade, etc.) occurs in which column of your input data. Change the column number (1,2,...) to the one you wish to process. The variable box allows you to select the name of the variable in the column you wish to process using the number associated with the commodity in the list at the bottom of the screen (Fig. 2). For example, if you are processing column one in your data, which is Ag grade, the column box should contain a "1" and the variable box a "10." If the commodity you wish to process does not occur in this list, refer to Figure 3 for the variable range applicable to commodities which are available and select one which contains a range of values similar to the commodity you are plotting. Be sure to modify the variable name in the resulting plot! Seven scales are available among the 20 commodities. Once the plot title, column number, and variable number are selected, check OK. The next screen should be a view of the upper 2/3 portion of the figure. No other part of the figure, at this point, can be viewed on the screen unless a 19" monitor or a program such as Stepping Out II is used. You can print a copy of the full figure by selecting Page Size on the Print Option menu, followed by Page Setup..., and Print... in the File menu. Note that the Color menu allows the color of the grade, axis, and background to be modified but it should be done prior to execution of the plot. If you are satisfied with the plot, select *Save as...* in the *File* menu and enter a file name; a MacDraw document will be created. Additional columns and variables can be processed from the currently selected data set by selecting *Plot* again from the *File* menu. If finished, exit the program via *Quit* on the *File* menu. You now need to open the figure which will open MacDraw to make the final changes.

MODIFICATIONS IN MacDraw

The MacDraw figure generated by the program shown in Figure 4 contains a number of letters in circles which have been added to assist in the following explanation. The figure is grouped twice so it should be ungrouped twice to make changes. "A" is the title which occurs in the Plot title box (Fig. 2). It can be changed in MacDraw as well. "B" is the number of observations (mineral deposits) in the data set and is taken from location "A1" in Table 1. The intersection of the horizontal line and vertical line at "C" is the 90th percentile of the cumulative lognormal (or normal for Fe, Ba, Mn, Cr, or Sb) curve fitted to 29 data points, one of which is identified by "D"; the value of the 90th percentile is 1.212 grams per tonne (g/tonne)and given at "G" along the X axis (the leading "1" is obscured partially in Fig. 5). It can be stated that 90 percent of deposits have a Au grade of 1.2 g/tonne or greater. The scale along the Y axis at "E" is the proportion of deposits; the variable title is located along the bottom of the figure at "K." The intersection of the horizontal line and vertical line at "F" is the 50th percentile of the cumulative curve and has a value of 6.155 g/tonne Au at "H." As before, it can be stated that 50 percent of the deposits have a Au grade of 6.2 g/tonne or greater. The intersection of the horizontal line and vertical line at "J" is the 10th percentile of the cumulative curve, the value for which is given along the X axis at "I." The scale along the X axis ,"M", is for gold grades in this example, the title for which is provided by the program (at "L"). Care should be taken, if other commodities are used, to change the title accordingly. Figure 5 gives what the final figure should resemble with modifications. The modifications are as follows: (1) The title at "K" is selected, rotated left and centered along the Y axis; (2) the vertical lines for "C" to "G", "F" to "H", and "J" to "I" are changed to arrows pointing to the X axis; (3) the values at "G", "H", and "I" are rounded to two significant digits, formatted as bold, and located to the right of each arrow (this change is more a matter of taste than clarity), and (4) an inverted "S-shaped" curve is fitted to both the data and the three points on the curve ("C", "F", and "J"). The fitted curve is created using the MacDraw polygon tool. After a

VARIABLE SELECTED

RANGE AND TITLE

T = 1

0.0016	0.0063	0.025	0.1	0.4	1.6	6.3	25	100	400	1,600

MILLION TONNES

Cu = 2
Ni = 3
Sn = 4
W = 5
Nb = 6
Hg = 15

0.032	0.056	0.1	0.18	0.32	0.56	1.0	1.8	3.2	5.6	10.0

COPPER GRADE IN PERCENT
NICKEL GRADE IN PERCENT
TIN GRADE IN PERCENT
TUNGSTEN GRADE IN PERCENT WO_3
NIOBIUM GRADE IN PERCENT Nb_2O_5
MERCURY GRADE IN PERCENT

Au = 7
Zn = 8
Pb = 9

0.01	0.025	0.063	0.16	0.4	1.0	2.5	6.3	16	40	100

GOLD GRADE IN GRAMS/TONNE
ZINC GRADE IN PERCENT
LEAD GRADE IN PERCENT

Ag = 10

0.4	1.0	2.5	6.3	16	40	100	250	630	1600	4000

SILVER GRADE IN GRAMS PER TONNE

Mo = 11

0.001	0.002	0.004	0.0079	0.016	0.032	0.063	0.13	0.25	0.5	1.0

MOLYBDENUM GRADE IN PERCENT Mo

Co = 12

0.01	0.016	0.025	0.04	0.063	0.1	0.16	0.25	0.4	0.63	1.0

COBALT GRADE IN PERCENT

Pt = 13
Pd = 14

1	2	4	8	16	32	63	130	250	500	1000

PLATINUM GRADE IN PARTS PER BILLION
PALLADIUM IN PARTS PER BILLION

Sb= 16
Fe = 17
Cr = 18
Mn = 19
Ba = 20

0	10	20	30	40	50	60	70	80	90	100

ANTIMONY GRADE IN PERCENT
IRON GRADE IN PERCENT
CHROMITE GRADE IN PERCENT Cr_2O_3
MANGANESE GRADE IN PERCENT
BARITE GRADE IN PERCENT

Figure 3.   Available default labels and scales for commodities along X axis for preparation of grade-tonnage models.

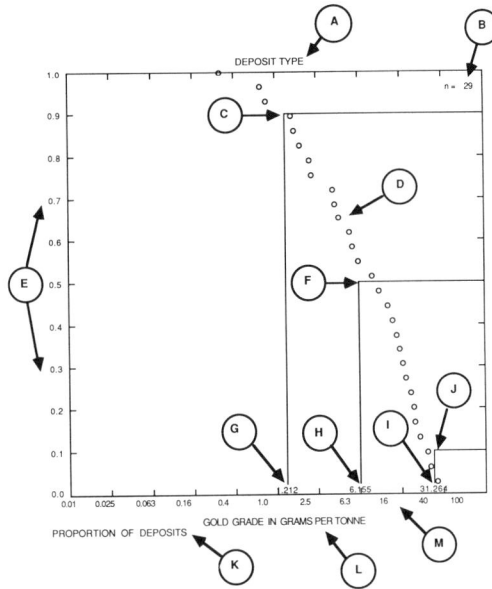

Figure 4. Output by Grade-Tonnage Plots prior to modifications made in MacDraw. Alphabetic codes in circles have been added to MacDraw figure for purposes of explanation of layout (see text).

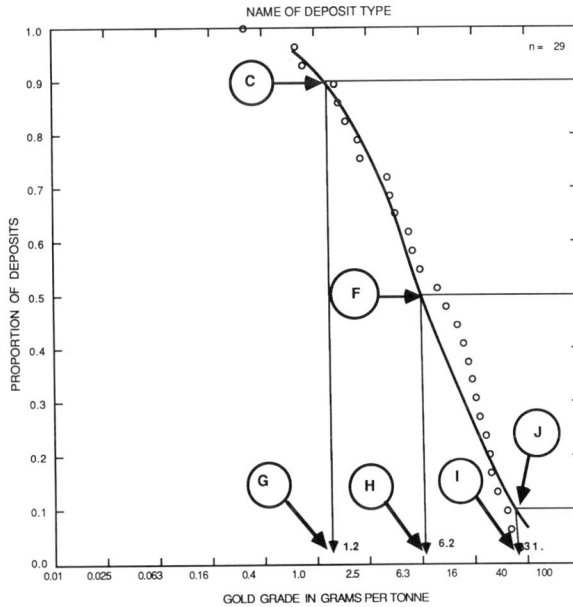

Figure 5. Figure after modifications made in MacDraw. Alphabetic codes in circles have been added to MacDraw figure for purposes of explanation of layout and are compatible with those in Figure 4.

rough fit is made by passing the line through the intersecting lines at "C", "F", and "J" with two inflection points about midway between the three, and starting and ending at about the 95th and 5th percentiles near the data, the curve is smoothed and reshaped via the *Edit* menu. An example of the tonnage model (Fig. 6) also shows how deposit names are posted. The figures created by this program will overlay those in Cox and Singer (1986) if they are printed with a 60 percent reduction in size.

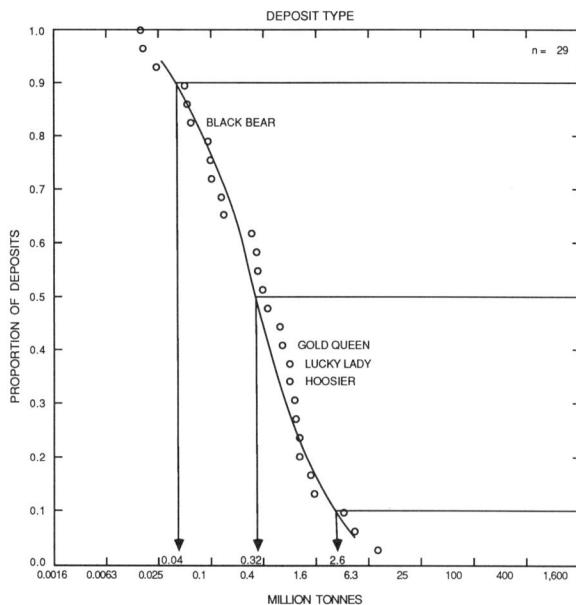

Figure 6. Output after modifications made in MacDraw.

SUMMARY

Grade-tonnage models are frequency distributions of deposit tonnage and grades of mineral deposits of a specific type. The program described here allows users to prepare standard figures comparable to those in Cox and Singer (1986). The program also displays the deposit name associated with each data point.

REFERENCES

Cox, D.P., and Singer, D.A., eds., 1986,, Mineral deposit models: U.S. Geological Survey Bull. 1693, 379 p.

Kelly, D., and Smith, D., 1988, Pascal procedures--Multifinder friendly MacDraw plotter: MacTutor, v. 4, no. 2, p. 16-38.

Singer, D. A., 1984, Mineral resource assessments of large regions: Now and in the future, *in* Geol. Survey of Japan, ed., U.S. – Japan Joint Seminar on Resources in the 1990's; June, 1984; publ. Earth Resources Satellite Data Analysis Center, v. 2, p. 31-40.

FILT-PC, A One-Dimensional Fourier Transform Program In FORTRAN For The PC

M.A. Sondergard
Wichita State University

J.E. Robinson
Syracuse University

D.F. Merriam
Wichita State University

ABSTRACT

FILT-PC is a computer program written in FORTRAN 77 to calculate one-dimensional Fast Fourier Transforms using the algorithm of Good as modified by Cooley-Tukey, Gentleman-Sande, and Robinson-Cohn. Options permit the selection of time or distance to frequency or the inverse transforms. The program runs on an IBM PC or compatible. A test sample is given as an example.

INTRODUCTION

Fourier transforms have been used for many years by geologists for analyzing and filtering sets of data. The analysis can be made in one or more dimensions depending on whether the data are along a string such as a profile or spatial such as map data or even volumetric. Thus any single-valued function recorded in time or distance can be represented as a series of sinusoidal waveforms, which can be characterized by wavelength, amplitude, and phase (Lee, 1960). Frequency components that suggest either cyclic processes or unique singular features can be isolated from other components by the application of selective filters (e.g. Jenkins and Watts, 1968). Filtering of the frequency domain data is accomplished by retaining the amplitudes of the desired components in their original form and making the amplitudes of all other equal to zero. Both positive and negative frequencies are filtered. The phase spectrum is not altered. The inverse transform completes the filtered display.

This program, which is an adaptation of FILTRAN (Robinson and Cohn, 1979), runs on an IBM microcomputer or compatible. It calculates a one-dimensional Fast Fourier Transform (FFT) based on the algorithm of Good (1958), which was modified by Cooley and Tukey (1965), and Gentleman and Sande (1966). The transform subroutines were written by G. Sande in 1965 for the University of Alberta. The original version of this complete program was presented by Robinson (1968), with an expanded version in Cohn (1975). Information on the spectral decomposition, frequency-domain filtering, and interpolation can be obtained from Robinson and Cohn (1979). Fourier transforms and filter theory are described in Lee (1960), Robinson and Treitel (1964), and Otnes and

Enochson (1972). A landmark publication is the *Measurement of Power Spectra* by Blackman and Tukey (1958).

PROGRAM DESCRIPTION

FILTRAN, first published in FORTRAN IV, has been rewritten in FORTRAN 77 for use with an IBM PC or compatible as FILT-PC. There are a few differences, however FILT-PC now is run interactively instead of in batch mode. There is no interface to a plotter; only character-generated plots are available. The data are limited to 2000 items per data set. The number of data points can be expanded easily by changing parameters within the program.

Requirements to run the program are as follows:

A data set containing:

        Line 1
        Cols. 1-80   Format, enclosed in parentheses and using standard
                        FORTRAN 77 syntax.

        Remaining Lines
        Cols. 1-80   The actual data set.

The data need to be regularly spaced in time or distance. Other "Fast Fourier" algorithms require the number of data points to be in powers of 2. This program, however, does not have this restriction, but there is a small tradeoff in speed. Data sets are limited to sequences that are factorable by prime numbers from 2 to 13.

After starting the program by entering the name of the program "FILT-PC", you will be asked for the filename containing the data set as mentioned, another filename for the output reports and graphs (a filename of "PRN" will send the output straight to the printer), and the number of data sets to process.

        Enter Filename of Input File
        TEST.DAT
        Enter Filename of Output File
        OUT.TXT
        Enter the Number of Data Sets to Process
        1

Then you will be asked for the number of data points in the first data set along with the data interval and the options for that set.

        Enter the Number of Data Points.

        Data Set#   1
        Number of Data Points: 99
        Data Interval: 1.0
        Enter Options

        Delete DC Component (Mean): 1
        Delete Zero Frequency: 1
        Amplitude & Phase Spectrum: 1
        3-Point Running Average (Amplitude): 1
        Graph Power Spectrum: 1
        3-Point Running Average (Power): 1

Inverse Prepare: 0
Inverse Transformation: 0

(Where a 1 enables the option and a 0 disables the option.)

Delete DC Component (Mean):	Deletes the mean from the data series.
Delete Zero Frequency:	Deletes the zero frequency from the graph display.
Amplitude & Phase Spectrum:	Graphs the Amplitude and Phase spectrum.
3-Point Running Average (Amplitude):	Graphs the 3-point running average of amplitude spectrum.
Graph Power Spectrum:	Graphs the power spectrum.
3-Point Running Average (Power):	Graphs the 3-point running average of power spectrum.
Inverse Prepare:	Writes frequency domain data to the output file.
Inverse Transformation:	Computes frequency to time transform, requires complex frequency values in input file.

The output to the screen are as follows:

Data Interval: 1.000 Nyquist Frequency = .500
Number of Data Points: 99
Frequency = J* .010101 Cycles per Unit Length

Average Value Deleted = 245.6913
Stop - Program terminated.

Pertaining to the first data set, and in this example only the data set, the filtered data, and graphs will be in the filename entered earlier (OUT.TXT).

EXAMPLE AND TEST RUN

The set is sample "boxcar" data made up as reported in Robinson and Cohn (1979). For instructions as to input see the section on Program Description in Robinson and Cohn (1979). Both the DC and zero frequency components of the input data were deleted before transformation.

Output from the program is given in Table 1 and Figure 1. Program output consists of the input data, reordered input, and one-dimensional Fourier Transform. The real R(J) and the imaginary Im(J) components for each of the frequencies (J) are listed with the amplitude power and phase. Note here that the power is defined as the amplitude squared. The phase (in degrees) denotes the displacement from the origin. Figure 1 consists of the amplitude spectrum, three-point running average of the amplitude spectrum, phase spectrum, power spectrum, and three-point running average power spectrum.

A listing of the program is available through COGS.

APPLICATIONS

There are numerous examples of applications of Fourier analysis to solving geologic problems. Some typical applications are water-level fluctuations (Cohn and Robinson, 1975), varves (Anderson and Koopmans, 1963), fold geometry (Beckman and Whitten,

Table 1. A, Boxcar data for testing and illustration; B, reordered data from A; C, one-
dimensional Fourier Transform. R(J) is real component; I(J) is imaginary
component; power, amplitude, and phase for each frequency J.

Input Data

```
.000    .000    .000    .000    .000    .000    .000    .000    .000    .000
.000    .000    .000    .000    .000    .000    .000    .000    .000    .000
.000    .000    .000    .000    .000    .000    .000    .000    .000    .000
.000    .000    .000    .000    .000    .000    .000    .000    .000    .000
.000    .000    .000    .000    .000   10.000  10.000  10.000  10.000  10.000
10.000  10.000  10.000  10.000    .000    .000    .000    .000    .000    .000
.000    .000    .000    .000    .000    .000    .000    .000    .000    .000
.000    .000    .000    .000    .000    .000    .000    .000    .000    .000
.000    .000    .000    .000    .000    .000    .000    .000    .000    .000
```
A
```
.000    .000    .000    .000    .000    .000    .000    .000    .000
```

Reordered Input
```
10.000  10.000  10.000  10.000  10.000    .000    .000    .000    .000    .000
.000    .000    .000    .000    .000    .000    .000    .000    .000    .000
.000    .000    .000    .000    .000    .000    .000    .000    .000    .000
.000    .000    .000    .000    .000    .000    .000    .000    .000    .000
.000    .000    .000    .000    .000    .000    .000    .000    .000    .000
.000    .000    .000    .000    .000    .000    .000    .000    .000    .000
.000    .000    .000    .000    .000    .000    .000    .000    .000    .000
.000    .000    .000    .000    .000    .000    .000    .000    .000    .000
.000    .000    .000    .000    .000    .000    .000    .000    .000    .000
```
B
```
.000    .000    .000    .000    .000   10.000  10.000  10.000  10.000
```

```
******************************************************
Data Interval:   1.000 Nyquist Frequency=   .500
Number of Data Points:  99
Frequency= J*  .010101 Cycles per Unit Length

        Average Value Deleted=   .9091
                          One-Dimensional Fourier Transform
                          **********************************
 J       R(J)            IM(J)           Power            Amplitude         Phase
 1   -.1907E-05       .0000E+00        .3638E-11        .1907E-05        .1800E+03
 2    .8880E+02      -.5841E-05        .7885E+04        .8880E+02       -.3769E-05
 3    .8524E+02      -.5126E-05        .7266E+04        .8524E+02       -.3445E-05
 4    .7951E+02      -.5603E-05        .6321E+04        .7951E+02       -.4038E-05
 5    .7186E+02      -.9060E-05        .5163E+04        .7186E+02       -.7224E-05
 6    .6265E+02      -.7629E-05        .3925E+04        .6265E+02       -.6978E-05
 7    .5230E+02      -.2146E-05        .2735E+04        .5230E+02       -.2351E-05
 8    .4129E+02      -.6199E-05        .1705E+04        .4129E+02       -.8602E-05
 9    .3009E+02      -.2384E-05        .9055E+03        .3009E+02       -.4540E-05
10    .1919E+02      -.2088E-05        .3683E+03        .1919E+02       -.6233E-05
11    .9029E+01      -.1181E-05        .8152E+02        .9029E+01       -.7497E-05
12    .8346E-05       .2865E-05        .7787E-10        .8824E-05        .1895E+02
13   -.7580E+01       .2354E-05        .5746E+02        .7580E+01        .1800E+03
14   -.1348E+02       .3373E-05        .1818E+03        .1348E+02        .1800E+03
15   -.1758E+02       .9223E-05        .3092E+03        .1758E+02        .1800E+03
16   -.1985E+02       .2942E-05        .3941E+03        .1985E+02        .1800E+03
17   -.2036E+02       .7628E-05        .4145E+03        .2036E+02        .1800E+03
18   -.1927E+02       .7852E-05        .3713E+03        .1927E+02        .1800E+03
19   -.1683E+02       .3712E-05        .2831E+03        .1683E+02        .1800E+03
20   -.1333E+02       .4723E-05        .1776E+03        .1333E+02        .1800E+03
21   -.9118E+01       .6367E-05        .8315E+02        .9118E+01        .1800E+03
22   -.4558E+01       .3397E-05        .2077E+02        .4558E+01        .1800E+03
23    .1982E-05       .2106E-05        .8363E-11        .2892E-05        .4674E+02
24    .4225E+01       .5922E-06        .1785E+02        .4225E+01        .8030E-05
25    .7834E+01       .4379E-06        .6138E+02        .7834E+01        .3202E-05
26    .1060E+02       .6034E-06        .1124E+03        .1060E+02        .3260E-05
27    .1238E+02      -.3550E-05        .1533E+03        .1238E+02       -.1643E-04
28    .1310E+02      -.3004E-06        .1715E+03        .1310E+02       -.1314E-05
29    .1275E+02      -.1854E-05        .1626E+03        .1275E+02       -.8329E-05
30    .1143E+02      -.1443E-05        .1307E+03        .1143E+02       -.7232E-05
```

31	.9278E+01	-.8563E-06	.8608E+02	.9278E+01	-.5288E-05
32	.6494E+01	-.4728E-07	.4217E+02	.6494E+01	-.4172E-06
33	.3316E+01	.1624E-05	.1099E+02	.3316E+01	.2807E-04
34	.1741E-05	.3506E-05	.1533E-10	.3915E-05	.6359E+02
35	-.3196E+01	.4936E-05	.1022E+02	.3196E+01	.1800E+03
36	-.6034E+01	.4368E-05	.3641E+02	.6034E+01	.1800E+03
37	-.8308E+01	.4084E-05	.6903E+02	.8308E+01	.1800E+03
38	-.9862E+01	.1186E-04	.9726E+02	.9862E+01	.1800E+03
39	-.1060E+02	.1244E-04	.1123E+03	.1060E+02	.1800E+03
40	-.1047E+02	.1581E-04	.1097E+03	.1047E+02	.1800E+03
41	-.9526E+01	.1720E-04	.9074E+02	.9526E+01	.1800E+03
42	-.7841E+01	.1760E-04	.6148E+02	.7841E+01	.1800E+03
43	-.5563E+01	.1713E-04	.3095E+02	.5563E+01	.1800E+03
44	-.2878E+01	.1742E-04	.8285E+01	.2878E+01	.1800E+03
45	.1894E-04	.1877E-04	.7110E-09	.2667E-04	.4474E+02
46	.2846E+01	.3372E-06	.8101E+01	.2846E+01	.6787E-05
47	.5440E+01	.6043E-05	.2959E+02	.5440E+01	.6365E-04
48	.7581E+01	.7545E-05	.5748E+02	.7581E+01	.5702E-04
49	.9107E+01	.7092E-05	.8293E+02	.9107E+01	.4462E-04
50	.9899E+01	.6807E-05	.9800E+02	.9899E+01	.3940E-04
51	.9899E+01	.7495E-05	.9800E+02	.9899E+01	.4338E-04
52	.9107E+01	.8029E-05	.8293E+02	.9107E+01	.5051E-04
53	.7581E+01	.9733E-05	.5748E+02	.7581E+01	.7355E-04
54	.5440E+01	.9830E-05	.2959E+02	.5440E+01	.1035E-03
55	.2846E+01	-.2862E-05	.8101E+01	.2846E+01	-.5762E-04
56	.1837E-04	-.9031E-05	.4192E-09	.2047E-04	-.2617E+02
57	-.2878E+01	-.6094E-05	.8285E+01	.2878E+01	-.1800E+03
58	-.5563E+01	-.3796E-05	.3095E+02	.5563E+01	-.1800E+03
59	-.7841E+01	.1589E-05	.6148E+02	.7841E+01	.1800E+03
60	-.9526E+01	.4393E-05	.9074E+02	.9526E+01	.1800E+03
61	-.1047E+02	.5389E-05	.1097E+03	.1047E+02	.1800E+03
62	-.1060E+02	.8837E-05	.1123E+03	.1060E+02	.1800E+03
63	-.9862E+01	.8768E-05	.9726E+02	.9862E+01	.1800E+03
64	-.8308E+01	.1313E-04	.6903E+02	.8308E+01	.1800E+03
65	-.6034E+01	-.4459E-05	.3641E+02	.6034E+01	-.1800E+03
66	-.3196E+01	-.7699E-05	.1022E+02	.3196E+01	-.1800E+03
67	-.2783E-05	-.7470E-05	.6354E-10	.7971E-05	-.1104E+03
68	.3316E+01	-.8957E-05	.1099E+02	.3316E+01	-.1548E+02
69	.6494E+01	-.1141E-04	.4217E+02	.6494E+01	-.1007E-03
70	.9278E+01	-.1295E-04	.8608E+02	.9278E+01	-.7997E-04
71	.1143E+02	-.1536E-04	.1307E+03	.1143E+02	-.7699E-04
72	.1275E+02	-.1627E-04	.1626E+03	.1275E+02	-.7311E-04
73	.1310E+02	-.8229E-05	.1715E+03	.1310E+02	-.3600E-04
74	.1238E+02	-.4289E-05	.1533E+03	.1238E+02	-.1984E-04
75	.1060E+02	-.3262E-05	.1124E+03	.1060E+02	-.1763E-04
76	.7834E+01	.2075E-05	.6138E+02	.7834E+01	.1517E-04
77	.4225E+01	.5780E-05	.1785E+02	.4225E+01	.7939E-04
78	-.1197E-04	.8473E-05	.2150E-09	.1466E-04	.1447E+03
79	-.4558E+01	.1173E-04	.2077E+02	.4558E+01	.1800E+03
80	-.9118E+01	.1749E-04	.8315E+02	.9118E+01	.1800E+03
81	-.1333E+02	.2070E-04	.1776E+03	.1333E+02	.1800E+03
82	-.1683E+02	.1612E-04	.2831E+03	.1683E+02	.1800E+03
83	-.1927E+02	.1167E-04	.3713E+03	.1927E+02	.1800E+03
84	-.2036E+02	.6306E-05	.4145E+03	.2036E+02	.1800E+03
85	-.1985E+02	.4845E-05	.3941E+03	.1985E+02	.1800E+03
86	-.1758E+02	.3911E-05	.3092E+03	.1758E+02	.1800E+03
87	-.1348E+02	-.3854E-05	.1818E+03	.1348E+02	-.1800E+03
88	-.7580E+01	-.1438E-04	.5746E+02	.7580E+01	-.1800E+03
89	-.3960E-04	-.1367E-04	.1755E-08	.4189E-04	-.1610E+03
90	.9029E+01	-.2677E-04	.8152E+02	.9029E+01	-.1699E+03
91	.1919E+02	-.1227E-04	.3683E+03	.1913E+02	-.3662E-04
92	.3009E+02	-.4724E-04	.9055E+03	.3009E+02	-.8995E-04
93	.4129E+02	-.6143E-04	.1705E+04	.4129E+02	-.8524E-04
94	.5230E+02	-.7128E-04	.2735E+04	.5230E+02	-.7808E-04
95	.6265E+02	-.8671E-04	.3925E+04	.6265E+02	-.7931E-04
96	.7186E+02	-.9780E-04	.5163E+04	.7186E+02	-.7798E-04
97	.7951E+02	-.9939E-04	.6321E+04	.7951E+02	-.7162E-04
98	.8524E+02	-.1107E-03	.7266E+04	.8524E+02	-.7443E-04
99	.8880E+02	-.1155E-03	.7885E+04	.8880E+02	-.7452E-04

C

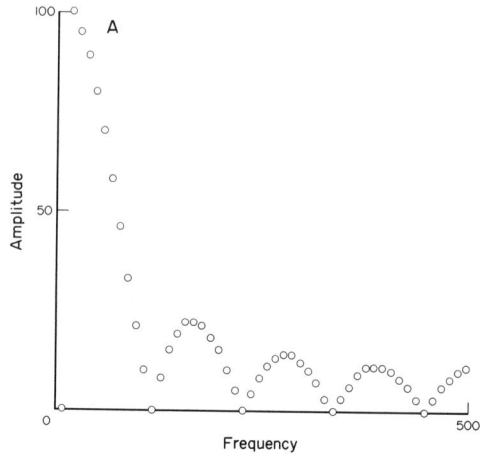

A

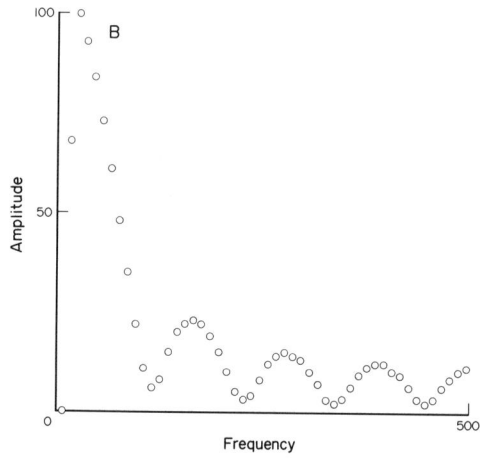

B

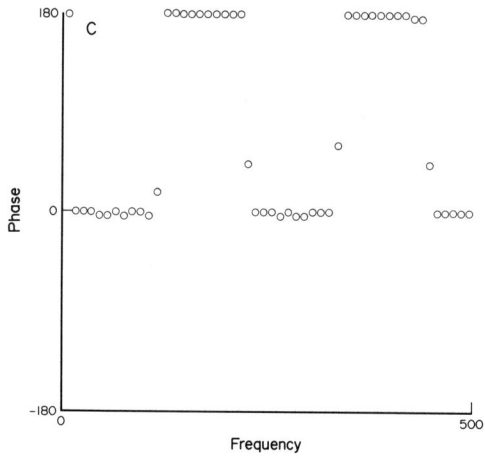

C

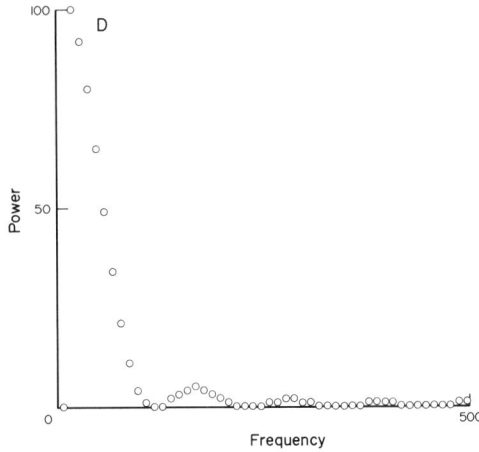

D

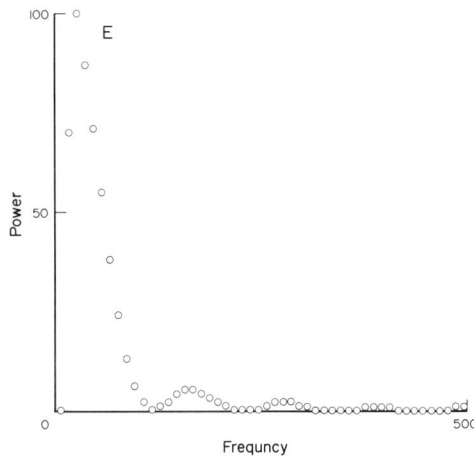

E

Figure 1. A, 3-point running average of amplitude spectrum; B, amplitude spectrum; C, phase spectrum; D, 3-point running average of power spectrum; and E, power spectrum.

1969), variations in reservoir rock properties (Bennion and Griffiths, 1966), magnetic changes (Bhattacharyya, 1965), turbidite correlation (Dean and Anderson, 1967), grain shape (Ehrlich and Weinberg, 1970), precipitation (Horn and Bryson, 1960), structure (Lustig, 1969), stratigraphic information (Merriam and Doria Medina, 1968), waves (Pierson, 1960), cyclic sediments (Preston and Henderson, 1964), volcanologic data (Reyment, 1969), x-ray peaks (Rothman and Cohen, 1968), paleontology (Stehle, McAlester, and Helsley, 1967), and microrelief (Stone and Dugundji, 1965).

REFERENCES

Anderson, R.Y., and Koopmans, L.H., 1963, Harmonic analysis of varve time series: Jour. Geophysical Res., v. 68, no. 3, p. 877-893.

Beckman, W. A., Jr., and Whitten, E.H.T., 1969, Three-dimensional variability of fold geometry in the Michigan Basin: Geol. Soc. America Bull., v. 80, no. 8, p. 1629-1634.

Bennion, D.W., and Griffiths, J.C., 1966, A stochastic model for predicting variations in reservoir rock properties: Jour. Soc. Pet. Eng., v. 6, no. 1, p. 9-16.

Bhattacharyya, B.K., 1965, Two-dimensional harmonic analysis as a tool for magnetic interpretation: Geophysics, v. 30, no. 5, p. 829-857.

Blackman, R.B., and Tukey, J.W., 1958, The measurement of power spectra: Dover Publ., New York, 190 p.

Cohn, B.P., 1975, A forecast model of Great Lakes water levels: unpubl. doctoral dissertation, Syracuse Univ., 235 p.

Cohn, B.P., and Robinson, J.E., 1975, Cyclic fluctuations of water levels in Lake Ontario: Computers & Geosciences, v. 1, no. 1/2, p. 1105-1111.

Cooley, J.W., and Tukey, J.W., 1965, An algorithm for the machine calculation of complex Fourier series: Mathematics of Computing, v. 19, p. 297-301.

Dean, W.E., Jr., and Anderson, R.Y., 1967, Correlation of turbidite strata in the Pennsylvanian Haymond Formation, Marathon Region, Texas: Jour. Geology, v. 75, no. 1, p. 59-75.

Ehrlich, R., and Weinberg, B., 1970, An exact method for characterization of grain shape: Jour. Sed. Pet., v. 40, no. 1, p. 205-212.

Gentleman, W.M., and Sande, G., 1966, Fast Fourier Transforms for fun and profit: AFIPS, Fall Joint Computer Conf., v. 29, p. 563-578.

Good, J., 1958, The interaction of algorithm and practical Fourier series: Jour. Roy. Stat. Soc., v. 20, ser. B, p. 361-372.

Horn, L.H., and Bryson, R.A., 1960, Harmonic analysis of the annual march of precipitation over the United States: Ann. Assoc. Amer. Geographers, v. 50, no. 2, p. 157-171.

Jenkins, F.M., and Watts, D.G., 1968, Spectral analysis and its applications: Holden-Day, San Francisco, 525 p.

Lee, Y.W., 1960, The statistical theory of communication: John Wiley & Sons, New York, 509 p.

Lustig, L.K., 1969, Trend surface analysis of the Basin and Range Province and some geomorphic implications: U.S. Geol. Survey Prof. Paper 500-D, p. D1-D70.

Merriam, D.F., and Doria Medina, J.H., 1968, Analysis of polynomial and Fourier trend surfaces applied to stratigraphic information: Boletin del Inst. Boliviano del Petroleo, v. 8, no. 1, p. 59-74.

Otnes, R.K., and Enochson, L., 1972, Digital time series analysis: John Wiley & Sons, New York, 467 p.

Pierson, W.J., Jr., 1960, The directional spectrum of a wind generated sea as determined from data obtained by the stereo wave observation project: Meteor. Papers, Coll. Eng. New York Univ., v. 2, no. 6, 88 p.

Preston, F.W., and Henderson, J.H., 1964, Fourier series characterization of cyclic sediments for stratigraphic correlation, *in* Symposium on cyclic sedimentation: Kansas Geol. Survey Bull. 169, v. 2, p. 415-425.

Reyment, R.A., 1969, Statistical analysis of some volcanologic data regarded as series of point events: Pure and Applied Geophysics, v. 74, no. 79, p. 57-77.

Robinson, J.E., 1968, Spatial filtering of structures: unpubl. doctoral dissertation, Univ. Alberta, 173 p.

Robinson, J.E., and Cohn, B.P., 1979, FILTRAN: a FORTRAN program for one-dimensional Fourier transforms: Computers & Geosciences, v. 5, no. 2, p. 231-249.

Robinson, E.A., and Treital, S., 19644, Principles of digital filtering: Geophysics, v. 29, no. 3, p. 395-404.

Rothman, R.L., and Cohen, J.B., 1968, A method of Fourier analysis of x-ray peak shapes using only first order peaks: Clearinghouse AD 665 784, 36 p.

Stehli, F.G., McAlester, A.L., and Helsley, C.E., 1967, Taxonomic diversity of Recent bivalves and some implications for geology: Geol. Soc. America Bull., v. 78, no. 4, p. 455-466.

Stone, R.O. and Dugundji, J., 1965, A study of microrelief - its mapping classification and quantification by means of a Fourier analysis: Eng. Geology, v. 1, no. 2, p. 89-187.

SIMULATION OF SEDIMENT-FLUID INTERACTION IN SUBSIDING BASINS

John C. Tipper
Australian National University

Richard Looi
Copland College

Tung Trinh
Lake Ginninderra College

ABSTRACT

The mechanism of fluid movement in a developing sedimentary basin drives most of the diagenetic change there. When movement rates are sufficiently high, initially small chemical differences can be amplified relatively easily, and this can result in the propagation of patterns of diagenetic inhomegeneity in the growing sediment pile. A discrete-time, discrete-space simulator is described here which allows those patterns to be generated and the mechanism's effects investigated. The simulator is implemented in a straightforward Pascal program.

INTRODUCTION

The fluids in actively subsiding sedimentary basins are not static: that has long been appreciated. Sediment compaction, for instance, always must tend to induce advective flow, and thermal instability in a growing sediment pile also can induce fluid convection. In general, although not inevitably, the fluid flow in basins prior to their invasion by meteoric waters will be centrifugal and lateral (Coustau and others, 1975), upward and outward, away from the basin center. For compaction-induced flow, the vertical flow component is thought to be more important in shallow sediments, the lateral component at depth (Bethke, 1985).

This moving fluid has two major effects on the sediments through which it passes, one thermal, the other chemical. The thermal effect is the transport of more heat upward and outward, more quickly, than is possible simply by conduction. This may result in diagenetic histories for basins with unrestricted fluid movement that are grossly different from those for basins in which fluid movement has been impeded. The chemical effect is to bring together at all stages of the basin's development combinations of sediment and fluid that can be chemically out of equilibrium with each other. The particular disequilibrium at any point in the basin will drive the diagenetic processes there, such as precipitation, dissolution and recrystallization.

261

Fluid flow, however, should not be seen as essential in order for thermal and chemical effects to be produced and for diagenetic processes to operate. Conductive heat transfer, after all, will occur along any temperature gradient, and ion diffusion through static pore fluids. Yet, where fluid flow is substantial, thermal and chemical effects will be magnified and the processes of diagenesis must be expected to operate rapidly.

Appreciating the full potential that fluid flow has to enhance the rates of many diagenetic processes is the key to understanding the variety of possibilities for basin diagenesis. This paper looks at one aspect of this potential, the way in which initially small chemical inhomogeneities can be generated, amplified, and propagated in a developing sediment pile. It does so by providing a straightforward model of the chemical interactions between a layer of sediment and the fluid moving through it. This model is implemented in an algorithm that enables the experimental generation of simple basin diagenesis patterns to be carried out on any microcomputer.

SEDIMENTATION, COMPACTION, AND FLUID-FLOW MODELING CONTINUITY AND DISCONTINUITY

In most modern environments sedimentation is manifestly a discontinuous process. Brief episodes of deposition and erosion usually punctuate longer periods in which effectively nothing happens at the surface. Below the surface, however, this fundamental discontinuity is less obvious. The time-constant of the compaction process, for instance, is so long in comparison with the duration of the typical episode of deposition or erosion that the discontinuities of the sedimentation process are "smeared out." Furthermore, as only episodes of deposition can stimulate compaction— episodes of erosion only having the effect of diminishing (or at most halting) ongoing porosity reduction that was actually a response to previous deposition— there is no clear one-to-one correspondence between the individual sedimentation episodes and the compaction history of the resulting sediment pile.

Just as sediment compaction then is surely a continuous process, even if not an entirely steady one, so the fluid flow that it drives must be continuous. And even more so the fluid flow driven by convection must be continuous, if only because of the inertia of any established convecting system. (In fact, the component of fluid flow resulting from compaction can be almost insignificant when compared to that resulting from convection. Bethke (1985) has commented on the low rates of compaction-induced flow in intracratonic basins, and suggested that its role in processes such as secondary petroleum migration— and, by implication, diagenesis— is limited.)

But does it follow that modeling the chemical interactions between the fluids moving in a sedimentary basin and the basin sediments also must be based in continuity? Is it essential that such models simply be solutions of the appropriate set of differential equations (Oran and Boris, 1987)? Must any discontinuity be the result of, for instance, the finite-difference or finite-element formulation of those equations?

In fact, a continuum approach may not be the most appropriate one to take. The most obvious reason for this is that most sedimentary successions, after all, are layered, either compositionally or texturally, and it is not unexpected that the chemistry of the fluids in adjacent layers can be distinct. Karlsen and Larter (1988), for instance, have documented fine-scale patterns of petroleum compositional layering in North Sea reservoirs. Lateral chemical heterogeneity exists too, reflecting both the nature and the geometry of lateral facies variations in any one layer, and this heterogeneity may be more or less distinct. Yet the existence of distinct chemical heterogeneity, lateral or vertical, is in itself no argument against a continuum approach to reactive fluid modeling, for such heterogeneity might be attributable simply to marked lateral or vertical permeability barriers. Any low vertical permeability, for instance, restricts flow between adjacent layers, and diffusion, a much slower process, then is all that can break down differences in the chemical composition of their pore fluids.

However, there are two more substantial arguments against the continuum approach. The first is that, when considered at the pore scale, the scale at which the chemical reactions between sediment and fluid actually take place, the fluid flow that accompanies those reactions is not a simple continuous phenomenon. Precipitation, for instance, anywhere in a porous network, will act at once to reorganize the flow everywhere in the network, and the resultant changes in pressure, temperature, and salinity gradients then can enhance locally that same precipitation. This feedback then can result rapidly in complete cementation in part of the pore network; Chadam (1987) has provided an analysis of the counter situation, dissolution. Pressure build-up in porous systems also can produce sudden flow discontinuities as fracturing takes place, and an interesting example of this, involving repeated episodes of overpressuring and fracturing, followed each time by dolomite cementation, has been documented in the Gippsland Basin (Bodard, Wall, and Cas, 1984).

The second substantial argument against the continuum approach is that the typical sediment-fluid reactions of diagenesis involve the propagation of reaction fronts through the porous medium (Giles, 1987; Guy, 1987). These can be incorporated into a continuum analysis (Walsh and others, 1984; Lichtner, 1985) but, depending on the sharpness of the front, a discrete model may be considerably more natural.

THE SEDIMENT-FLUID "CONVEYOR BELT" MODEL

The approach taken here is to adopt a discrete-time, discrete-space formulation for a developing sediment pile. As each layer of sediment is buried farther, the fluids that pass through it are deemed to do so as distinct packets. If the fluid movement pathways are known, the sediment-fluid interactions can be modeled as a set of "conveyor belts" (Fig. 1).

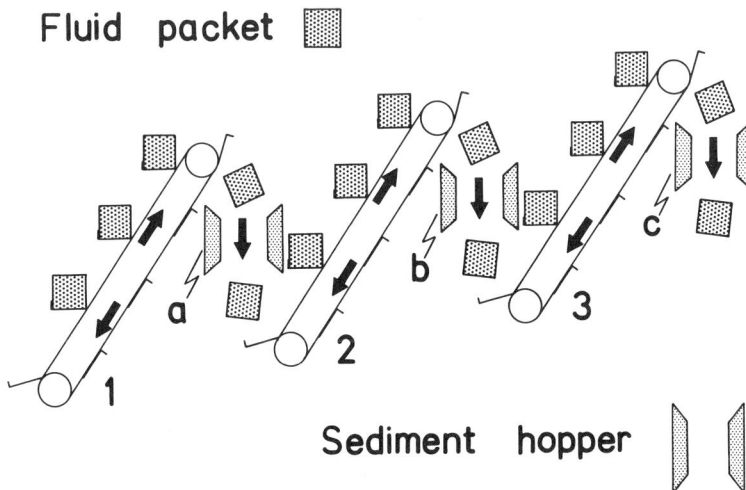

Figure 1. Sediment-fluid interactions along one flow path through developing sediment pile, modeled using set of moving "conveyor belts." Individual sediment layers of pile are "hoppers" a, b, and c, and each fluid "packet" is passed, in turn, through these by conveyor belts 1, 2, and 3. Any diagenetic reactions between sediment and fluid occur as fluid packet passes through sediment hopper.

To appreciate the simplicity of this model's structure, consider a single layer of sediment that has been "tagged" as it was deposited, and then followed down. This layer receives, in successive time steps, the fluids from the layers immediately upstream of it. These fluids are packets on the input belt, and their chemistry, of course, will have been determined by each packet's own history. During each time step the sediment of the tagged layer will react with the fluid that is then in contact with it, and its chemistry and that of that fluid will both change. The nature and extent of these changes will depend on the ambient physical conditions and on the reaction rates relative to the duration of the time step. Finally, at the end of the time step, the fluid will be passed to the layer that is immediately downstream of the tagged one. This fluid will be a packet on the output belt.

The heart of the model is the actual set of reactions between sediment and fluid at the intersection of each input and output belt. The equilibria are controlled by the thermodynamics of the particular chemical system and the reaction rates are controlled by its kinetics. Although, in principle, it is possible to specify what will happen as any packet of fluid from the input belt reacts with a layer of sediment, that is to predict the chemistry both of the fluid packet sent to the output belt and of the sediment left behind, in practice this rarely will be feasible other than for the simplest of sediment-fluid combinations. For the purpose of this paper, no such analytical approach will be attempted. The reactions will be treated as taking place in a "black box" with the input-output mapping specified empirically.

The detailed structure of the conveyor belt model for a single layer is illustrated in Figure 2.

AN IMPLEMENTATION FOR A SINGLE COMPONENT SYSTEM

Implementing the model requires just three specifications: (1) the composition of the fluid packets on the input belt, (2) the initial composition of the tagged sediment layer, and (3) the input-output mapping of the reaction "black box." To illustrate how even the simplest of specifications can potentially produce complex results, a chemical system that has just a single reactive component will be used. For convenience the concentration of this component will be taken to be in the range 0-1.

(1) Three alternatives are used for the compositions of fluid packets on the input belt:
(a) X_i = constant, for all i;
(b) X_i is drawn independently from B(p,q);
(c) X_i is the output of a Gauss-Markov, discrete-time, signal model (Anderson and Moore, 1979). This provides autocorrelation between packets on the input belt.

(2) The initial sediment composition, Y_1, is selected using a uniform distribution on 0-1.

(3) The input-output mapping is implemented using the geometric arrangement shown in Figure 3. The composition of the output fluid packet, Z_i, is related to that of the input fluid packet, X_i, and that of the sediment, Y_i, by a single continuous, smooth surface in the coordinate system (X_i, Y_i, Z_i). To provide a computationally convenient form for this surface, while at the same time offering a wealth of possible behavior for the sediment-fluid reaction, a Coons surface formulation is used (Tipper, 1979; Fig. 3).

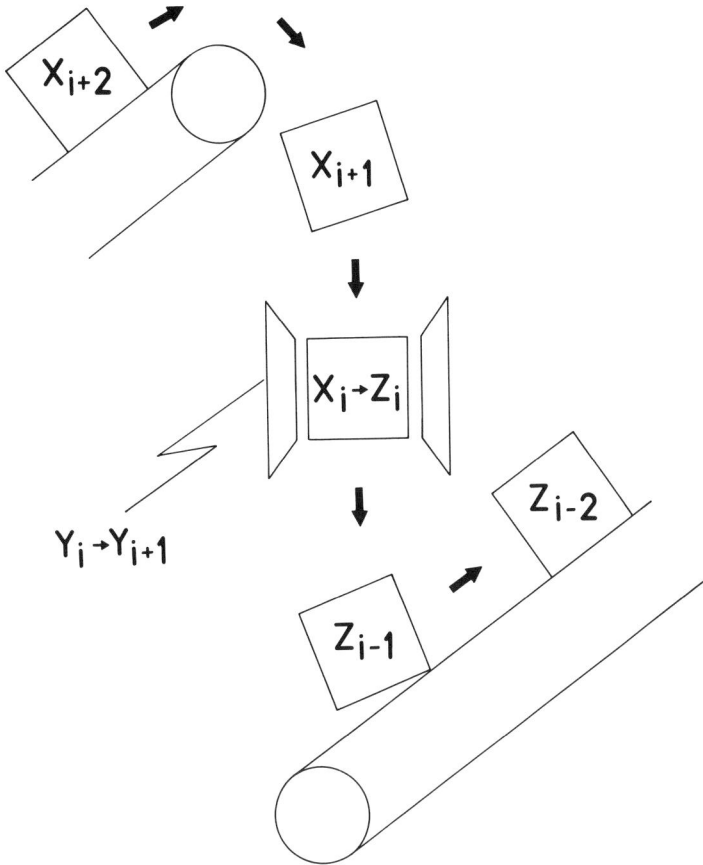

Figure 2. Detailed structure of conveyor belt model. Fluid packets, X, from input belt react in turn with sediment, Y, in hopper, changing both its own composition and their own: $X_i + Y_i$ $Z_i + Y_{i+1}$. Fluid packets, Z, move away on output belt.

The Coons surface is

$$Z_i = [f_0(X_i) \, f_1(X_i)] \begin{bmatrix} Z(0, Y_i) \\ Z(1, Y_i) \end{bmatrix}$$

$$+ [Z(X_i, 0) \quad Z(X_i, 1)] \begin{bmatrix} f_0(Y_i) \\ f_1(Y_i) \end{bmatrix}$$

$$- [f_0(X_i) \, f_1(X_i)] \begin{bmatrix} 0 & Z(0,1) \\ Z(1,0) & 1 \end{bmatrix} \begin{bmatrix} f_0(Y_i) \\ f_1(Y_i) \end{bmatrix}$$

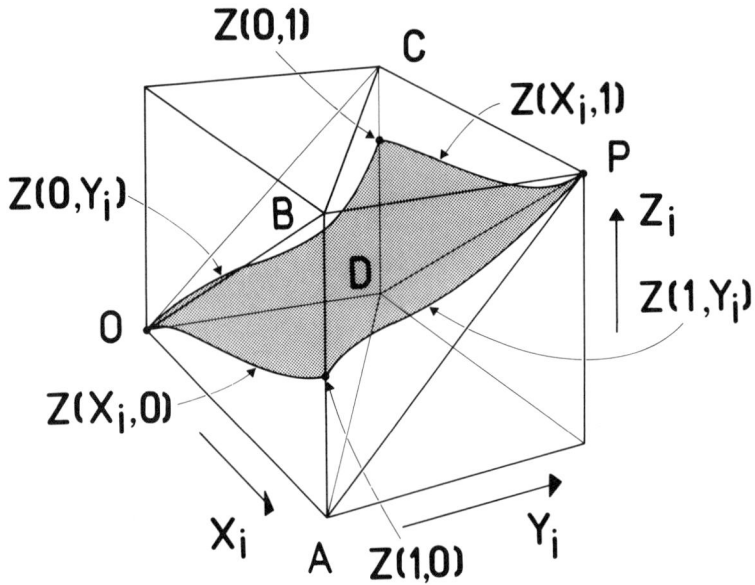

Figure 3. Input-output mapping surface for Z_i, output fluid composition. X_i and Y_i are compositions of input fluid and sediment, respectively. Edge curves, $\{Z(X_i,0)$, $Z(X_i,1)$, $Z(0,Y_i)$, $Z(1,Y_i)\}$ are blended together in Coons surface formulation. Surface lies within polyhedron OABPCD: its corner points are O, P, Z(0,1), and Z(1,0).

The functions $f_0(\ )$ and $f_1(\ )$ serve to blend together the edge curves and corner points. For a linear blend they are:

$$f_0(X_i) = 1 - X_i$$
$$f_1(X_i) = X_i$$

The new composition of the sediment, Y_{i+1}, is given by:

$$Y_{i+1} = X_i + Y_i - Z_i$$

A versatile form for the surface that is useful for experimentation employs cubics as the blended edge curves:

$$Z(0,Y_i) = a_y Y_i^3 + b_y Y_i^2 + c_y Y_i$$
$$Z(1,Y_i) = A_y Y_i^3 + B_y Y_i^2 + C_y Y_i + a_x + b_x + c_x$$
$$Z(X_i, 0) = a_x X_i^3 + b_x X_i^2 + c_x X_i$$
$$Z(X_i, 1) = A_x X_i^3 + B_x X_i^2 + C_x X_i + a_y + b_y + c_y$$

The corner points are:

$$Z(0,1) = a_y + b_y + c_y$$
$$Z(1,0) = a_x + b_x + c_x$$

The restriction that the component's concentration always lie in 0-1 requires that the surface for Z_i lie within the polyhedron OABPCD (Fig.3). To achieve this, the 10 free parameters that describe the Coons surface, $(a_x, b_x, c_x, a_y, b_y, c_y, A_x, B_x, A_y, B_y)$, are somewhat restricted (see Appendix). Within this restriction they can be adjusted conveniently to change the behavior of the sediment-fluid reaction, for instance to vary the nature of the equilibrium conditions and the ways in which they can be approached. The other parameters are related by:

$$C_x = 1 - (A_x + B_x) - (a_y + b_y + c_y)$$
$$C_y = 1 - (A_y + B_y) - (a_x + b_x + c_x).$$

THE MODEL IN OPERATION

To show how the conveyor belt model can both amplify and suppress patterns of chemical heterogeneity, five examples will be given of the single component version in operation. These use different parameters for the Coons surface and different patterns of input fluid composition: they result in substantially different composition patterns for the output fluid.

(1)   The simplest, most symmetric configurations for the input-output mapping surface typically result in the smoothing out of any fluctuations in the input fluid composition (Fig. 4A). The autocorrelation of the fluid compositions is enhanced as they pass through each sediment layer, and inhomogeneities in the fluid continuum thus can be seen only at greater and greater scales as the fluid moves farther away from its initial position as the sediment pile develops. Fluid inhomogeneities will be expected to be seen at the smallest scales in the most recent sediments, and at the largest scales in the oldest: this is intuitively reasonable.

(2)   A similar smoothing effect can be predicted by supplying input fluid of constant composition to a layer that is out of equilibrium with it (Fig. 4B). The result is the classic convergent-exponential approach to equilibrium. This can be expected to occur in basins in which fluid flow has been blocked temporarily, and then suddenly reestablished: the episodic sequence of cementation, overpressuring, and fracturing quoted earlier from the Gippsland Basin (Bodard, Wall, and Cas, 1984) would be an example of such a setting.

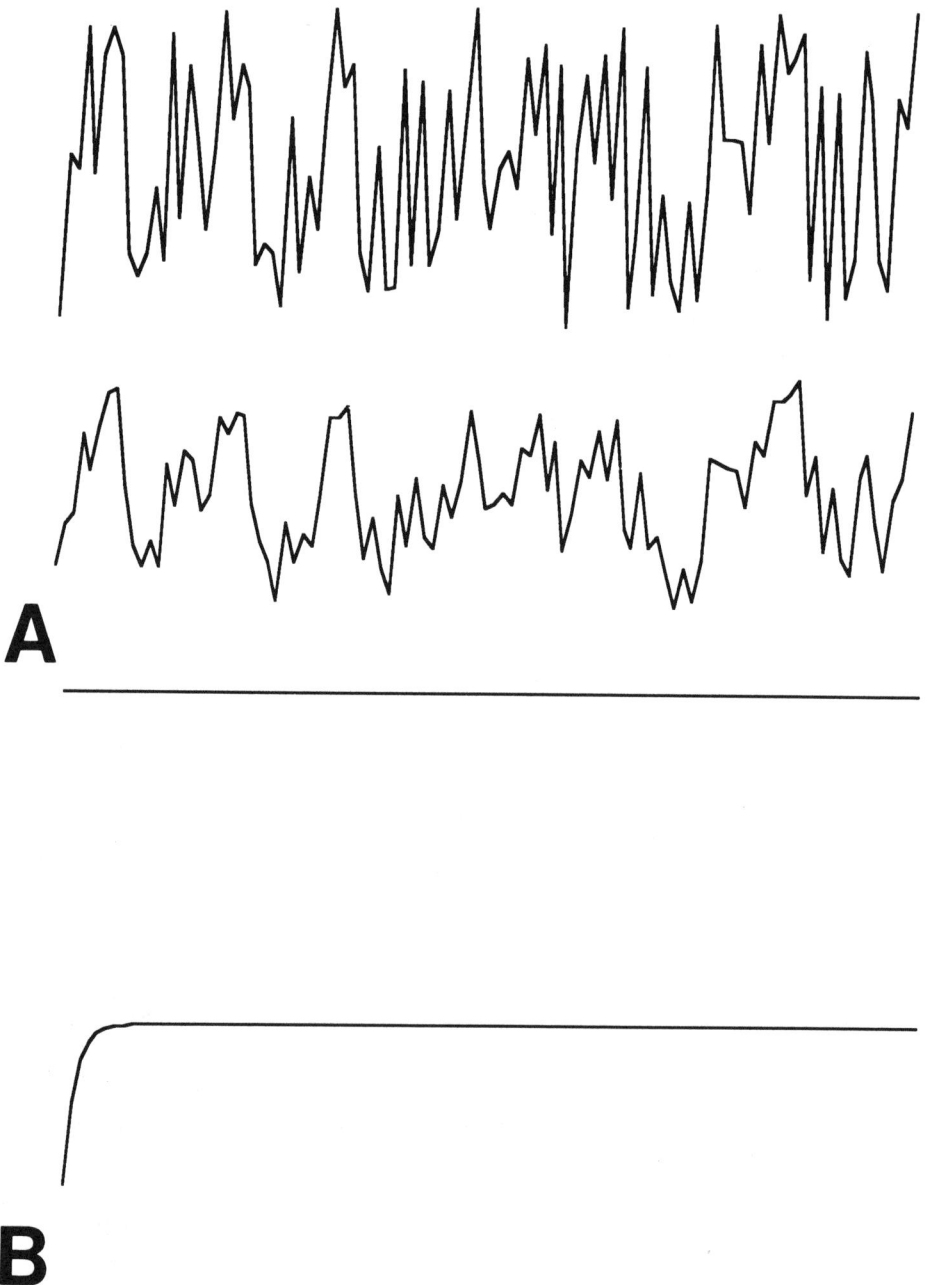

Figure 4. Input and output time-series (upper and lower diagrams, respectively) for fluid
           compositions. Each time-series is of 100 points. A, Input is from symmetric
           beta distribution: output is smoothed version of input. Lag 1 autocorrelations
           are 0.022 (input series) and 0.544 (output series). B, Constant input, out of
           equilibrium with initial sediment composition. Output shows convergent-
           exponential approach to equilibrium.

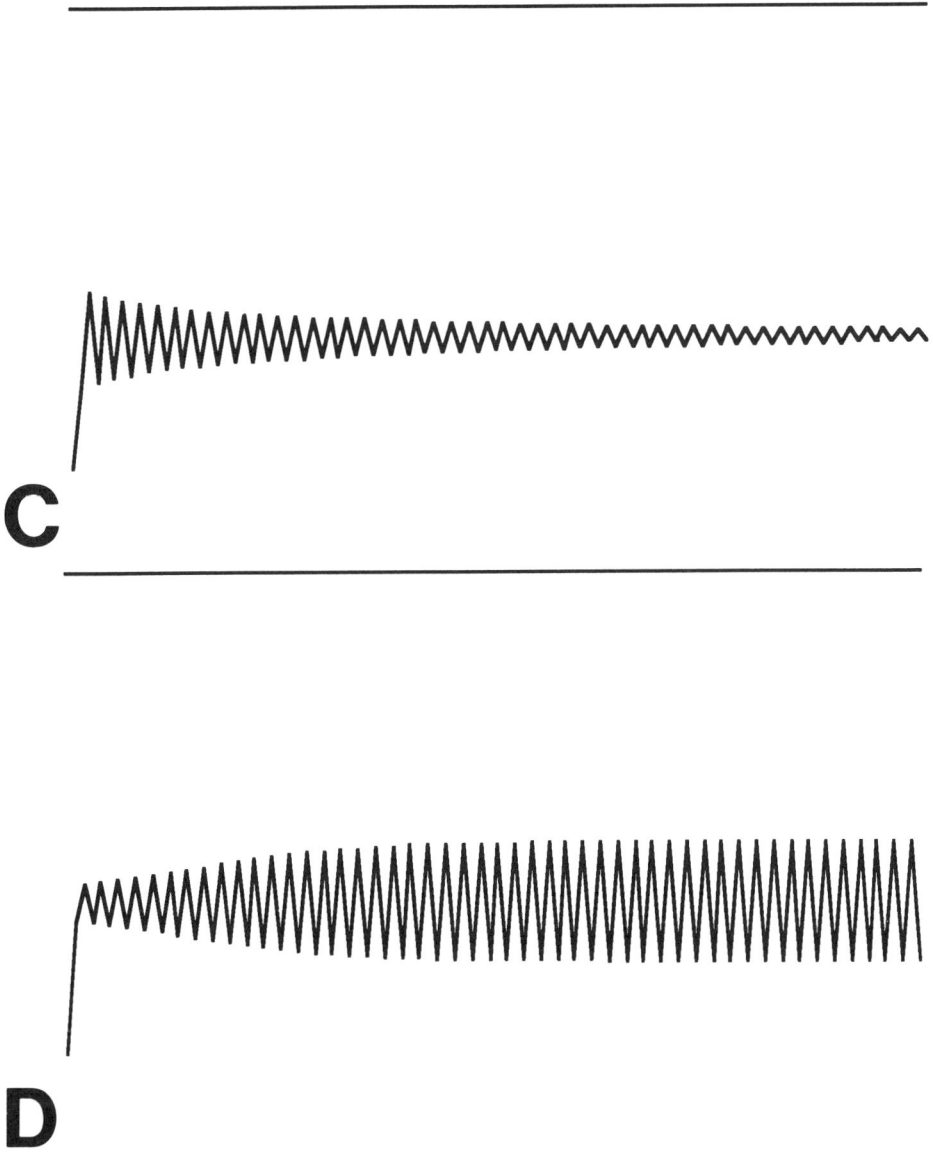

Figure 4. (cont.) C, Input as for B. Output shows convergent-oscillatory approach to equilibrium. D, Input as for B. Output is divergent oscillation, stabilizing as limit-cycle.

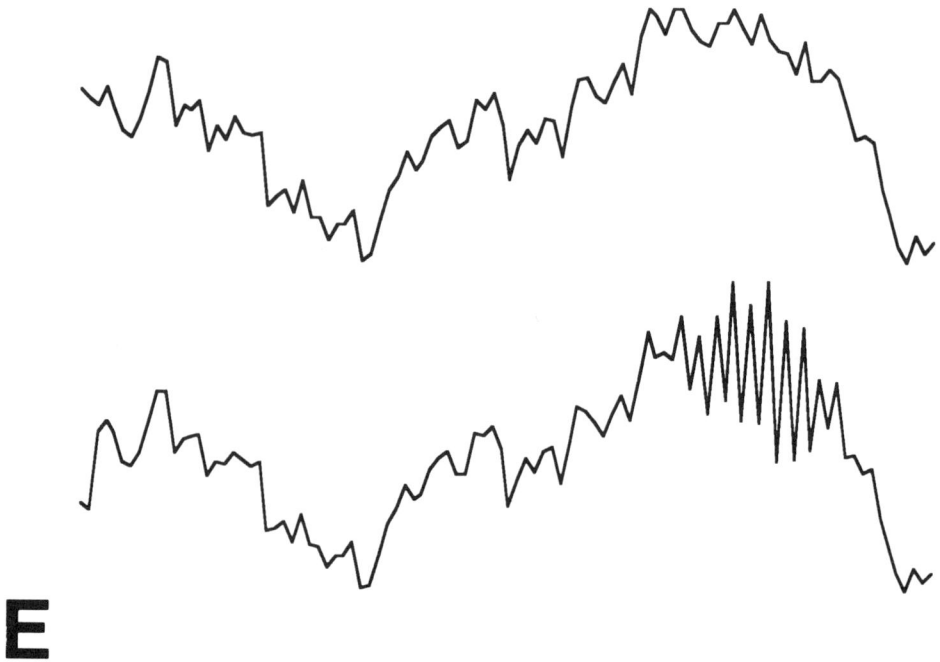

E

Figure 4. (cont.)   E, Input is from Gauss-Markov, discrete-time, signal generator. Nature
of reaction changes across range of sediment and fluid compositions, and
output is in places, smoother version of input, in others, oscillation.

(3,4) Different forms of the input-output mapping surface can represent reactions that
are less straightforward in effect. For instance, these may be reactions which
are far from their equilibrium, a condition that is likely to be usual in basins for
which flow rates are relatively high. Hayes (1979) has suggested, for instance,
that in many basins: "...equilibrium is continually thwarted by the dynamic
transfer of dissolved material around a basin by bulk fluid flow" (Hayes, 1979,
p. 133). With a constant source of input fluid, oscillation in the output fluid
composition is relatively easy to produce. This may be either a convergent-
oscillatory approach to a single equilibrium point (Fig 4C), or divergent
oscillation that stabilizes as the alternation of two equilibria, in effect a limit-
cycle (Fig. 4D).

(5) Finally, the combination of (a) reaction behavior that ranges in its nature across
the range of sediment and fluid compositions with (b) input fluid compositions
that differ in time in a reasonable way (for instance because they have
themselves been subject to smoothing as in example #1) can produce predictably
complex effects (Fig. 4E). In this example the output fluid composition is
effectively a smoothed version of the input fluid composition except when, by
chance, the input fluid composition makes an excursion into a region for which
the behavior is far-from-equilibrium. The output fluid composition then

oscillates, and this oscillation is propagated, even back into the region of convergent-exponential behavior.

It should be clear, although an example of it is not given here, that the superimposition of effects, such as have been illustrated by the passing of fluid from conveyor belt to conveyor belt (Fig. 1), must be capable of producing an extremely wide range of chemical patterns, both in the fluids and in the sediments through which they move. Disequilibrium can be produced readily, emphasized, and then, just as readily, be suppressed. Heterogeneities can arise, be changed in scale, then lose their coherence.

DISCUSSION

The particular model that has been used here for demonstration, the single component system with a reaction "black box", is in itself no more than an attractive toy. Yet the effects that it produces are both reasonable for many basin situations and not unreasonable for others. It provides an excellent instance of what seems more-and-more to be appreciated nowadays in geology— that simple models may produce extremely complex results. Much may be achieved without any undue multiplication of parameters, and without appeal to special causes or circumstances. William of Ockham would approve.

Of course, there are many obvious shortcomings to the particular model used here. Some are relatively easy to identify and remove, and of these the clearest is that the actual nature of the sediment-fluid reaction in any layer must be linked to the physical conditions in that layer at the time of the reaction. The form of the input-output mapping surface must change from time step to time step. If the model is implemented in a real basin situation, this can be done by linking it with conventional forward basin modeling (see, for instance, Bethke, 1985; and Hutchison, 1985).

There also may be a feedback effect that should not be ignored. If, for instance, the sediment-fluid reaction involves any change in the relative volume of those phases (for instance, by precipitation or dissolution), then the flow rate of the fluid through the sediment will alter from time step to time step. This will affect both the physical conditions under which the reactions take place (by changing the thermal gradient), and the reaction rate (which in this model is implicitly scaled by the flow rate). Work by Wood and Hewitt (1982, 1984) points to how this feedback effect operates, and in principle, it can be accommodated simply .

One other shortcoming to the conveyor-belt model, however, far less easy to remove, makes problematic its use in any real basin setting. This concerns the way that the model represents the sediment-fluid reaction kinetics only as an empirically defined input-output mapping surface. This is unfortunate, but inevitable: our knowledge of the reaction kinetics of realistic diagenetic systems is too limited to do otherwise. In fact, it is doubtful if in many situations we can specify correctly the components actually involved in a particular diagenetic system, let alone the kinetics of the reactions within the system assuming it were specified correctly. Organic acids, for instance, have a major influence on many critical diagenetic changes (Surdam and Crossey, 1987), but it rarely will be possible to do other than guess at their actual concentration in any one part of a basin at any one time. Even their presence may be difficult to detect. Perhaps the only light here is that the problem of correctly identifying the components in a real diagenetic system, of estimating their individual concentrations, and of utilizing the proper reaction kinetics, is shared by this conveyor-belt model with even the most sophisticated of continuum approaches.

The philosophy behind this approach to modeling sediment-fluid interactions is that there seems to be little reason why the modeling of basin diagenesis should not benefit from the rather powerful techniques that already are well used elsewhere in the study of other complex systems. The state-space formulation, for instance, that has been employed here in such a simple way allows the interaction of sediment and fluid to be treated as nothing more than a filtering process— the sediment acts as a filter for the composition of the fluid passing through it, and the fluid at the same time acts as a filter for the composition of the sediment through which it passes. Representing the diagenetic system in this way, a direct analog of the "composition space" with which petrologists have worked for years (see, e.g., Thompson, 1982) but incorporating the idea of time explicitly, allows a vast body of results to be transferred across from fields such as systems engineering, in which filtering operations are routine (Anderson and Moore, 1979).

Finally, a comment on modeling systems in discrete time. It is important to appreciate that even the simplest of discrete-time systems can behave readily in ways that would demand far greater structural complexity from a continuous-time counterpart; May (1976) provides excellent examples of this (but see Ortoleva, 1987, for examples of self-organization in continuum geochemical systems). Yet although this may have its drawbacks (for at times that behavior may simply not be realistic), it is on balance an advantage. Too often, in looking at sedimentary basins, we are bemused by their complexity and led to model that complexity with models that are inherently complex in structure. Sedimentary basins are certainly complex things, but that complexity can be produced equally well by models that are of inherently simple structure— provided that the appropriate types of model are used. The conveyor-belt model described here is just one such model: straightforward, with a richness of behavior, and ideally suited to experimentation using microcomputer systems.

ACKNOWLEDGMENTS

This work was carried out partly under the auspices of the CSIRO Student Research Scheme.

REFERENCES

Anderson, B.D.O., and Moore, J.B., 1979, Optimal filtering: Prentice-Hall, Englewood Cliffs, New Jersey, 357 p.

Bethke, C.M., 1985, A numerical model of compaction-driven groundwater flow and heat transfer and its application to the paleohydrology of intracratonic sedimentary basins: Jour. Geophysical Res., v. 90, no. 8, p. 6817-6828.

Bodard, J.M., Wall, V.J., and Cas, R.A.F., 1984, Diagenesis and the evolution of Gippsland Basin reservoirs: Australian Petroleum Expl. Assoc. Jour., v. 24, pt. 1, p. 314-335.

Chadam, J., 1987, Reaction-percolation instability, in Nicolis, C., and Nicolis, G., eds., Irreversible phenomena and dynamical systems analysis in geosciences: D. Reidel Publ. Co., Dordrecht, p. 523-532.

Coustau, J., Rumeau, J.L., Sourisse, C., Chiarelli, A., and Tison, J., 1975, Classification hydrodynamique des bassins sedimentaires, utilisation combinee avec d'autres methodes pour rationaliser l'exploration dans des bassins non-productifs: Ninth World Petroleum Congress, Proc., v. 2, Applied Science Publ., London, p. 105-119.

Giles, M.R., 1987, Mass transfer and problems of secondary porosity creation in deeply buried hydrocarbon reservoirs: Marine and Petroleum Geology, v. 4, no. 3, p. 188-204.

Guy, B., 1987, Nonlinear convection problems in geology, *in* Nicolis, C., and Nicolis, G., eds., Irreversible phenomena and dynamical systems analysis in geosciences: D. Reidel Publ. Co., Dordrecht, p. 511-521.

Hayes, J.B., 1979, Sandstone diagenesis – the hole truth, *in* Scholle, P.A., and Schluger, P. R., eds., Aspects of diagenesis: Soc. Economic Paleontologists and Mineralogists, Tulsa, Oklahoma, p. 127-139.

Hutchison, I., 1985, The effects of sedimentation and compaction on oceanic heat-flow: Royal Astronomical Soc., Geophysical Jour., v. 82, p. 439-459.

Karlsen, D.A., and Larter, S., 1988, A rapid correlation method for petroleum population mapping within individual petroleum reservoirs – applications to petroleum reservoir description (abst.): Abstracts of meeting, "Correlation in Hydrocarbon Exploration", Bergen, Norway. Norsk Petroleumsforening.

Lichtner, P.C., 1985, Continuum model for simultaneous chemical reactions and mass transport in hydrothermal systems: Geochimica et Cosmochimica Acta, v. 49, no. 3, p. 779-800.

May, R.M., 1976, Simple mathematical models with very complicated dynamics: Nature, v. 261, no. 5560, p. 459-467.

Oran, E.S., and Boris, J.P., 1987, Numerical simulation of reactive flow: Elsevier, New York, 601 p.

Ortoleva, P.J., 1987, Modeling geochemical self organisation, *in* Nicolis, C., and Nicolis, G., eds., Irreversible phenomena and dynamical systems analysis in geosciences: D. Reidel Publ. Co., Dordrecht, p. 493-510.

Surdam, R.C., and Crossey, L.J., 1987, Integrated diagenetic modeling: a process-oriented approach for clastic systems: Annual Review of Earth and Planetary Sciences, v. 15, p. 141-170.

Thompson, J.B, Jr., 1982, Composition space: an algebraic and geometric approach: Reviews in Mineralogy, v. 10, p. 1-31.

Tipper, J.C. 1979, Surface modelling techniques: Kansas Geol. Survey, Series on Spatial Analysis, v. 4, 108 p.

Walsh, M.P., Bryant, S.L., Schechter, R.S, and Lake, L.W., 1984, Precipitation and dissolution of solids attending flow through porous media: American Inst. Chemical Eng. Jour., v. 30, no. 2, p. 317-328.

Wood, J. R., and Hewitt, T.A., 1982, Fluid convection and mass transfer in porous sandstones - a theoretical model: Geochimica et Cosmochimica Acta, v. 46, no. 10, p. 1707-1713.

Wood, J.R., and Hewitt, T.A., 1984, Reservoir diagenesis and convective fluid flow: Am. Assoc. Petroleum Geologists Mem. 37, p. 99-110.

APPENDIX

Assigning Parameter Values to the Coons Surface

Because of the restriction that the input-output mapping surface lie within the
polyhedron OABPCD (Fig. 3), the Coons surface that is used here to represent it has just 10
degrees of freedom. These can be filled conveniently in the following way.

(1)   Assign values in the range 0-1 to the slopes of the edge curves at O and P. These
      are:

$$\left.\frac{dZ(X_i,0)}{dX_i}\right|_0 = c_x$$

$$\left.\frac{dZ(0,Y_i)}{dY_i}\right|_0 = c_y$$

$$\left.\frac{dZ(X_i,1)}{dX_i}\right|_P = 3A_x + 2B_x + C_x$$

$$\left.\frac{dZ(1,Y_i)}{dY_i}\right|_P = 3A_y + 2B_y + C_y$$

(2)   Assign values in the range 0-1 to the abscissae that correspond to the edge curve
      inflection points. These are:

$-b_x/3a_x$	for	$Z(X_i,0)$
$-b_y/3a_y$	for	$Z(0,Y_i)$
$-B_x/3A_x$	for	$Z(X_i,1)$
$-B_y/3A_y$	for	$Z(1,Y_i)$

(3)   Assign values in the range 0-1 to the Z-values of the surface as it cuts AB and
      DC. These are:

$$Z(1,0) = a_x + b_x + c_x$$
$$Z(0,1) = a_y + b_y + c_y$$

(4)   Solve these equations to give the values of the free parameters (ax, bx, cx, ay,
      by, cy, Ax, Bx, Ay, By).

Porosity Advisor – An Expert System Used As An Aid In Interpreting The Origin Of Porosity In Carbonate Rocks

W. Lynn Watney, James E. Anderson, and Jan-Chung Wong
Kansas Geological Survey

ABSTRACT

Porosity Advisor was developed to assist in interpreting the origin of porosity in carbonate rocks and to evaluate inexpensive rule-based expert-system shells in solving practical geologic problems. The shells are programs that contain an inference engine without a knowledge base. These programs facilitate development of a knowledge base in addition to a method of linking rules and drawing conclusions.

The facts in this program are pore and cement types. The rules relate their stages of formation and origin in carbonate reservoirs of the Upper Pennsylvanian Lansing and Kansas City Groups in a study area in northwestern Kansas and southwestern Nebraska. Information on types and abundances of pores and cements is requested by the system which are operated on by the rules. Inquiries can be made during the course of the session to determine more about the pores and cement or why a question is being asked. Output consists of a database suited for use in a spreadsheet that tabulates the results and interpretations.

Porosity advisor assists decision-making in the field. It also can provide a means to catalogue information systematically on porosity and cement with the goal of refining existing interpretations. Facilitating both tasks can have a considerable impact on exploration strategy.

New features in expert-system shells have increased their versatility considerably. The cost and size of programs make them well suited for use on small systems in remote locations where access to basic forms of expertise suited to these programs can be of considerable use.

INTRODUCTION

Porosity Advisor is a rule-based expert system used to assist in description of the nature, origin, and timing of porosity development and destruction in Upper Pennsylvanian carbonate rocks in western Kansas. **VP-Expert** (Sawyer, 1987), the software used to implement Porosity Advisor, is a PC-based, expert-system shell offering considerable versatility. It is comparable to many shells now available.

A goal of Porosity Advisor is to reduce the search space of information related to porosity distribution to a manageable size. Another goal is to provide the field personnel, such as an independent petroleum geologist, with the ability to access expertise in establishing relationships between porosity and cement and their interpretations.

The information presented in this paper includes: (1) an overview of expert systems; (2) a summary of how Porosity Advisor was constructed; (3) a review of the knowledge domain used in Porosity Advisor; and (4) the results of a session using Porosity Advisor.

OVERVIEW OF EXPERT SYSTEMS

Expert-system shells are computer software that provide programming tools facilitating the development of a knowledge base and an inference program to link rules and draw conclusions. The programs generally have less flexibility than symbolics program languages such as PROLOG or LISP. The later are languages usually used in expert-systems development. The shells simplify programming steps and therefore expedite programming and facilitate program modification. However, shells also are limited in the technical functions performed and accordingly limit the scope of what ultimately can be accomplished.

In general, a knowledge-based expert system is focused on the accumulation, representation, and use of knowledge specific to a particular task or domain. An objective is to solve a problem by employing inference strategy that experts actually use to solve problems. The expert system can give advice comparable to that of a human expert. Conclusions are provided with justification based on facts and rules that are invoked.

Expert systems have been used to:

interpret	monitor
predict	debug
diagnose	repair
design	instruct
plan	control.

However, limitations exist in applying expert systems (Waterman, 1986). In general, the tasks should:

(1)   require only cognitive skills (involve logical decision making),
(2)   not require common sense (solution should not depend on intuition),
(3)   include those for which experts can articulate their methods (can be clearly and simply stated),
(4)   include those for which genuine experts exist (a unique domain of knowledge exists which is well understood),
(5)   those where experts agree on solutions (facts in the knowledge base are well constrained and the boundary conditions are known adequately), and
(6)   not be too difficult.

Nevertheless, expert systems can be useful to address certain ill-defined or complex problems. This is where the real utility of expert systems will be in the future.

An expert system consists of two basic components:

(1)   a domain knowledge base containing descriptions of the domain and facts about the domain, and

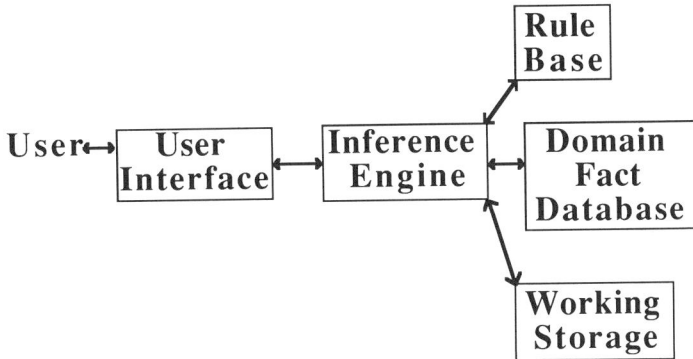

Figure 1. Structure of expert system.

(2)   an inference engine for inferring new conclusions from the knowledge base.

Figure 1 depicts the basic structure of a rule-based expert system. An expert-system shell is an expert system without the domain knowledge. The domain knowledge consists of facts and rules describing a specific area of knowledge such as porosity and cementation in carbonate rocks. The domain knowledge is encoded into the shell as rules. In **VP-Expert**, the domain facts are encoded in the programming language as variables. Values of these variables are stored in a database, as rules, or obtained from a keyboard response to a query.

VP Expert also has a facility to permit the description of the knowledge domain and to ask why queries are being made during various stages of the consultation. This feature is invoked through a user interface. The user interface familiarizes the user with the knowledge domain and inferences being made. This step is critical in transferring knowledge to the person using the system even though it is not being used directly to make inferences. The user interface also can tailor the program to seem as a logical extension of the thought process of the user making the interaction proceed more smoothly.

The rule system used by **VP-Expert** is termed a production rule system. Its structure is, IF (predicate), THEN (conclusion), for example, IF pore name = interparticle, THEN stage of formation = Ø. Inference is the process of verifying the truth of some statement, for example, stage of pore formation based on the truth of the other statements, such as pore name = interparticle. Expert systems can induce only from facts and their relationships. They cannot deduce from a theory.

An inference engine, the heart of the system, operates on or verifies:

(1)   rules provided in the rule base,
(2)   facts present in the domain fact database, and
(3)   facts present in working storage acquired through keyboard
       entry as the session is being conducted.

The inference engine creates a sequence or chain of inference leading to a conclusion. An obvious control problem is encountered in deciding what order to apply the rules. The application of the same sequence of rule selection and evaluation is a procedure termed forward chaining.

VP-Expert also can backward chain where prior inferences made from previously evaluated rules are recalled to evaluate a certain fact. Thus rules can be appended easily to the program and easily accessed during consultation. In backward chaining the inference engine identifies a goal variable and then moves through a sequence of rules until it determines a value that can help it assign a value to the goal variable. Once a value is known, the inference engine can retrace its steps and test the rules that provided its original path. Thus, the facts not addressed in the initial rules eventually are addressed indirectly.

Uncertainty of information is a feature addressed by expert systems. Uncertainty can result from several sources (KU Computing Services Seminar, 1988) including:

(1)   probabilistic knowledge,
(2)   level of belief,
(3)   imprecision of information,
(4)   vague terminology,
(5)   information may be quantized without precision, and
(6)   multiple sources of information leading to contradictory assumptions.

Uncertainty is an important aspect of expert systems and is certainly an important consideration regarding data evaluation in geology. A probabilistic scheme normally is used to quantify uncertainty including the use of:

(1)   confidence factors,
(2)   Bayesian probability,
(3)   fuzzy logic, and
(4)   belief theory.

Confidence factors are used by **VP-Expert**. Uncertainty can be assigned to both rules and facts. Confidence factors are included as multivalued logic. Invoking confidence factors affects the way the truth of a premise is determined. Porosity Advisor in its present stage of development assumes that the answers are definite.

CONSTRUCTING THE KNOWLEDGE BASE

Building this expert system followed several standard steps (Waterman, 1986):
1. *Identification of problem* (acknowledging constraints)
2. *Conceptualization* (basic understanding)
3. *Formalization* (inception and development of the logic used in the expert system)
4. *Implementation* (using **VP-Expert**)
5. *Testing* (working examples to test logic and ease of use)

Identification Of The Problem

The focus of Porosity Advisor is the assessment of the origin and timing of porosity associated with Upper Pennsylvanian Lansing and Kansas City Groups carbonate petroleum reservoirs in northwestern Kansas (Fig. 2). The Upper Pennsylvanian cyclic carbonate rocks were deposited on broad, carbonate-dominated shelf in northwestern Kansas and southwestern Nebraska 250 km north of the Anadarko Basin (Fig. 3). Correlative, subaerially exposed bounding surfaces separate depositional sequences that average 20 meters in thickness. Evidence for episodic emergence on the upper, northern shelf is pervasive including mixed-clast carbonate conglomerates, solution fissures and microkarst, calcrete, and red oxidized paleosols with rhizoliths and ped surfaces (Watney and Ebanks, 1978; Watney, 1980). The formation and later occlusion of pores associated

LITHOLOGY	UNIT	AGE	SEQUENCE
	Pierre Shale		
	Niobrara Formation		Zuni
	Dakota Sandstone		
		Jurassic-Triassic	
	Nippewalla and Sumner Group	Permian	
	Chase Group Council Grove and Admire Groups		Absaroka
	Wabaunsee, Shawnee and Douglas Groups	Virgilian	
	Lansing, Kansas City, and Pleasanton Groups	Missourian	Pennsylvanian
	Marmaton and Cherokee Groups sub-Pennsylvanian unconsolidation	Desmoinesian	
	"chat"	Mississippian	Kaskaskia
	Chattanooga Shale	Devonian-Mississippian	
	"Hunton" Formation	Silurian-Devonian	Tippecanoe
	Maquoketa Shale Viola Simpson Group Limestone	Middle-Upper Ordovician	
	Arbuckle Group	Cambrian-Ordovician	Sauk
	Reagan Sandstone	Precambrian	

approximate scale 500 m

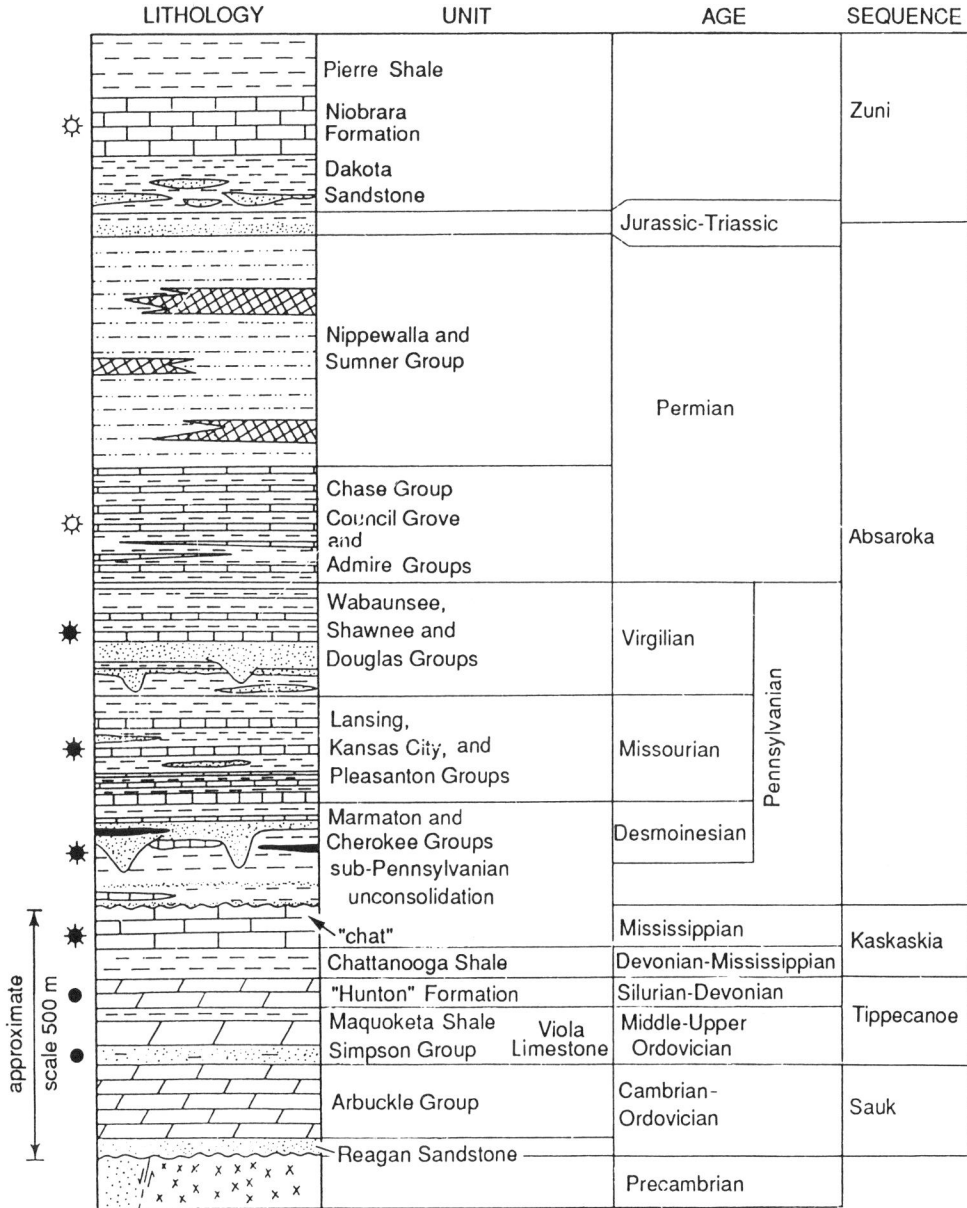

Figure 2. Stratigraphic column of Kansas identifying location of Upper Pennsylvanian Lansing and Kansas City Groups. Solid dots indicate intervals with oil production, dots with radiating spokes indicate oil and gas, and circle with spokes indicates gas-producing interval.

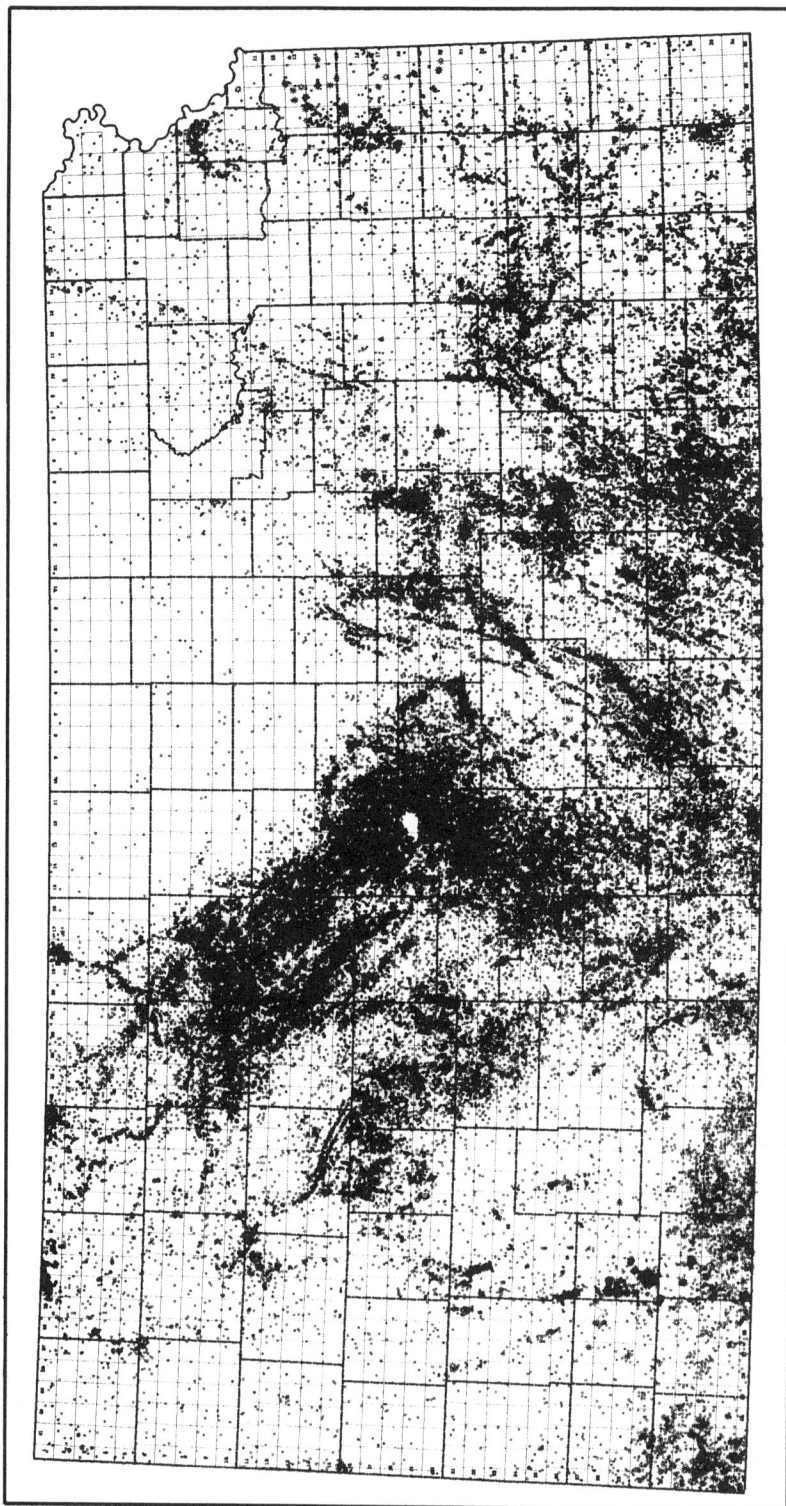

Figure 3. Map of Kansas showing wells penetrating Lansing and Kansas City Groups. Open circles indicate nonproductive penetrations and larger solid dots represent oil-producing wells. Gas wells are larger solid dots with radiating spokes. Smaller rectangles are townships that are 6 miles (9 km) square.

with these rocks is complex and is related only in part to these events of pronounced subaerial exposure.

The assessment of the origin and timing of the pores and cement is a *primary problem* of the independent geologist working in this region. A *secondary problem* is developing an inventory of observations to determine dominant characteristics of open porosity observed in reservoir quality rocks in this area. Together these are important problems in exploration play analysis.

Porosity Advisor is devised as a tool to assist the independent explorationist in developing a computer-based inventory of characteristics of pores and cements and as a consultant to aid in establishing the origin and timing of the pores and cement. Porosity Advisor in its present form does not predict porosity, but rather this is a long-range objective of the research project.

The knowledge base on porosity and cement origin and timing incorporated in Porosity Advisor was obtained through expertise developed in previous work. In particular, a petrographic study was conducted recently by Anderson (1989) in which cement-pore paragenetic sequence was established after looking at 1950 feet of core and 158 thin sections representing seven sedimentary cycles from 35 wells in a study area covering 8500 square miles.

Several facts and observations described here are conveyed to the user during consultation, but were not operated on by the inference engine. This information aids in explaining the importance of certain pores and cements, their interrelationships, and significance.

An example of this knowledge available for inquiry during a session include the facts and background on the fact that significant primary and secondary precementation porosity occurs in nearly all nonargillaceous carbonate rocks regardless of carbonate texture (Fig. 4). Furthermore, open pores in nonargillaceous carbonates are present in nearly equal amounts among various carbonate rock textures ranging from grain supported to mud

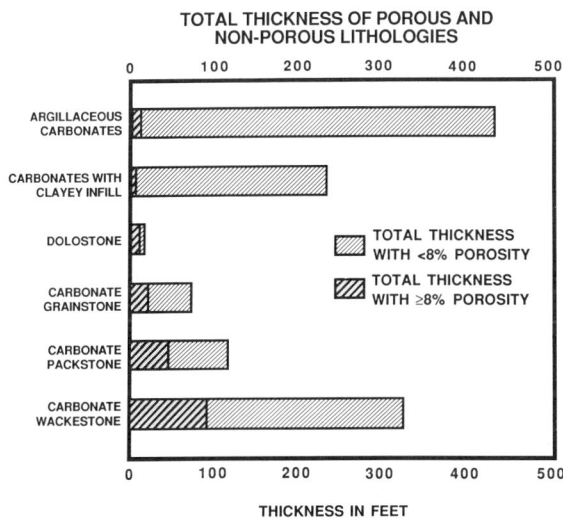

Figure 4. Total thickness of porous and nonporous lithologies. Argillaceous carbonates contain abundant clay seams. Clayey infill fills channel and vuggy porosity and is attributed to subaerial exposure occurring at tops of each shallowing upward carbonate unit.

AVERAGE POROSITY AND CEMENT FILLING

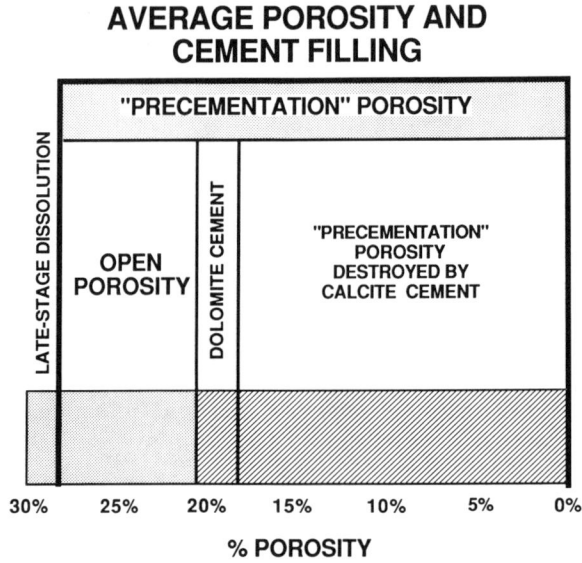

Figure 5. Average porosity and cement filling based on petrographic analysis.

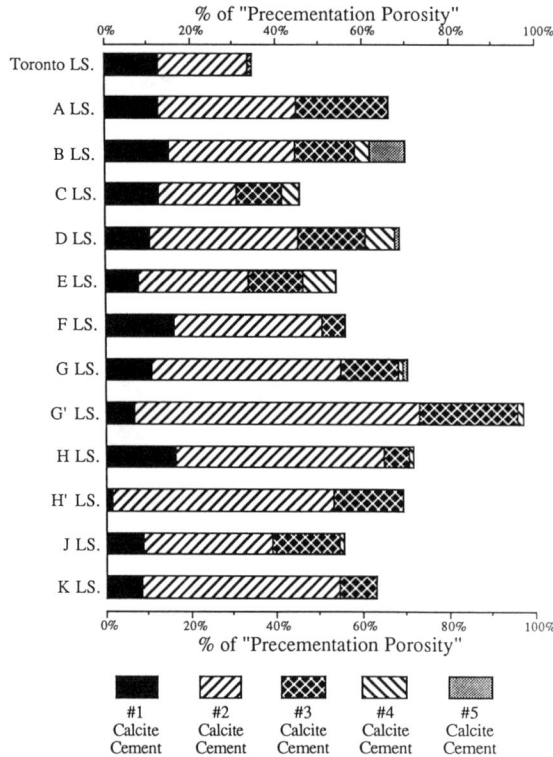

Figure 6. Average percent filling of precement porosity by equant calcite cement in individual limestone zones.

supported. Average precementation and existing porosity and the types of cement filling of pores are charted in Figure 5. "Precementation" porosity, including primary pores and molds, is pervasive in nonargillaceous carbonates. Although the causes are not well understood, this information is important to convey to the user. This information ultimately will be useful in understanding the processes and eventually in predicting porosity.

Other knowledge important in assessing porosity evolution is that the timing of most cement is late and widespread. Equant calcite spar and precementation pores are both nearly equally distributed through multiple stratigraphic zones in the area of study as summarized in Figure 6. Also equant calcite spar which consists of five distinct zones distinguished through cathodoluminescence whose abundance seems to correlate spatially with the presence or absence of overlying Lower Permian evaporites.

Moreover, a considerable portion of the secondary vuggy and moldic porosity is post-compactional. Some of the porosity is derived from late-stage dissolution (LSD), usually cross-cutting burial cements. Figure 7 is a map of the distribution of late-stage dissolution reported as a percentage of existing open porosity. LSD porosity seems to be associated with structural trends.

Although information related to burial history and paleohydrogeology are important, the data are not used directly in the inference strategy because of the hypothetical nature and limited data availability. Rather, this knowledge is conveyed to the user through their query during the consultation. The intent is to make this information part of the working knowledge of the user and encourage its testing. Access to, and consideration of, this knowledge at appropriate times may impact significantly the user's strategy in search for porosity in the area.

Conceptualization

Porosity evaluation requires objective appraisal of pore and cement types and their inferred timing and origin. Figure 8 is a chart of the paragenetic sequence of pores and cement and sediment infilling based on the study of Anderson (1989). The chart is based

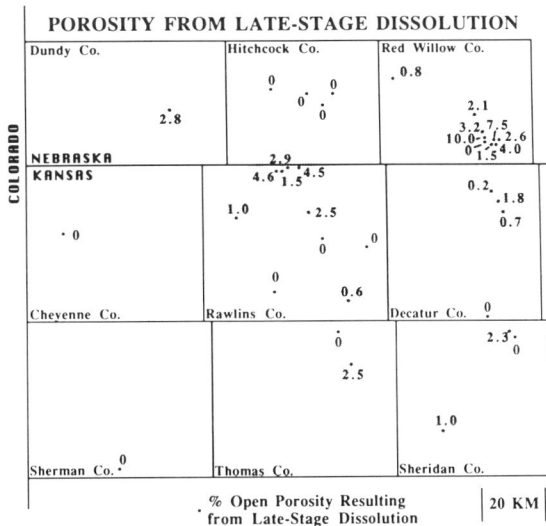

Figure 7. Map of percent open space resulting from late-stage dissolution based on Anderson (1989) study area in northwestern Kansas and southwestern Nebraska.

PARAGENETIC STAGES

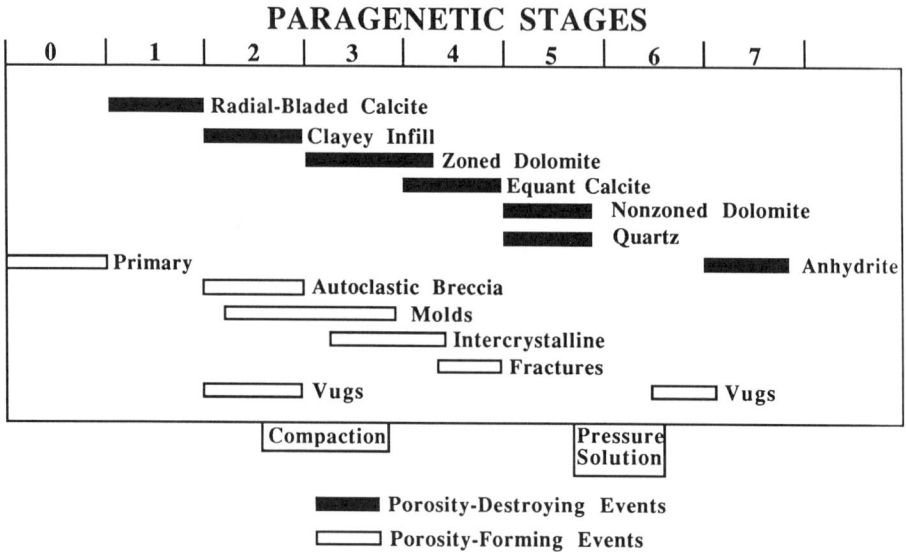

Figure 8. Stages of paragenesis assigned for porosity-destroying and porosity-forming events.

PORE NAME	PORE TYPE	(ORIGIN) PORE FORMATION	(TIMING) STAGE OF FORMATION
Interparticle	Primary	Depositional	0
Intraparticle	Primary	Pre-depositional	0
Shelter	Primary	Depositional	0
Mold	Secondary	{ Pre-compaction / Other }	{ 2 / 3 }
Intercrystalline	Secondary	Recrystallization	4
Breccia	Secondary	Dissolution	1
Vug	Secondary	Dissolution	?
Fracture	Secondary	Structural	?

Figure 9. Pore decision table.

primarily on petrographic information and the use of fluid inclusion and cathodoluminescence. Porosity Advisor restricts the query to petrographic criteria usually available to the independent geologist. Supporting data such as interpretations of timing based on compaction were determined to be subject to some disagreement by the "Experts." Consequently, the timing based on compaction was not considered a significant criteria that could be evaluated with certainty.

CEMENT TYPE	(DIAGENETIC STAGE) PORE FILLING	CEMENT ORIGIN
Radial-bladed Calcite	1	Marine Phreatic
{ Meniscus form Pendant form Internal sediment }	2	Meteoric Vadose
Zoned Dolomite	3	Early Mesogenetic
Equant Calcite	4	Early- to Mid-Mesogenetic
{ Non-zoned Dolomite Baroque Dolomite Quartz }	5	Mid-Mesogenetic
Anhydrite	6	Late-Mesogenetic
No Cement	7	Late-Mesogenetic

Figure 10. Cement decision table.

Formalization

The inception and development of Porosity Advisor is best illustrated through a decision table that lists the facts and interpretations which were used to construct the rules in Porosity Advisor (Figs. 9 and 10). The major categories include classification of porosity type, abundance, and interpretation of porosity and cement origin and timing. Timing of vugs and fractures is constrained by the stage of cement filling as an upper time limit and any cement that these pores cross cut as the lower time limit. Direct assignment of the timing of other pores was made based on empirical results gained in the petrographic study.

Implementation

VP-Expert is implemented as a program with three parts:

(1)   an ACTION block that defines goals and manages the solution process,
(2)   a set of production rules, and
(3)   a set of statements that define variable types and templates for questions to be directed to the user.

The ACTION block contains clauses to be processed in sequential order and seems to be a normal program. Rules contain the knowledge in the knowledge base and conclusions that can be executed if the premise is true, for example, assign variables (facts), display something, or locate additional variable(s) to initiate backward chaining. Statements define variable types and input/output templates for interaction with the user. For example, an ASK statement defines a prompt to seek the value of a specified variable (fact).

VP-Expert also can generate rules automatically. The program can read decision tables from a database that describes a knowledge domain and generate rule sets from it. This is done through the INDUCE command. A database consisting of only the variables that make up the rules makes the expert system flexible and permits rapid modification of the knowledge base.

```
HELP-Select definition
'stop' to exit:                    Brown to dark brown micritic cements include:

  Interparticle               1) meniscus-types which form at grain contacts in
  Intraparticle                  interparticle pores,
  Shelter                     2) pendant forms which extend down from grains
  Mold                           into pores,
  Intercrystallize            3) and irregular linings along walls of channel
  Breccia                        and irregular vugs.
  Vug
  Fracture                    The cements are identical to micritic cement
  Radial bladed calite     associated with calcrete formed at or near the tops
  Micritic cements         of the upper limestones(Watney, 1980). Some micritic
  Internal sediment        cements are actually contiguous with the calcretes,
  Zoned dolomite           especially those lining channel vugs. They seldom
  Equant calcite spar      contribute appreciable infilling, but provide a
  Non zoned dolomite       distinctive means to time the development of reduced
  Quartz                   pores. The stage assigned is 2.
  Anhydrite
  No cement
  Stop
```

Enter to select   END to complete   /Q to Quit   ? for Unknown

Figure 11. Menu of types of porosity and cements from Porosity Advisor.

```
POROSITY ADVISOR -- REPORT
-----------------------------------------------------------------------------
WELL NAME:    Gulf #1-22 Hughes
  LOCATION:   Sec22 T9 29W  C SWSW          ZONE: G          DEPTH: 3969.5
-----------------------------------------------------------------------------
The # 1 pore: Interparticle    Type: primary    Origin :   depositional

    The # 1 cement:  Equant calite spar    Abundance: 12    Stage: 4
    The # 2 cement:  Baroque dolomite    Abundance: 3    Stage: 5

---Remaining abundance of the PORE is : 5    Stage:  0-0  Pre-Porosity:  20

  Which pore exists in the sample (select one at time, ? to stop)?

    Interparticle           Intraparticle           Shelter
    Mold                    Intercrystallize        Breccia
    Vug                     Fracture

  Does cross cutting exist ?
    yes                          no
```

Enter to select   END to complete   /Q to Quit   ? for Unknown

Figure 12. Example of query.

The session begins with a description of Porosity Advisor. This is followed by a menu
listing the types of pores and cement present in this domain (Fig. 11). These can be selected
and queried at will before the rule base is accessed in order to learn more about the nature
and importance of the pores and cement observed in this study area. Information then is
requested to identify the well and sample being described.

Data requested in the program include pore and cement type, percentage of each pore and
cement type, and information on cross-cutting relationships.

```
POROSITY ADVISOR -- REPORT
--------------------------------------------------------------------------
WELL NAME:    Gulf #1-22 Hughes
  LOCATION:   Sec22 T9 29W  C SWSW         ZONE: G            DEPTH: 3969.5
--------------------------------------------------------------------------
The # 1 pore: Interparticle   Type: primary   Origin :  depositional

     The # 1 cement:  Equant calite spar   Abundance: 12   Stage: 4
     The # 2 cement:  Baroque dolomite    Abundance: 3   Stage: 5

---Remaining abundance of the PORE is : 5    Stage: Ø-Ø  Pre-Porosity:  20

The # 2 pore: Intraparticle   Type: primary   Origin :  pre depositional

     The # 1 cement:  Equant calite spar   Abundance: 6.5   Stage: 4
     The # 2 cement:  Baroque dolomite    Abundance: 2   Stage: 5

---Remaining abundance of the PORE is : 1.5ØØØØØ   Stage: Ø-Ø  Pre-Porosity:  10

The # 3 pore: Mold   Type: secondary   Origin :  Dissolution

     The # 1 cement:  Equant calite spar   Abundance: 12   Stage: 4
     The # 2 cement:  Baroque dolomite    Abundance: 3   Stage: 5

---Remaining abundance of the PORE is : 5    Stage: 3-3  Pre-Porosity:  20

The # 4 pore: Vug   Type: secondary   Origin :  dissolution

     The latest crosscut cement: Equant calite spar  Abundance: Ø  Stage: 4
     The youngest fill cement:  Baroque dolomite  Abundance: Ø  Stage: 5

---Remaining abundance of the PORE is : 10   Stage: 4-5  Pre-Porosity:  10

--------------------------------------------------------------------------
     The dominant pore is : Vug  Origin: dissolution  Stage: 4-5
              Pre-porosity: 10      Remaining Porosity: 10
```

Figure 13. Example of output.

Once the consultation begins, questions appear to first select the type of pores identified in the rock. This is followed by a menu from which to select each pore noted in the sample. Next comes the assignment of pore abundance, then the initial cement type reducing the particular pore and its abundance (Fig. 12). Later cements further reducing this particular pore are identified. In addition, vugs and fractures require information as to cross-cutting relationships with preexisting pores. If so, the question is asked as to the type of the latest pore-reducing cement in the pores which are cross cut. This is used to provide the minimum age for the vug or fracture. The questioning continues until all of the descriptive information is provided on the pores and cement. If the user desires to determine why a question is being asked, a key can be hit to bring up a window that will provide an answer.

The results are displayed in the background as the session proceeds. The questions appear as windows in front of the results. At the conclusion of the session the results appear on the full screen (Fig. 13). The output of **VP-Expert** is flexible with results written to either the screen or database. In Porosity Advisor, the output begins with a line indicating the well name, location, zone, and depth being examined. This is followed by a table listing each pore and associated cements that are present, the relative amounts of each, the remaining pore space, and interpretation of their timing and origin (Fig. 13).

A trace can be made while the consultation is being made. This can be printed at the conclusion of the session to examine the exact path of the program. Windows can be added at the bottom of the screen when the consultation is being conducted to examine the particular rules being activated.

Also, what-if strategies can be invoked to determine variations in the outcome with changes in facts. One can move back and forth between a particular rule and the conclusion during the consultation session to modify the response to the rule and examine the result.

The structure of Porosity Advisor in **VP-Expert** is designed for flexibility. More data can be added easily to the database. Graphics could be included to help describe the pores or cement. Other programs written in other languages can be called and run to perform specific functions to evaluate data and create more facts to aid in the evaluation.

CONCLUSIONS

Porosity Advisor assists in interpreting pores and cements present in carbonate rocks in northwestern Kansas. It assists in developing a database from which additional analyses can be conducted. Porosity Advisor in its present form does not attempt to predict porosity, but this is an ultimate objective.

Expert-system shells have a large, but only lightly tapped potential for use in decision making and transferring knowledge in geoscience. They are becoming increasingly user-friendly and suited for use by the researcher who has limited time to develop an expert system. They no longer have to be an end in themselves in research.

Nevertheless, the expert system can provide a way for an individual to learn from an expert when the expert is not available. Programs now are designed such that knowledge bases can be modified easily. They can grow as the boundary conditions and constraints of the variables are defined better. The program can lead an individual through the process of inductive reasoning, step-by-step, giving reasons for questions, provide a sense of how to handle the problems, and aid in organizing the knowledge domain and results.

REFERENCES

Anderson, J.E., 1989, Diagenesis of the Lansing and Kansas City Groups (Upper Pennsylvanian), northwestern Kansas and southwestern Nebraska: Geol. Soc. America, South-Central Section, Abstracts With Programs, p. 1-2.

Kansas University Computing Services, 1988, Seminar on expert systems.

Sawyer, B., 1987, VP-Expert: rule-based expert system development tool: Paperback Software International, Berkeley, California, unpaginated.

Waterman, D. A., 1986, A guide to expert systems: Addison-Wesley Publ. Company, Reading, Massachusetts, 419 p.

Watney, W. L., 1980, Cyclic sedimentation of the Lansing and Kansas City Groups (Missourian) in northwestern Kansas and southwestern Nebraska—A guide for petroleum exploration: Kansas Geol. Survey Bull. 220, 72 p.

Watney, W. L., and Ebanks, W. J., 1978, Early subaerial exposure and freshwater diagenesis of the Upper Pennsylvanian cyclic sediments in northern Kansas and southern Nebraska (abst.): Am. Assoc. Petroleum Geologists Bull., v. 62, no. 3, p. 570-571.

A Frame-Based Expert System to Identify Minerals in Thin Section

D. Wright
Shelton State Community College

D. Stanley
U.S. Bureau of Mines

H. C. Chen
A. W. Shultz
J. H. Fang
The University of Alabama

ABSTRACT

A computer program to aid in identifying minerals in thin section, XMIN-S, was built using a commercial expert-system shell and is designed to run on IBM-compatible microcomputer systems. The program uses a menu-driven method of data input to avoid syntax errors typical of free-form user input. The menu also allows the user to input any number of optical properties of a mineral in any order. The program will match the user input with the optical properties of the minerals in the knowledge base. Frames are used to store the characteristics of each mineral in the knowledge base and a B-tree structure is used to index the minerals on secondary storage. The use of a B-tree structure permits the use of a large database of minerals on microcomputer systems with reasonable access times. Approximate reasoning is used to accommodate imprecision of measurements and uncertainty of observations. Help facilities are available to provide definition of terms or to explain the procedure to be used for determining a particular optical property.

INTRODUCTION

Mineral identification in thin section requires the use of a broad range of optical properties, which can be categorized easily by an experienced petrographer, but which can be difficult for a novice. Each optical property must be measured or estimated and the process can be tedious and frustrating for novices, especially given that there may be hundreds or even thousands of candidate minerals from which to select. The process can be even more difficult if the observations are incomplete or imprecise. This paper will discuss the development of an expert system which makes the identification of minerals simpler for the novice user.

BACKGROUND

Expert Systems

Since the 1950's, researchers have been attempting to develop computer systems whose behavior can be classified as intelligent, with one of the primary fields being expert systems. Expert systems are computer programs that are characterized by their ability to capture and apply the knowledge of human experts. Facts alone do not constitute knowledge; knowledge is reached when the relationships between facts are understood and analyzed to arrive at conclusions. Knowledge of factual information will be referred to in this article as declarative knowledge, whereas understanding of relationships or methods will be referred to as procedural knowledge (Mishkoff, 1985).

Until recently, expert systems were designed and coded by computer programmers working with domain experts. Some of these expert systems were emptied of their domain specific knowledge and used as shells or templates for expert systems in other areas of knowledge. Computer capabilities and software development have progressed rapidly and these shells have evolved into development tools, usable on personal computers. Access to personal computers and availability of the new development tools allow the expert to play a larger role in the development of expert systems.

Two examples of expert systems which have been developed for optical identification of minerals are MICA and XMIN. Hart, McQueen, and Newmarch (1988) reported on the expert system termed MICA which identifies minerals in hand specimen, thin section, and polished section. MICA has two major drawbacks: (1) The system is designed to run on a SUN workstation using a local version of the programming language Prolog, which is a configuration not present in many teaching situations. (2) MICA is relatively slow to identify candidate minerals, requiring up to four minutes to identify a mineral. Donahoe, Green, and Fang (1989) developed an expert system, XMIN, for identification of minerals in thin section. XMIN is relatively fast in identifying minerals and is implemented on an IBM-compatible computer. However, XMIN requires the user to enter text information, and is prone to errors related to mismatches in spelling, punctuation, or formatting. This article will discuss the development of an expert system which is designed to rectify the problem areas in both of these previous programs.

Terminology and Definitions

Expert systems usually have the following components: (1) a user interface, (2) an inference engine, and (3) a knowledge base. The inference engine is the part of the expert system which simulates the reasoning of the expert, specifically which search techniques or rules which are to be used for a particular problem. The knowledge base is the store of knowledge which contains the declarative and procedural knowledge related to the domain of the expert. The user interface is a bridge between the system and the user and is responsible for question and answer sessions, explanation of reasoning processes, tutorial information, and assignment of confidence factors by the user (Fig. 1).

An expert's declarative and procedural knowledge must be represented in a form the machine can use to simulate the techniques used by an expert. Procedural information is concerned with what courses of action are appropriate based upon the situation and available information and can be represented by a series of rules which are equivalent to the IF-THEN construction in other programming languages. Declarative knowledge can be thought of as factual information, which can be stored as single pieces of information, or can be grouped if the information is related to a single name or attribute. Grouped factual information usually is referred to in expert-system literature as frame structure. The programming language Prolog and the expert-system shell used in the development of this system refer to single pieces of information as facts.

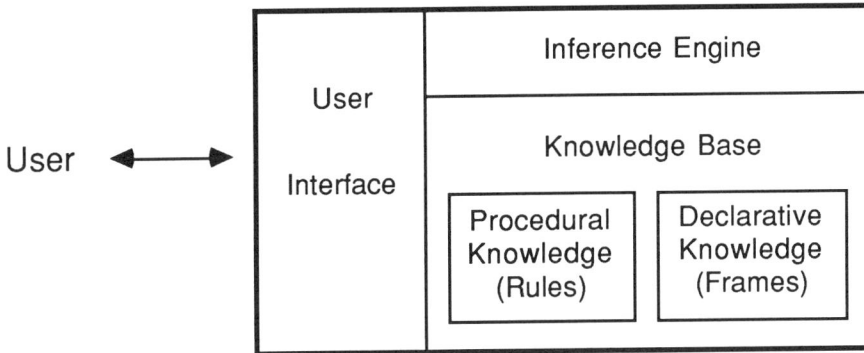

Figure 1. Expert-system architecture.

Structures

Rules

Rules are used to represent methods of reasoning used by the expert in reaching a
conclusion and can be forward-chaining or backward-chaining. Forward-chaining rules
take the available facts or premises and follow a logical pattern from the facts to arrive at
a conclusion. Backward-chaining rules follow a different logical pattern by taking a
conclusion as an idea or hypothesis and attempting to test all the premises which are
needed to prove the conclusion. Backward-chaining rules have the advantage of working
on one specific conclusion to prove or disprove, whereas forward-chaining rules must
attempt to derive the conclusion from the information available.

Forward-chaining rules have the format:

```
If
   premise
Then
   conclusion
```

Backward-chaining rules have the format:

```
   conclusion
If
   premise
```

An example of the use of a backward-chaining rule is in answering the question "Is X the
grandparent of Y?" which is written in the following way:

```
   grandparent 'X' , 'Y'
if 'X' is_parent_of 'Z' and
   'Z' is_parent_of 'Y'
```

Facts

A fact, for purposes of this paper, is a single piece of declarative information which
defines a particular relationship between a single item (or person) and a value. For

example, a person named George has a parent named Sam. A fact representing this might look like:

Sam is_parent_of George

where "is_parent_of " is a relational phrase connecting Sam as a parent to George.

Frames

Another method of storing declarative knowledge is the frame structure, which is used in many expert systems. The frame structure is a set of related pieces of factual information with a common index or name. The advantage of this method of representing information versus single facts is that the common index or name need not be repeated for each piece of information. A frame for a person might appear as follows:

                    Frame: George
                    Parent: Sam
                    Parent: Sue
                    Slot: Age   Value: 30
                    Slot: Sex   Value: Male
                    Slot: Eye Color   Value: blue
                    Slot: Hair Color   Value: blond

Frames can have parents (in this example Sam and Sue are the parents of George) which in turn are also frames.

USE OF EXPERT-SYSTEM SHELLS TO DEVELOP EXPERT SYSTEMS

Why Use Expert-System Shells?

Expert-system shells are becoming more widely used to develop expert systems. Shells are designed to provide a template of the structures usually used in expert systems, although not all shells provide all of the structures described in the expert-system literature. The primary reason for using expert-system shells is that a developer does not have to design these structures as would be required when using a procedural programming language, but can simply make use of the structures provided by the shell. This allows the developer to concentrate on the logic needed to implement the expert system, rather than on the design and verification of low-level details.

Intelligence Compiler, a commercial expert system shell, was selected for this project. Intelligence Compiler uses a set of Prolog-like rules and facts and provides a frame structure for storage of the knowledge base. The shell provides forward-chaining and backward-chaining rules, inexact rules, and facts (exact and inexact).

Rules in Intelligence Compiler use the format of the rule structure typical of expert systems and can be forward-chaining or backward-chaining. The rules can be either exact or inexact with assigned confidence values. Rules in Intelligence Compiler can be used to generate new information during execution, as well as to access existing information. The rules in this mineral identification program are composed entirely of backward-chaining rules.

Facts in Intelligence Compiler are used to store single pieces of declarative knowledge, such as the fact that the mineral epidote can be colorless. The format for facts is:

    epidote color colorless

where epidote is a mineral and color is a verb linking epidote with the value colorless. Each mineral can have multiple colors or a single color associated with it. However, each color of a mineral is stored as an individual fact. For example, if a mineral can be yellow or green, the representation using facts would appear as follows:

> epidote color green
> epidote color yellow

Frames in Intelligence Compiler are used to store related factual information using a common attribute or name. An example of a frame used in a mineral-identification system is shown next. The mineral is aegirine and its parent mineral group is pyroxene. The slots contain the optical properties and the value is the associated text or numerical value for each category. C1, C2, and C3 slots are text lines which explain distinguishing features of the mineral.

> Frame: aegirine
> Parent: pyroxene
> Slot: Birefringence Low range Value: 0.0370
> Slot: Birefringence High range Value: 0.0590
> Slot: TwoVLow Value: 60
> Slot: TwoVHigh Value: 66
> Slot: Elongation Value: length fast
> Slot: Extinction Value: 2 to 10 degrees
> Slot: Twinning Value: no
> Slot: Refractive Index Low Value: 1.745
> Slot: Refractive Index High Value: 1.777
> Slot: Pleochroism Value: strong
> Slot: Optic Group Value: biaxial negative
> Slot: Cleavage Value: at 87, 93 degrees
> Slot: C1 Value: "Long prismatic crystals. Often bladed with"
> Slot: C2 Value: "typical 4-8 sided cross-section. 100 better"
> Slot: C3 Value: "developed than 010."

B-trees in Intelligence Compiler

B-trees are not a knowledge structure of expert systems, but instead are a method of indexing frames and are concerned with actual storage problems and accessing methods of the knowledge base. Each member of the B-tree structure is equivalent to the frame structure, except that it is maintained on secondary storage and only part of the file is loaded into memory. A B-tree structure permits access to the frames sequentially or in a random-access method. The frames in a B-tree structure are stored in alphabetical order (by name) so that new frames can be inserted into or deleted from the file as desired. The frames also are indexed by a series of pointers into the knowledge base which allow the inference engine to locate frames in a direct-access method. The developer does not have be concerned with the implementation details of the B-tree structure because the indexing methods are maintained by the system.

IMPLEMENTATION

Development of a Mineral-Identification System

XMIN-S, a thin-section mineral-identification system, is a revision of a earlier program, XMIN, developed at the University of Alabama. Changes include improvements to the user interface and new search techniques for selecting candidate minerals. XMIN was developed using Pascal, whereas XMIN-S was developed using the expert-system shell,

Intelligence Compiler. XMIN-S was developed using Intelligence Compiler instead of Pascal to compare the development process of the two methods.

User Interface

The user interface of XMIN-S is divided into a hierarchical series of menus, tutorials, and help screens. User input, wherever possible, has been limited to a single-letter selection from a menu. The only exceptions are when the user keys in text involving associated minerals or requests to view the data contained on a specific mineral. The program displays help screens upon request during the session and uses a two-level menu to implement the user interface. When the user types a letter from the menu of the optical properties, the optical property selected appears and displays the possible values.

Adjustments for User Input by XMIN-S

Because users of this program are not necessarily experts, problems may occur during the mineral-identification process, specifically with precision and certainty of the user input. Approximate reasoning is used to accommodate imprecision of measurements and uncertainty of observations. Three problem areas where approximate reasoning is required are (1) incomplete information, (2) uncertainty of information, and (3) imprecision of user input.

Incomplete information is dealt with by allowing the user to enter only those values for properties which the user can identify, rather than for all of the fields which the program makes available. The program actually can search with no information at all, although this will result in every mineral in the database being displayed. The user does not have to enter the information in any order and can change a field value by simply entering the letter of the category. All optical properties for which the user does not enter a value are ignored in the mineral-selection process.

Uncertainty is present when the user is not sure of the exact answer to enter for an optical property. Uncertainty of user input can be dealt with most easily in quantifiable optical properties by presenting a menu from which group values can be selected. Uncertainty can be accommodated further for such properties by widening the ranges of the various groups. For example, each range of the optical property birefringence numerically overlaps the adjacent ranges to compensate for user uncertainty or imprecision. Some of the optical properties which have text values also can compensate for uncertainty of user input by allowing the user to search for yes or no values. For example, cleavage can have many specific values, such as perfect, rhombohedral, among others, which the user may able to match to the particular mineral being examined. However, if the user only knows that cleavage is present, then the system locates all minerals for which the cleavage has some value other than "no."

Imprecision occurs when a user is not careful about the observations of optical properties. Careless measurements can be accommodated by the same techniques used for uncertainty and therefore this tends to be an area of overlap between the problem areas. Imprecision of format (errors such as spacing, punctuation, etc.) is avoided completely by the use of menu-driven user input.

Search Techniques

The search methodology of XMIN-S differs greatly from that of XMIN. XMIN makes a complete search of every mineral in the database to identify matching minerals. XMIN-S, however, eliminates groups of minerals by comparing the characteristics that the user has entered with the shared characteristics of each mineral group. The characteristics of the mineral groups are matched with the user input and if any of the characteristics of the mineral group do not match the user input, then all of the minerals of that group are dropped from the list of candidate minerals to be searched. After all of the minerals groups have been examined, each mineral which is a viable candidate is examined to see if it should be selected as a match with the user input.

XMIN-S uses the relationships between each mineral and its parent mineral group to select candidate minerals during its search phase. Each mineral has a parent mineral group, and each mineral group is stored in a frame structure with its distinguishing characteristics. If the search process determines that a mineral group has not been eliminated as a source of candidate minerals, then each of the minerals in the group is examined to compare its properties with those reported by the user. However, if a mineral group has a characteristic which conflicts with the user input, then all of the minerals in the group are eliminated from the search process. The feldspar mineral group and its distinguishing characteristics are shown here.

```
Frame: feldspar
Parent: mineral groups
Slot: Twinning          Value: yes
Slot: Relief            Value: low relief
Slot: Optic Group       Value: biaxial
```

Considerations Made in Knowledge Representation

The methods used in mineral identification and the volume of mineral data to be stored in the knowledge base presented some problems in deciding what structures were to be used for storing the knowledge base. Most of the optical properties of each mineral have either a single value or range of values which can be stored easily in a frame structure, such as optic group or twinning. Other properties, such as color and associated minerals, can have multiple values which are difficult to store and access in a fixed-length structure such as a B-tree structure. However, the size of the knowledge base precluded using a fact-based system for all of the optical properties. Because of the differing requirements of the various optical values represented, a combination of the B-tree, frames, and facts were needed to store the knowledge base. The B-tree structure is the main method of storing the bulk of the knowledge base because the number of minerals is anticipated to expand in the future and the amount of data which would be accumulated probably would be too large to load in memory. However, facts were used to store two optical properties, colors and associated minerals, because many of the minerals in the knowledge base have multiple values for color and associated minerals.

USING XMIN-S

The user is assumed to be a novice in terms of knowledge of specific minerals, but must be familiar with the use of the polarizing microscope. The interface is menu-driven and normally will need only a numerical value or a one-letter selection by the user, with help screens available when appropriate. An example illustrating the use of the program will be concerned with two of the characteristics of quartz, specifically birefringence (which is weak to moderate) and optical group (which is uniaxial positive). The user selects optical properties one at a time and enters all available information before starting the search process. It is recommended that the user enter as many of the optical properties as can be determined so as to restrict the number of candidate minerals from the 131 minerals in the current knowledge base to as small a set as possible. For example, if the user enters only the color and the color selected is colorless, there will be many candidate minerals which can have this characteristic. If the user provides at least a few optical properties, then the list of candidate minerals is reduced to usually less than 10. If the user selects only the optical group and enters the value "uniaxial positive", then a list of 5 minerals is developed for the user to consider. If the user selects the birefringence option only and enters weak-moderate, a list of 40 minerals is developed. If the user selects both of the properties before beginning the search process, then quartz is identified as the only candidate mineral.

The main menu presents the user with the screen in Figure 2.

University of Alabama Mineral Identification Program

Input Instructions
Enter letter of optical property of interest and press return.
A - Name
B - Optical Group
C - Cleavage
D - Twinning
E - Color
F - Pleochroism
G - Relief
H - 2V
I   - Birefringence
J - Optic Angle
K - Elongation
L - Associated Minerals
M - Clear all values
N - Display Mineral Groups
O - Display Minerals
X - Execute program
Q - Exit Program
? - Help

Figure 2. Main menu of XMIN-S

The user selects the letter of the option desired. If the user selects the birefringence option, then the submenu for birefringence is displayed in Figure 3.

If the user enters D for weak-moderate, then any mineral which can have a value in this range will be considered to be a candidate mineral if all other values entered by the user also match. Once the search process is started, matching of a particular mineral according to the birefringence range will be carried out by the birefringence rule in the program. If the range covered by the user-input value (which is weak-moderate in this example) overlaps any part of the range that a candidate mineral can encompass, then the rule will return a true result; otherwise, it will return a false result.

At this point, the main menu (Fig. 2) will appear again, and the user selects option B for optical group. The menu for optical group is displayed in Figure 4.

The user enters the letter A for uniaxial positive and presses return. The main menu (Fig. 2) will appear and the user enters X to start the search process. The program develops a set of candidate minerals and lists the number of matches which are exact and the number of matches which are considered to be close but not exact. At this point the

Birefringence Menu

Term	Value	Order	Colors
A-nil	.000	--	black
B-very weak	.000-.003	1	dark gray
C-weak	.002-.008	1	gray, white
D-weak-moderate	.007-.012	1	white, yellow
E-moderate	.010-.030	1-2	yellow, red, blue, green
F-moderate-strong	.025-.040	2-3	green, yellow, red, violet
G-strong	.035-.055	3	violet, green, yellow, pink
H-very strong	.050-.080	4-5	pink, green
I-extreme	> .075	5 & up	pearly white

Figure 3. Birefringence menu.

What is the optical group ?

A - uniaxial positive
B - uniaxial negative
C - uniaxial (can be negative or positive)
D - biaxial positive
E - biaxial negative
F - biaxial (can be negative or positive)
G - isotropic

Figure 4. Optical group menu.

program will print full descriptions of all of the exact matches in alphabetical order, followed by the inexact matches if requested by the user. Quartz, as expected, is a candidate mineral and is displayed in Figure 5.

SYSTEM REQUIREMENTS AND AVAILABILITY

This system is designed to run on IBM PC-compatible systems. An AT-class machine with a hard disk is recommended strongly. The use of the B-tree structure indicates that many disk I/O accesses are performed in the execution of the program and substantially improved access times are achieved by using a hard disk.

XMIN-S on diskette, together with a user's manual, is available through the COGS public-domain software library.

Mineral is quartz
Birefringence Range   - Low is 0.009 High is 0.009
TwoV Range   - Low is 0   High is 0
Elongation is length slow
Extinction is parallel/symmetrical
Twinning is none
Refractive Index range   - Low is 1.544 High is 1.553
Pleochroism is none
Optic Group is uniaxial positive
Cleavage is none
Colors can be colorless
Associated minerals are feldspar muscovite

Distinguishing features are
Generally unaltered. Basal sections are dark in all
positions. Wavy extinction due to strain common.
--

100% match on all inputs.

Press return to continue.

Figure 5.   Program output for quartz.

CONCLUSIONS

The primary purpose of this research project was to compare an expert-system shell with that of a traditional programming language and to examine the advantages and disadvantages of the two implementation methods. A reasonable comparison could be made between the two approaches because XMIN-S and XMIN are both designed for optical mineral identification of thin sections. The minerals in the two programs are the same and the user interface has the same basic format. Tradeoffs are present when using an expert-system shell versus a traditional programming language, specifically in the areas of speed and efficiency of design. A system developed using a shell may execute more slowly than one written from scratch because the developer has little control over the actual implementation details of the system. Also, the provided template structures may not fit the characteristics of the problem as well as those that could be custom-designed using a traditional programming language. This is true especially for data structures.

The major advantage of using an expert-system shell is that the developers do not have to verify that the underlying structures of knowledge representation have been designed and tested correctly. Exactly how the rules, facts, and frames are stored, retrieved ,and accessed are no longer the concern of the developer, who then can concentrate on the correctness of the logic in the system. Some understanding of programming is required at this time but the lower-level implementation details of the structures have been taken care of by the expert-system shell. The use of expert-system shells is anticipated to increase as shells become more user-friendly and require decreasing amounts of programming knowledge.

REFERENCES

Donahoe, J.L., Green, N.L., and Fang, J.H, 1989, An expert system for the identification of minerals in thin section: Jour. Geol. Education, v. 37 no. 1, p. 4-6.

Hart, A.B., McQueen, K.G., and Newmarch, J.D. 1988, A computer program which uses an expert systems approach to identifying minerals: Jour. Geol. Education, v. 36, no. 1, p. 30-33.

Mishkoff, H. C., 1985, Understanding artificial intelligence: Texas Instruments, 258 p.

Plates 1, 2. Hall

FIGURE 1. First oblique scenes produced in early 1988 by Kevin J. Hussey at Jet Propulsion Laboratory's Digital Image Analysis Laboratory (DIAL). Satellite image consists of 10 m pixels interpolated by KRS (Kodak Remote Sensing) from LANDSAT 5 TM scene from 18 January 1987 according to variations seen in 10-m resolution panchromatic image from French SPOT-1 satellite. Pixel elevations are interpolated from 25-m grid of DTM data from Jerusalem, Ramallah, Kallia, and Jericho 1:50,000 topographic sheets. Images were produced on JPL-DIAL MicroVAX image-processing system using methodology described by Hussey, Hall, and Mortensen (1986). Vertical exaggeration is about 250%, and images are 1024 × 1024 pixels except as noted.

FIGURE 1A. Computer-generated view showing whole test region from considerably higher altitude than Figures 1B-D. Viewpoint is in Jordan Valley over Tomer, looking southwest across northern Judea and southern Samaria. Jericho lies at center left, whereas Jerusalem and Ramallah are in middle background.

FIGURE 1B. Computer-generated view from point near Jordan River northeast of Jericho, looking southwest toward Jerusalem. Mount of Temptation lies in center towering some 250 m above Jericho, world's oldest city. Note Wadi Kelt at left center where it disgorges into Jordan Valley, and oval form of Jericho racecourse in left foreground.

FIGURE 1C. Computer-generated view to east toward Ramallah and Atarot (Jerusalem Airport). Arab village of Beit I'nan is in foreground. with Jewish settlements of Mevo Horon and Givat Ze'ev in intermediate background. Note banding of topography resulting from geologic outcrops of gently folded sediments and artificial agricultural terracing. Image is 2048 × 2048 pixels, but shows only slight improvement in resolution for same scene at 1024 × 1024 pixels.

FIGURE 1D. Computer-generated view to southwest across upper parts of Judean desert some 10 km southwest of view in Figure 1B. Ma'ale Mukhmas is krill-shaped settlement in central foreground. Wadi in middle intermediate foreground is upper part of Wadi Kelt, through which Mikhmas River flows to Jericho. Jerusalem lies in center in far background, whereas Ramallah is just beyond wadi to right. Relatively heavy vegetation can be attributed to high rainfall in winter of 1986–87.

Plate 3. Hittleman

FIGURE 2. By creating graphic representations of data, users can locate data errors rapidly. In this view of topography one can identify two areas in vicinity of Cape Cod where bathymetric data were processed using wrong unit-of-measure. Areas are depicted by rectangular holes in continental shelf.

FIGURE 3. Color images of various geophysical data can be viewed rapidly. This image of isostatic gravity data highlights Snake River Plain in Idaho.

FIGURE 4. Topography of coterminous United States can be represented with color selections defined by user. Overlays of political boundaries, or contours of geophysical data can be displayed.

FIGURE 5. In this topographic view of San Francisco region, oceanic colors were extended to include land elevations up to 25 feet. Current coastline data were represented as overlay. Much of San Joaquin Valley becomes submerged in this scenario.

Plate 4. Roberts

FIGURE 5. Example of normal plot of climate variables. This plot illustrates solutions for January using sea-surface temperatures characteristic of modern conditions.

Plate 1

FIGURE 1A

FIGURE 1B

Plate 2

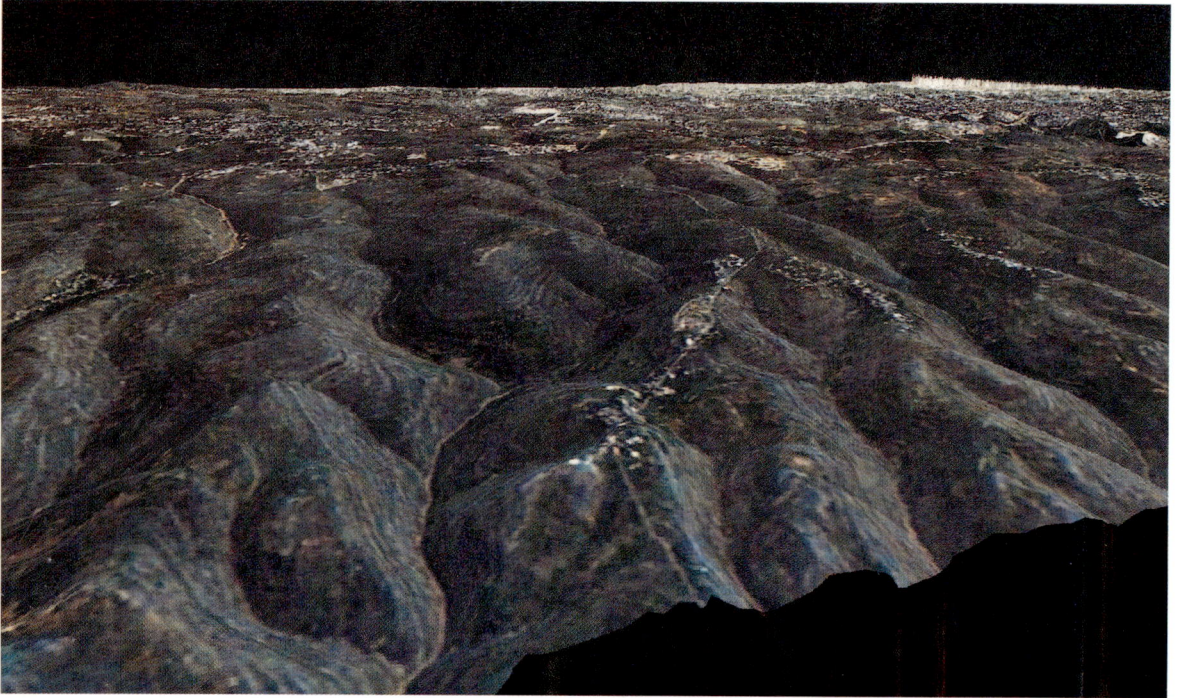

FIGURE 1C

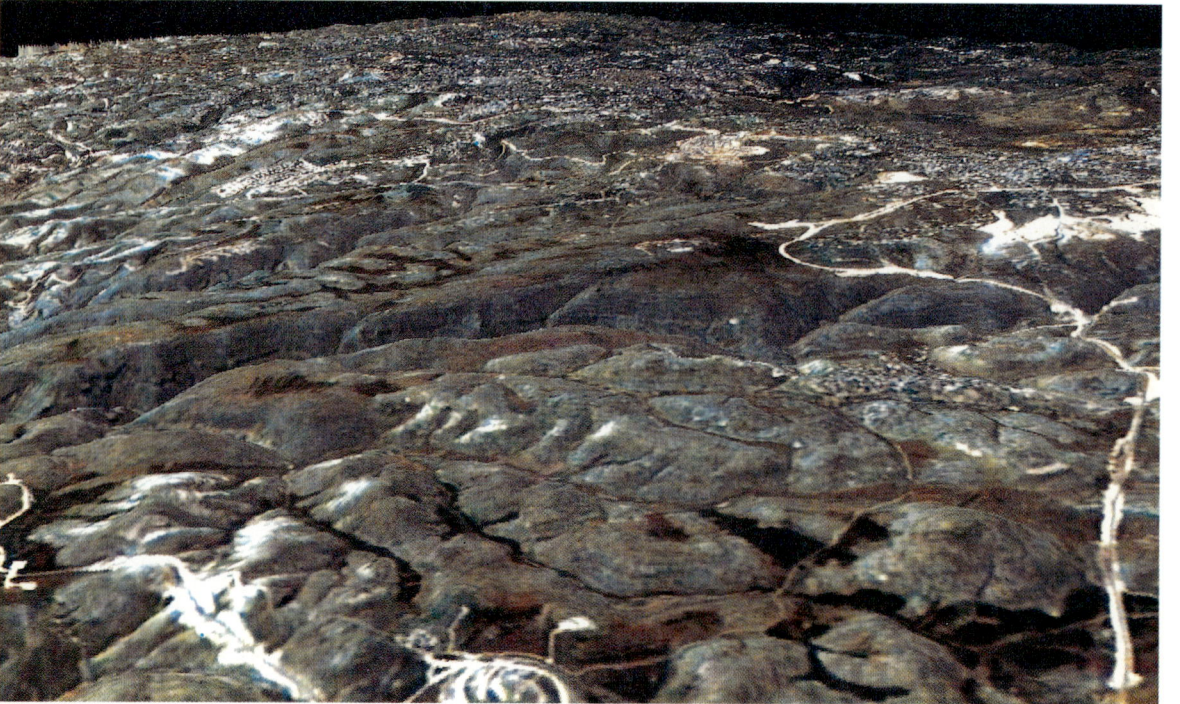

FIGURE 1D

Plate 3

FIGURE 2

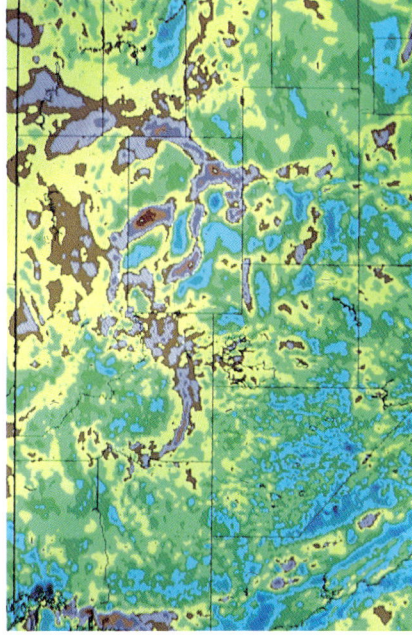

FIGURE 3

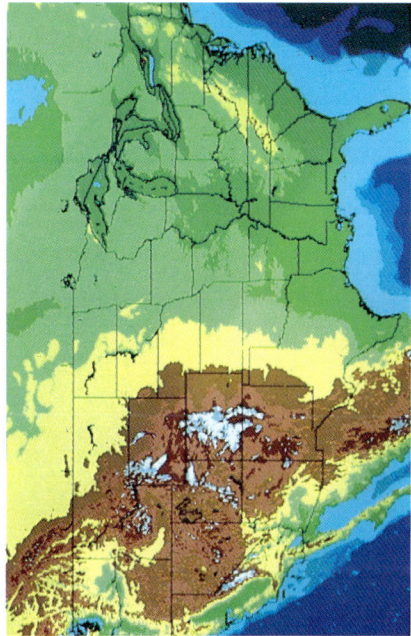

FIGURE 4

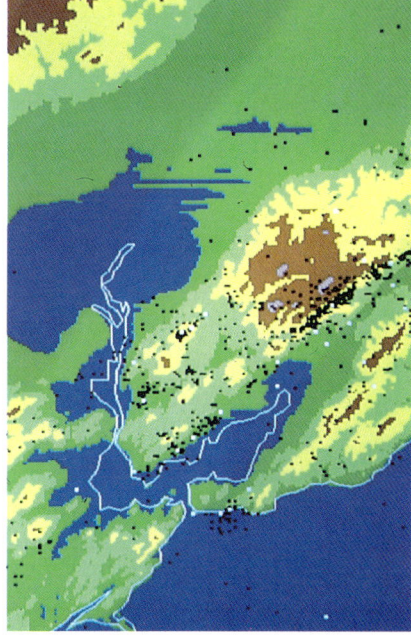

FIGURE 5

Plate 4

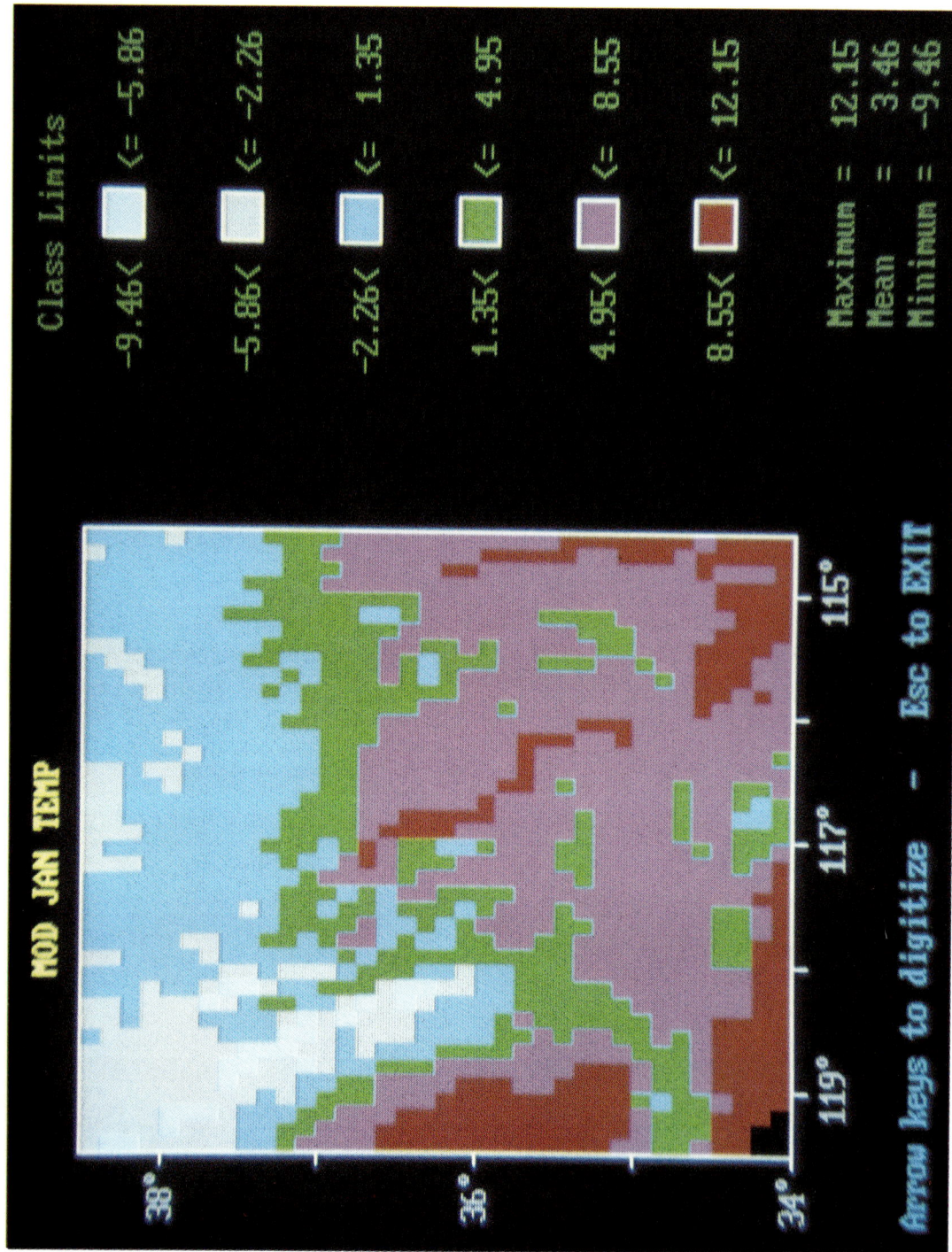

Class Limits

-9.46< <= -5.86
-5.86< <= -2.26
-2.26< <= 1.35
1.35< <= 4.95
4.95< <= 8.55
8.55< <= 12.15

Maximum = 12.15
Mean   = 3.46
Minimum = -9.46

MOD JAN TEMP

38°
36°
34°

119°   117°   115°

Arrow keys to digitize - Esc to EXIT

FIGURE 5

Index